Methods in Molecular Biology

For further volumes:
http://www.springer.com/series/7651

For over 35 years, biological scientists have come to rely on the research protocols and methodologies in the critically acclaimed *Methods in Molecular Biology* series. The series was the first to introduce the step-by-step protocols approach that has become the standard in all biomedical protocol publishing. Each protocol is provided in readily-reproducible step-by-step fashion, opening with an introductory overview, a list of the materials and reagents needed to complete the experiment, and followed by a detailed procedure that is supported with a helpful notes section offering tips and tricks of the trade as well as troubleshooting advice. These hallmark features were introduced by series editor Dr. John Walker and constitute the key ingredient in each and every volume of the *Methods in Molecular Biology* series. Tested and trusted, comprehensive and reliable, all protocols from the series are indexed in PubMed.

Pollen and Pollen Tube Biology

Methods and Protocols

Edited by

Anja Geitmann

Department of Plant Science, McGill University, Montreal, QC, Canada

Humana Press

Editor
Anja Geitmann
Department of Plant Science
McGill University
Montreal, QC, Canada

ISSN 1064-3745 ISSN 1940-6029 (electronic)
ISBN 978-1-0716-0671-1 ISBN 978-1-0716-0672-8 (eBook)
https://doi.org/10.1007/978-1-0716-0672-8

This Humana imprint is published by the registered company Springer Science+Business Media, LLC, part of Springer Nature.
The registered company address is: 1 New York Plaza, New York, NY 10004, U.S.A.

Preface

In plants, the life cycle alternates between a haploid gametophytic and a diploid sporophytic generation. The male gametophyte in seed plants is the pollen grain and its extension, the pollen tube. To accomplish its function, the pollen must first travel between plants and then deliver its cargo, the sperm cells precisely to its counterpart, the female gametophyte. The pollen tube is a crucial mobile vector for the immotile sperm cells; as in seed plants, they need assistance in reaching the female gametes, which are located in an ovule nestled deeply within the ovary. This second phase of the process requires an invasive penetration of the sporophytic pistil tissues to reach a receptive ovule—an accomplishment that involves chemical communication and physical interaction between the mating partners. This book represents a collection of experimental protocols designed to study various aspects of pollen development and function, including the mechanism it employs to reach the ovule and deliver the sperm cells. The protocols range from basic in vitro cell culture methodology to sophisticated experimental designs involving single cell micromanipulation, Lab-on-a-Chip technology, or high-end imaging.

The development of pollen occurs in the anther where microspore mother cells undergo meiosis to give rise to tetrads of haploid microspores. These tetrads typically separate and each of the four microspores develop into mature pollen grains involving one or two mitotic cell divisions generating bicellular or tricellular pollen grains, respectively. The number of viable pollen grains is critical for breeding because reduced yield due to pollen abortion can be induced by diverse abiotic stresses, such as excessively low or high temperature, salt stress, or osmotic shock. A basic assay to determine one of the key parameters in the reproductive potential of a plant is, therefore, the quantification of viable pollen formed by flowers. Pollen grains can be counted microscopically, but a quicker method based on an electric field multichannel cell counting system is provided in Chapter 1 of this book. This method translates the change in voltage upon the passage of a suspended pollen grain through a pore into biologically interpretable values including total number of pollen grains, percentage of viable grains, and grain size.

The mature pollen grain consists of a vegetative cell and, depending on the species, either a generative cell or two sperm cells. In the former, the second mitotic division generating the sperm cells occurs after the grain has germinated, and the generative cell has entered the elongating pollen tube. The pollen tube is a cellular extension of the pollen grain that is formed upon contact with a receptive stigma in vivo or in vitro under suitable growth conditions. It functions akin to a catheter and transports the sperm cells to an ovule over distances that correspond to several thousand folds that of the diameter of the grain. The pollen tube elongates through tip growth, a highly polar cellular elongation process that is fueled by a continuous supply of vesicles delivering cell wall material to the growing tip. Pollen tube elongation is the fastest known cell growth process in the plant kingdom.

Pollen and pollen tubes are excellent single cell experimental model systems for multiple reasons. The haploid genome affords the opportunity to explore gene expression unencumbered by dominance. On the flip side, unless the pollen is generated from a homozygous pure line, the pollen population, even when harvested from a single anther, displays segregation. The resulting variability in cellular behavior is thus an inherent phenomenon that is reflected in the variability of experimental data. Other experimental advantages of the pollen system is its easy storage between harvest and experimentation (dried and frozen), the ability of the pollen grain to successfully germinate and form a functional pollen tube in absence of its usual host, the pistil, and the extreme speed of the pollen tube growth process which ascertains quick responses to environmental stimuli allowing for short experimentation times.

Pollen tubes from many plants are perfectly able to grow in vitro and in the absence of a pistil, but intriguingly, in situ they have to interact with many different cell types. On their way from the stigma to a receptive female gametophyte, the pollen tubes enter in contact with sporophytic tissues (stigmatic papillae, style, transmitting tissue, septum, funiculus, ovule integuments, and nucellus) and with cells of the female gametophyte (synergids). This offers the opportunity for the investigation of a kaleidoscope of cell-cell interactions and signaling processes that guide the pollen tube on its path and probe its compatibility with the female partner at every stage and with different cell types listed above. Methods to assess pollen-stigma interactions and the identification of new players regulating these processes are described in Chapters 2, 3, 4, 8, and 9. The detailed characterization of pollen tube behavior within the pistil requires their visualization when traversing the female tissues (Chapters 2, 5, and 8). All of these methods are crucial for the characterization of mutants (Chapters 3, 5, and 8). In Chapter 11, a dual-staining protocol is presented that allows comparison of pollen/pollen tube behavior of two different genotypes/species in vivo in the same pistil. The characterization of mutants can be complemented by a simple and straightforward assessment of fertility based on seed set (Chapters 7 and 8).

The rapid growth process of the pollen tube allows researchers to efficiently investigate the fundamental underpinnings of polar plant cell growth, such as cell wall assembly, intracellular transport, regulation of polar growth, and the response to external guidance cues. In vitro growth of pollen tubes in experimental platforms facilitates high resolution imaging and micromanipulation. The in vitro cultivation of pollen is simple in some species but requires careful control of the chemical composition of the growth medium as well as environmental conditions in others. Pollen of the model plant *Arabidopsis*, unfortunately, falls in the latter category. Many chapters in this book provide protocols that have been carefully optimized to overcome the challenges associated with in vitro germination of *Arabidopsis* pollen (Chapters 9, 10, 11, 12, 17, 20, and 21). One crucial advancement towards successful in vitro culture of *Arabidopsis* pollen has been the development of semi-in vivo techniques (Chapters 4, 6, and 9). Here, pollen grains are applied to stigmas excised from the pistil. The initial growth through the stigmatic tissue provides pollen tubes with signals and/or nutrients that are required for optimal elongation and ovule targeting, a process called capacitation. In the semi-in vivo setup, the elongating pollen tubes emerge from the cut surface of the excised stigma and can now be observed microscopically and/or exposed to chemical gradients, structured microenvironments, or isolated ovules (Chapter 6). Alternatively, proteins and compounds secreted by the elongating pollen tubes can be collected and analyzed (Chapter 4).

While the semi-in vivo assay is desirable for certain experimental approaches, others require pollen to be devoid of any pistil tissues and must therefore be performed completely in vitro. These include, for example, testing for pollen viability using flow cytometry (Chapter 11), quantifying the germination rate (Chapter 9), or performing the extraction of protein, RNA, or cell wall components (Chapter 10). Pollen or in vitro growing pollen tubes can also be chemically fixed for immunohistochemistry (Chapter 10), whereas the investigation of dynamic cellular processes requires live cell imaging (Chapters 13–21). A straightforward parameter to determine pollen tube performance is the quantification of the pollen tube elongation rate, which can be performed in vivo, semi-in vivo, or in vitro (Chapter 9).

A crucial feature of the pollen tube is its ability to respond to guidance cues by changing its growth direction towards an attractant or away from a repulsing agent. Pollen tube guidance can be investigated with a variety of strategies that include in vivo (Chapter 8) or their exposure to excised ovules in vitro (Chapter 6) or triggering their reorientation through the application of electrical fields (Chapter 13). These experimental approaches can be optimized through the use of in vitro cultivation chambers that allow for easy exchange of solutions or incorporation of microenvironments (Chapters 15, 19, and 20).

In vitro or semi-in vivo growth of pollen tubes is ideal for the detailed analysis of intracellular processes underlying the rapid growth of this cell. Given the small size of the cell and its subcellular structures, live cell image occurs beyond the limit of the resolution power of the optical microscope and the development of sophisticated image analysis methods has been helpful in augmenting the significance of the extracted information (Chapters 14 and 20). Advanced imaging allows, for example, the spatio-temporal detection of intracellular ion gradients including that of cytosolic calcium by way of stable or transient expression of calcium indicators (Chapters 14, 16, and 20) or other FRET (Förster resonance energy transfer)-based nanosensors able to detect specific substrates (Chapter 19). The transformation of pollen grains with genes encoding fluorescent markers through particle bombardment offers a rapid way to express such constructs (Chapters 16 and 22) and is, despite the inherently low transformation efficiency, a viable alternative for the more time-consuming production of stably transformed lines (Chapter 19).

Assembly of the cell wall is the main metabolic activity of the short-lived pollen tube. The characterization of the polysaccharidic composition of the pollen tube cell wall often relies on antibody-based labeling techniques which require fixation, but several live dyes are now available to stain both pectin and cellulose (Chapter 17). Cell wall assembly is fueled by the delivery of new cell wall material to the pollen tube tip, and exocytosis is a crucial process governing the speed and efficiency of cell growth. A new labeling strategy, corrected fluorescence recovery after photoconversion (cFRAPc), allows measuring the exocytosis rate by monitoring both exocytosis-dependent and exocytosis-independent events (Chapter 21). Exocytosis and endocytosis are precisely coordinated in the pollen tube, and the two processes are thought to occur in distinct plasma membrane domains at and near the tip. The subcellular sites of exocytosis and endocytosis are characterized by the accumulation of specific sets of proteins and lipids, which determine the properties of these plasma membrane domains. Phosphatidic acid plays an important role in this context as it has signaling functions, and Chapter 22 describes how proteins associating with this lipid can be identified.

The cytomechanical characteristics of pollen tubes are intriguing since the cell is able to drill its way through the pistil tissues against the obstacle of the apoplast. It does so while following geometrical and chemical guidance cues, and the investigation of these features has propelled the development of sophisticated experimental devices based on microfluidics (Chapter 20) as well as ingenious technical assays to quantify the pollen tube's force generation that endows it with the ability to penetrate obstacles (Chapters 18 and 20). These approaches demonstrate the phenomenal potential of Lab-on-a-Chip (LOC) and Microelectromechanical system (MEMS) technologies in pollen tube research.

Decades of research on the male gametophyte in the flowering plants have resulted in an important body of work straddling fundamentals in cell biology, genetics, biochemistry, signal transduction, and cell mechanics and leading to breeding applications. Our improved understanding of plant evolution, our ability to perform molecular breeding for the adaptation to climate change, the development of novel trait combinations, and sustainable agricultural practices will greatly benefit from continued work in this field. For this, the present book provides the technical know-how and detailed experimental protocols to serve the plant reproduction research community.

Montreal, QC, Canada *Anja Geitmann*

Contents

Chapter 1

Pollen Grain Counting Using a Cell Counter

Hiroyuki Kakui, Misako Yamazaki, Naoto-Benjamin Hamaya, and Kentaro K. Shimizu

Abstract

The number of pollen grains is a critical part of the reproductive strategies in plants and varies greatly between and within species. In agriculture, pollen viability is important for crop breeding. It is a laborious work to count pollen tubes using a counting chamber under a microscope. Here, we present a method of counting the number of pollen grains using a cell counter. In this method, the counting step is shortened to 3 min per flower, which, in our setting, is more than five times faster than the counting chamber method. This technique is applicable to species with a lower and higher number of pollen grains, as it can count particles in a wide range, from 0 to 20,000 particles, in one measurement. The cell counter also estimates the size of the particles together with the number. Because aborted pollen shows abnormal membrane characteristics and/or a distorted or smaller shape, a cell counter can quantify the number of normal and aborted pollen separately. We explain how to count the number of pollen grains and measure pollen size in *Arabidopsis thaliana*, *Arabidopsis kamchatica*, and wheat (*Triticum aestivum*).

Key words Pollen number, Pollen size, Pollen viability, Cell counter (CASY cell counter), *Arabidopsis thaliana*, *Arabidopsis kamchatica*, Wheat (*Triticum aestivum*)

1 Introduction

Pollen grain number has long been studied from agricultural and evolutionary viewpoints [1–3]. For example, the number of pollen grains was shown to have decreased during rice domestication [4]. The number of viable pollen grains is critical for breeding because decreased yield due to pollen abortion can be induced by diverse abiotic stresses, such as low/high temperature, salt stress, and osmotic shock [5, 6]. To control pollen mass is a major challenge to increase the efficiency of hybrid breeding in wheat [7]. In an evolutionary context, shifts from outcrossing to selfing are one of the most frequent evolutionary transitions in angiosperms [8]. Selfing plants tend to show a set of traits termed selfing syndrome, such as smaller flower size, reduced scent and nectar, and a reduced pollen grain number (or low pollen–ovule ratio). The

Anja Geitmann (ed.), *Pollen and Pollen Tube Biology: Methods and Protocols*, Methods in Molecular Biology, vol. 2160, https://doi.org/10.1007/978-1-0716-0672-8_1,

selfing syndrome is regarded as adaptive evolution because selfing plants do not need an excess amount of pollen to assure mating success and, instead of producing pollen, use that resource for other purposes.

Recently, we identified a pollen-grain-number controlling gene in *Arabidopsis thaliana* (*A. thaliana*) by a genome-wide association study. We named this gene *REDUCED POLLEN NUMBER 1* (*RDP1* [9]). We found that *rdp1* null mutants produced about half the number of pollen grains per flower compared with wild type. In our study, we also found that pollen grain number per flower varied greatly between accessions (an average of 2000–8000 pollen grains from 144 accessions) and within samples from the same individual (1000–5000 pollen grains in Col-0). Because of this large variation, many samples (~50 samples per genotype) had to be counted to obtain a reliable estimate of pollen grain number from an accession and a genotype. Traditionally, pollen grain number is counted under a microscope using a counting chamber (hemocytometer) [10–12], but this method is time-consuming and laborious. A more recent approach is based on automated image analysis. Researchers can apply this method to the images obtained using counting chambers or conventional glass slides. Its advantage is that the analytical software is available for free and saves time compared with manual counting [13]. However, this method has difficulties with handling samples that contain aborted pollen and/or debris, because these can adversely affect the identification of the normal pollen. Another drawback of the counting chamber method arises when measuring pollen size. A small difference in microscope handling (i.e., focus, light intensity, position in the optical field, etc.) may lead to a different region being recognized as a pollen and may cause a large difference in size calculation.

Recently, cell counters (e.g., CASY cell counter (OLS OMNI Life Science [14]) or Ampha Z32 (Amphasys [15])) provided convenient and reliable methods for the counting of pollen grain number, the measurement of the size of pollen grains, and examination of pollen viability. Here, we describe a pollen grain counting method that uses a CASY cell counter. Cell counters measure the number and the size of cells based on electrical impulses. For example, the CASY Model TT uses an established method called "Electrical Sensing Zone" paired with "Pulse Area Analysis" [14]. Pollen is suspended in an electrolyte (CASYton) that is aspirated through a precision measuring pore. The electrolyte then moves through the pore, which has a defined electrical resistance, at a constant flow speed. Each pollen grain that passes through the measuring pore creates a change in voltage and is recorded as a cell count. The amplitude of the signal corresponds to the volume of each pollen grain, as it displaces as much electrolyte as its own volume. Moreover, the shape and time course of the electrical impulse are recorded. From that, the cell counter can translate the

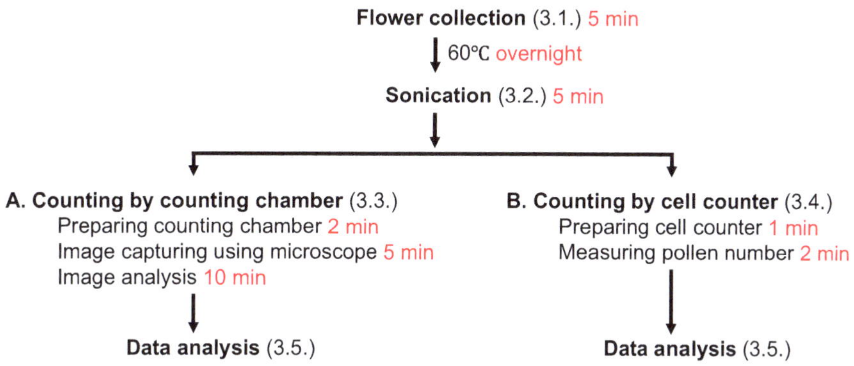

Fig. 1 Overview of pollen grain counting using a counting chamber (**A**) or a cell counter (**B**). The numbers in brackets refer to the subheadings. The numbers in red indicate the estimated time required for the counting of a single sample in our setting

change in voltage into biologically interpretable values, such as cell diameter, because it calculates the integral of the electrical impulse. The sampling rate of the cell counter is 1 MHz, which means that the measuring pore performs one million measurements per second. Furthermore, it is possible to distinguish between viable and dead cells. Given that viable cells are electrical isolators because of their cell membrane, their volume can be determined precisely. Conversely, the cell membrane of dead cells may not be intact, leading to the detection of a reduced signal compared with viable cells [14].

Figure 1 shows a comparative overview of the two counting methods, one (A) using a counting chamber and the other (B) using a cell counter. Both methods follow the same procedures for flower collection (in Subheading 3.1), incubation, and sonication (in Subheading 3.2). The next step in the counting chamber method requires image capturing and image analysis (in Subheading 3.3), which are time-consuming (in total, 17 min per sample in our setting). The cell counter method needs sample preparation before measurement (in Subheading 3.4); including this preparation time, measurement requires in total 3 min per sample, meaning that the cell counter method is more than five times faster than the counting chamber method.

2 Materials

2.1 Preparation of Plant Materials

1. Plants with flower buds (i.e., *A. thaliana, A. kamchatica* or wheat (*Triticum aestivum*); *see* **Note 1**).
2. Fine forceps.
3. 0.5 ml tube.
4. Needle (18 G size).
5. 60 °C incubator.

2.2 Sonication

1. Sonicator (Bioruptor Plus, Diagenode, Belgium).
2. 5% (w/w) Tween-20 (SIGMA, St. Louis, MO, USA).

2.3 Pollen Counting

1. Counting chamber (Neubauer, Paul Marienfeld, Germany).
2. Light microscope.
3. Fiji [16].
4. Cell counter (model: CASY TT, OLS OMNI Life Science, Germany).
5. Isotonic buffer for CASY (CASYton, OLS OMNI Life Science).
6. Measuring vessel for CASY (CASYcups, OLS OMNI Life Science).
7. Software for CASY (CASYworX 1.26, OLS OMNI Life Science).

2.4 Data Analysis

R software [17].

3 Methods

3.1 Flower Collection

1. Choose closed anthers with fully matured pollen. Mature anthers are typically yellow, whereas immature anthers are green. Select *A. thaliana* or *A. kamchatica* flowers from late flower stage 12 to early stage 13 [18, 19]. Flowers from the main stem tend to produce more pollen grains compared with those from the side stems and side branches of the main inflorescence [9]. Because of this characteristic and the limitation of the number of flowers on the apical tip of one branch, take samples from the side stems or side branches of the main inflorescence, to collect more flowers on the same day. Avoid the first and second flowers on each branch because they tend to have an abnormal flower shape and pollen number (for wheat, *see* **Note 2**).
2. Open a flower with fine forceps and place it in a 0.5 ml tube. Close the tube, make a hole in it using a needle, and incubate it overnight at 60 °C, to dry and open the anthers (*see* **Note 3**).

3.2 Sonication

Add 30 μl of 5% Tween-20 and centrifuge at 100 × *g* for 1 min (for wheat, *see* **Note 4**). Place the tubes into the sonicator and start sonication at 4 °C for five cycles of sonication-ON for 30 s and sonication-OFF for 30 s, to release the pollen grains from the anthers (the majority of pollen grains are released into the Tween-20 (Fig. 2; *see* **Note 5**)).

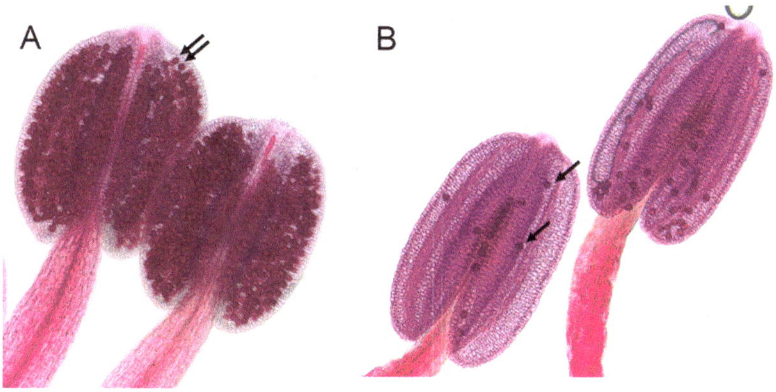

Fig. 2 Anthers before (**A**) and after (**B**) sonication. Only a few pollen grains remained after sonication (**B**). Anthers from *A. thaliana* were stained by Alexander staining [21]. The arrows indicate pollen grains

3.3 Pollen Counting Using a Counting Chamber

1. Apply a cover slip to the counting chamber and add 10 μl of well-mixed sample solution which is prepared in Subheading 3.2. Capture nine images per sample using a light microscope to obtain robust data (Fig. 3; *see* **Note 6**).
2. The number of pollen grains per image is counted using the particle counting implemented in Fiji. The total number of pollen grains per flower is estimated based on the image size and the total volume.

$$\textit{Calculated total pollen number} = \frac{\textit{counted pollen number} \times \textit{sample volume}\ (30\ \mu\text{l})}{\textit{image area}\ (\mu\text{m}^2) \times \textit{height of the chamber}\ (100\ \mu\text{m})}$$

3.4 Pollen Counting Using a Cell Counter (CASY Cell Counter)

1. Set the parameters on the CASY cell counter for *A. thaliana* as described in Table 1 (*see* **Table 2** for *A. kamchatica* and **Table 3** for wheat).
2. Fill the CASYcup with 10 ml of CASYton. Take 0.3 ml of CASYton from the CASYcup and mix the solution in a 0.5 ml tube after sonication. Transfer the entire liquid to the CASYcup (without the flower sample). Put a cap on the CASYcup and mix gently. Measure the sample immediately using the CASY cell counter (*see* **Note 7**).
3. A result screen appears after the measurement. A few examples are displayed in Fig. 4. The measurement data are transferred automatically to CASYworX.

3.5 Data Analysis

Import pollen grain number (or size) data from CASYworX to R, to generate a violin plot (Fig. 5). In *A. thaliana*, the pollen grain number is highly variable, even between flowers taken from the same individual (e.g., around 2000–6000 pollen grains from *A. thaliana*; Fig. 5A). Therefore, to obtain a robust result, the collection of more than 50 samples per genotype is recommended. *A.*

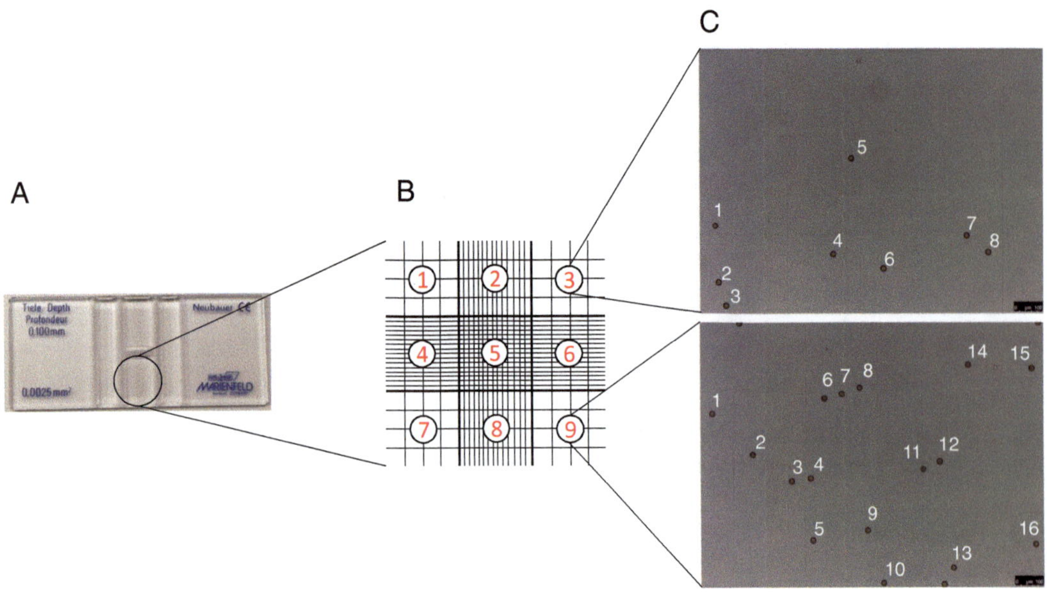

Fig. 3 The principle of the counting chamber. (**A**) Counting chamber. (**B**) Graphical illustration of the counting chamber grid, separated into nine fields. One image is captured for each of the nine fields because of the variation of pollen grain number between fields (**C**, top; $n = 8$ and bottom; $n = 16$). Scale bar: 100μm

Table 1
Parameter setup used for counting the number of *A. thaliana* pollen grains

Capillary	150 μm
Size scale	50 μm
Left norm. cursor	7.50 μm
Right norm. cursor	25.00 μm
Left eval. cursor	12.50 μm
Right eval. cursor	25.00 μm
Sample volume	3 × 400 μl
Dilution	1.00*E*+01
Aggregation corr.	0.000*E*+00 fl

The number of particles detected between the “Left Norm. Cursor” and “Right Norm. Cursor” represents the total amount of pollen (viable and aborted pollen). The number of particles detected between the “Left Eval. Cursor” and “Right Eval. Cursor” represents viable pollen. The dilution factor is used to calculate the original concentration (“Viable cells/ml” in Fig. 5). If the dilution factor is entered as the same value as the volume (ml) used in the CASYcup, the value of the “Viable cells/ml” can be interpreted as the number of normal pollen grains of the sample

kamchatica produced a great quantity of pollen (Figs. 4C and 5A), and wheat produced larger pollen grains (Figs. 4D and 5B) compared with the other two species.

Table 2
Parameter setup used for counting the number of *A. kamchatica* pollen grains

Capillary	150 μm
Size scale	50 μm
Left norm. cursor	7.50 μm
Right norm. cursor	30.00 μm
Left eval. cursor	15.00 μm
Right eval. cursor	30.00 μm
Sample volume	3 × 400 μl
Dilution	1.00*E*+01
Aggregation corr.	0.000*E*+00 fl

Table 3
Parameter setup used for counting the number of wheat pollen grains

Capillary	150 μm
Size scale	80 μm
Left norm. cursor	20.00 μm
Right norm. cursor	80.00 μm
Left eval. cursor	35.00 μm
Right eval. cursor	60.00 μm
Sample volume	3 × 400 μl
Dilution	1.00*E*+01
Aggregation corr.	0.000*E*+00 fl

4 Notes

1. For *A. thaliana*, plants flower typically 10 days after bolting. For wheat, choose mature ears with flowers in some spikelets. To obtain reproducible results, it is important to determine the anther stage.
2. For wheat anther sampling, open spikelets and identify closed (yellow) anthers (stage 13–14 [20]). Hold the ear in your hand and wait until the anther filament elongates (a few minutes). Collect elongated anthers before they release pollen.
3. Flower samples can be stored in the 60 °C incubator for about 1 week.

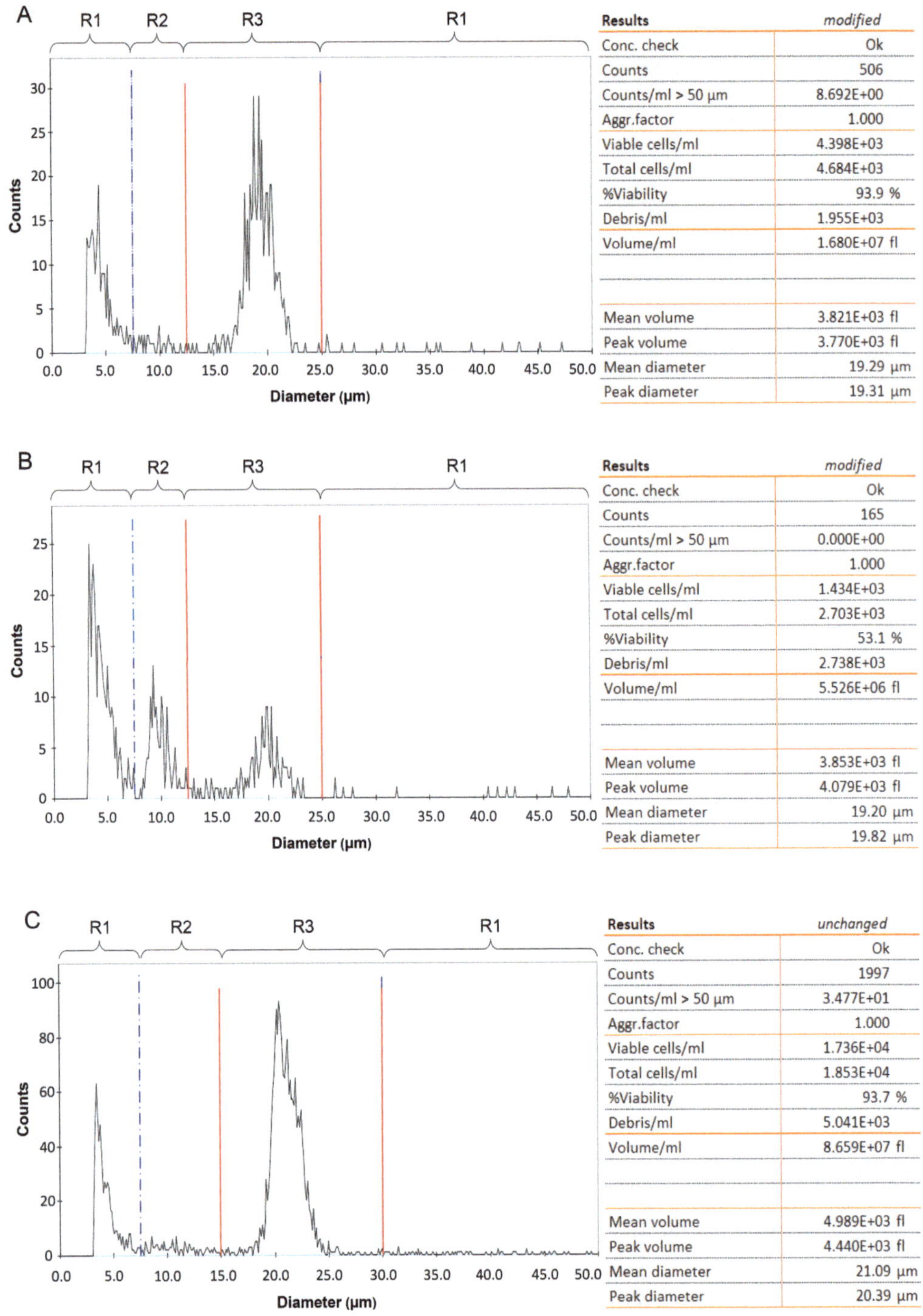

Fig. 4 Result screen of the CASY cell counter. Plots were created and analyses were performed using the CASYworX 1.26 software (modified). The X-axis indicates the particle size and the Y-axis indicates the particle counts. Debris (R1), aborted pollen (R2), and normal pollen (R3) are indicated. "Counts" indicate the measured number of normal pollen grain, "Viable cells/ml" indicate the calculated number of normal pollen grains per sample (calculated by "Counts", dilution factor, and machine bias), "Total cells/ml" indicate the total number of pollen grains (including aborted pollen), and "Mean diameter" indicates the average size of normal pollen. (**A**) Wild-type pollen distribution in *A. thaliana*. There is no peak in the region of aborted pollen (R2). (**B**) Flower with aborted pollen from *A. thaliana*. There is a clear peak in the aborted pollen region (R2). (**C**) Pollen measurement example from *A. kamchatica*, which produced around four times as many pollen grains as did *A. thaliana* (1.736×10^4 pollen grains per flower). (**D**) Pollen measurement in wheat. The size of its pollen is more than twice that of *A. thaliana* (~50 µm in diameter)

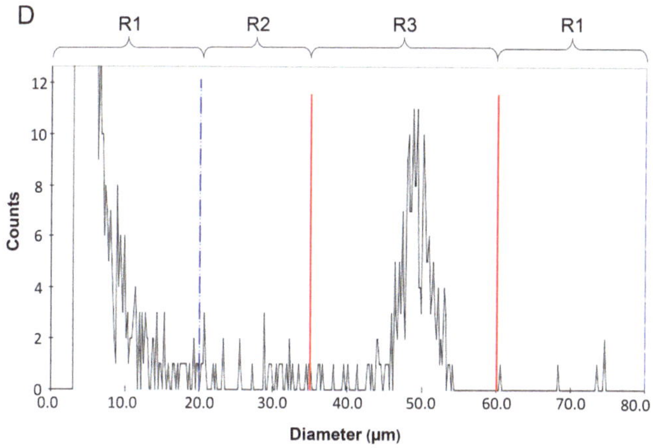

Results	unchanged
Conc. check	Ok
Counts	205
Counts/ml > 80 μm	1.043E+02
Aggr.factor	1.000
Viable cells/ml	1.782E+03
Total cells/ml	2.095E+03
%Viability	85.1 %
Debris/ml	8.640E+03
Volume/ml	1.084E+08 fl
Mean volume	6.084E+04 fl
Peak volume	6.043E+04 fl
Mean diameter	48.61 μm
Peak diameter	48.69 μm

Fig. 4 (continued)

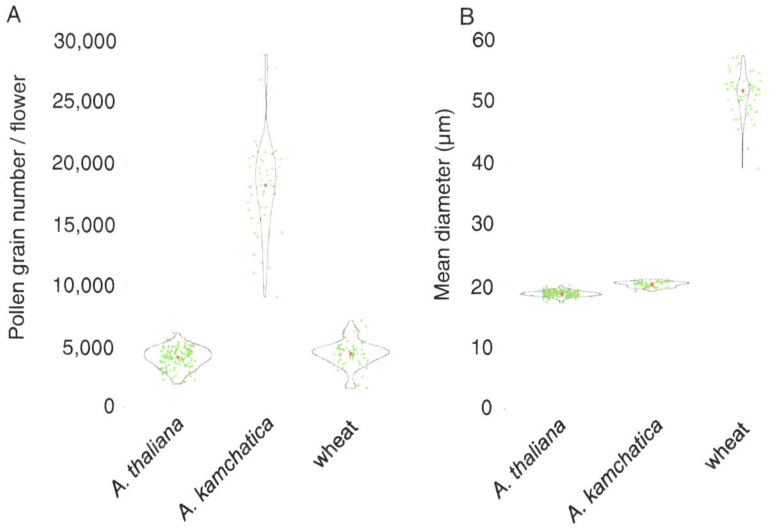

Fig. 5 Violin plot for pollen grain number (**A**) and size (**B**) from *A. thaliana* (Col-0, $n = 122$), *A. kamchatica* (Hukiinoura [22], $n = 51$) and wheat (Arina*LrFor* [23], $n = 60$). Green dots represent samples and red dots indicate medians

4. Add 5% Tween-20 to the sample and cut both tips of the anther using a syringe to help release the pollen after sonication (Fig. 6). To obtain a reliable result, it is essential to confirm the setup to release almost all pollen grains from the anther. The observation of several anthers after sonication using Alexander staining is recommended [21].
5. At the sonication step, use a blank tube if your sample has an empty slot in the sonication holder, to sonicate each sample homogeneously.

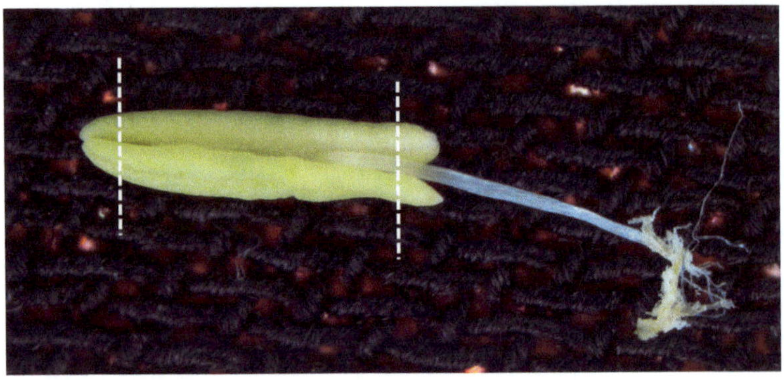

Fig. 6 Wheat anther. The dashed lines represent cutting lines

6. At the image analyzing step, pollen grains at the edge of the image should not be counted (Fig. 3c bottom).
7. At the counting step, if error messages appear, clean the capillary three times using CASYton (blank sample) by pushing the "clean" button.

Acknowledgements

This study was supported by the Swiss National Science Foundation 31003A_182318, a JST CREST Grant Number JPMJCR16O3 and MEXT KAKENHI Grant Number 16H06469 to KKS, and a MEXT KAKENHI Grant Number 19K05976 to H.K. We thank Beat Keller for plant material and Reiko Akiyama and Toshiaki Tameshige for their valuable suggestions.

References

1. Shimizu KK, Tsuchimatsu T (2015) Evolution of selfing: recurrent patterns in molecular adaptation. Annu Rev Ecol Evol Syst 46:593–622
2. Darwin C (1877) The different forms of flowers on plants of the same species. John Murray, London
3. Sicard A, Lenhard M (2011) The selfing syndrome: a model for studying the genetic and evolutionary basis of morphological adaptation in plants. Ann Bot 107:1433–1443
4. Oka H-I, Morishima H (1967) Variations in the breeding systems of a wild rice, *Oryza perennis*. Evolution 21:249–258
5. De Storme N, Geelen D (2014) The impact of environmental stress on male reproductive development in plants: biological processes and molecular mechanisms. Plant Cell Environ 37:1–18
6. Dolferus R, Ji X, Richards RA (2011) Abiotic stress and control of grain number in cereals. Plant Sci 181:331–341
7. Boeven PHG, Longin CFH, Leiser WL et al (2013) Genetic architecture of male floral traits required for hybrid wheat breeding. Theor Appl Genet 129:2343–2357
8. Wright SI, Kalisz S, Slotte T (2013) Evolutionary consequences of self-fertilization in plants. Proc R Soc B 280:20130133
9. Tsuchimatsu T, Kakui H, Yamazaki M et al (2020) Adaptive reduction of male gamete number in the selfing plant Arabidopsis thaliana. Nat Commun (provisionally accepted)

10. Willis JH (1999) The contribution of male-sterility mutations to inbreeding depression in *Mimulus guttatus*. Heredity 83:337–346
11. De Vries AP (1974) Some aspects of cross-pollination in wheat (*Triticum aestivum* L.). 3. Anther length and number of pollen grains per anther. Euphytica 23:11–19
12. Ortega R, Aresti M, Pereira I (2011) Implementation and evaluation of an image analysis system for determining viability of pollen grains in temperate rice. Chil J Agr Res 71:16–22
13. Costa CM, Yang S (2009) Counting pollen grains using readily available, free image processing and analysis software. Ann Bot 104:1005–1010
14. OLS OMNI Life Science (2015) CASY TT cell counter & analyzer; operators guide. https://cellcounting.de/downloads/OLS_CASY_TT-OperatorsGuide_2015.pdf
15. Heidmann I, Di Berardino M (2017) Impedance flow cytometry as a tool to analyze microspore and pollen quality. In: Schmidt A (ed) Plant germline development. Springer, New York, NY
16. Schindelin J, Arganda-Carreras I, Frise E et al (2012) Fiji: an open-source platform for biological-image analysis. Nat Methods 9:676–682
17. R Core Team (2013) R foundation for statistical computing
18. Smyth DR, Bowman JL, Meyerowitz EM (1990) Early flower development in *Arabidopsis*. Plant Cell 2:755–767
19. Yew C-L, Kakui H, Shimizu KK (2017) *Agrobacterium*-mediated floral dip transformation of the model polyploid species *Arabidopsis kamchatica*. J Plant Res 9:1–10
20. Browne RG, Iacuone S, Li SF et al (2018) Anther morphological development and stage determination in *Triticum aestivum*. Front Plant Sci 9:1–13
21. Alexander MP (1969) Differential staining of aborted and nonaborted pollen. Stain Technol 44:117–122
22. Paape T, Briskine RV, Halstead-Nussloch G et al (2018) Patterns of polymorphism and selection in the subgenomes of the allopolyploid *Arabidopsis kamchatica*. Nat Commun 9:3909
23. Singla J, Lüthi L, Wicker T et al (2017) Characterization of *Lr75*: a partial, broad-spectrum leaf rust resistance gene in wheat. Theor Appl Genet 130:1–12

Chapter 2

A Toolkit for Teasing Apart the Early Stages of Pollen–Stigma Interactions in *Arabidopsis thaliana*

Hyun Kyung Lee, Stuart Macgregor, and Daphne R. Goring

Abstract

In hermaphroditic flowering plants, the female pistil serves as the main gatekeeper of mate acceptance as several mechanisms are present to prevent fertilization by unsuitable pollen. The characteristic Brassicaceae dry stigma at the top of pistil represents the first layer that requires pollen recognition to elicit appropriate physiological responses from the pistil. Successful pollen–stigma interactions then lead to pollen hydration, pollen germination, and pollen tube entry into the stigmatic surface. To assess these early stages in detail, our lab has used three experimental procedures to quantitatively and qualitatively characterize the outcome of compatible pollen–stigma interactions that would ultimately lead to the successful fertilization. These assays are also useful for assessing self-incompatible pollinations and mutations that affect these pathways. The model organism, *Arabidopsis thaliana*, offers an excellent platform for these investigations as loss-of-function or gain-of-function mutants can be easily generated using CRISPR/Cas9 technology, existing T-DNA insertion mutant collections, and heterologous expression constructs, respectively. Here, we provide a detailed description of the methods for these inexpensive assays that can be reliably used to assess pollen–stigma interactions and used to identify new players regulating these processes.

Key words Pollen, Pistil, Stigma, Brassicaceae, Self-incompatible, Compatible, Pollen hydration, Pollen adhesion, Pollen tube, Aniline blue stain

1 Introduction

Understanding plant fertility at the molecular level in the context of pollen–pistil interactions has garnered considerable attention due to its direct link to agricultural seed production. Determining the function of genes and proteins involved in this process is critical to fully understand the basic processes involved and how they are related to infertility and seed production. Many laboratory techniques have been developed to characterize pollen tube growth through the pistil. Both in vitro and semi-in vivo approaches have been honed to observe pollen tube behavior and guidance

Hyun Kyung Lee and Stuart Macgregor contributed equally to this work.

Anja Geitmann (ed.), *Pollen and Pollen Tube Biology: Methods and Protocols*, Methods in Molecular Biology, vol. 2160, https://doi.org/10.1007/978-1-0716-0672-8_2, © Springer Science+Business Media, LLC, part of Springer Nature 2020

[1, 2]. Moreover, tagging of actin and microtubule with fluorescent proteins or transiently expressing fluorophores by using particle bombardment has allowed for more dynamic observations in the living cells [3, 4]. Despite these accomplishments, restricting our views in assessing pollen behavior explains only one side of the story as the pistil plays a significant role in plant sexual reproduction.

Many classified angiosperm families, including the Brassicaceae, have a physical characteristic of a dry stigma which serves as the first contact point at top of the pistil for the incoming pollen grains [5–7]. This trait allows for a tight regulation of mate acceptance by electing to provide resources to the desiccated pollen grain for metabolic activation [8–12]. The provision of water and nutrients by the stigmatic papillae to the pollen has been found to be a critical step in facilitating successful fertilization, as pollen germination and viability depend heavily on stigmatic resources [5–7, 13]. Additionally, the pollen tube's ability to locate an ovule is dependent on pollen–pistil interactions, as pollen germinated in vitro lacks the guidance competency compared to the pollen tubes grown through the stigma/style tissues [2]. Together, these steps point to a complex series of pollen–pistil interactions where both work together to ensure reproductive success [5–9].

Following pollen contact and capture, the surface pollen coat and the proteinaceous pellicle of the stigmatic papillae intermix, leading to the formation of a "pollen foot" [14–17]. The pollen foot further strengthens pollen adhesion via stigma papillar glycoproteins and their putative pollen peptide ligands [18–21]. Through the exchange of information between the pollen grain and the stigmatic papilla, the Brassicaceae stigma then accepts or rejects the pollen grain. Upon acceptance, several physiological steps occur chronologically: pollen hydration, pollen germination, and pollen tube entry into the stigma, which would then eventually lead to pollen tube growth to an ovule for fertilization [6, 9, 22]. Recent ultrastructural studies performed on the pollen–papillar interface supports the importance of the stigma, as polarized secretory activities and vacuolar expansion were observed in *Brassica* and *Arabidopsis* stigmatic papillae [14, 23–28]. Delivery of resources to the pollen was identified to be facilitated by the hetero-octameric exocyst complex, conserved within eukaryotes, where its role is to tether vesicles to the plasma membrane for exocytosis [29–31]. Among the exocyst subunits, the EXO70A1 subunit was found to be a target for degradation in the self-incompatibility pathway by the E3 ubiquitin ligase, ARC1, supporting the notion that when stigmatic resources are halted, the pollen is no longer able to hydrate and germinate [30].

In this chapter, we describe three different techniques that can be used to quantitatively and qualitatively assess three different physiological stages following pollinations: pollen hydration, pollen germination, and pollen tube entry into the stigma. The assays are designed to follow these distinct post-pollination stages, occurring from 10 min to 2 h, and offer a semi-in vivo output without the requirement of expensive lab reagents. The pollen hydration assay measures the rate of pollen hydration by recording the pollen grain diameter following stigmatic papilla contact. The pollen adhesion assay measures the number of pollen grains that successfully germinated on the stigmatic surface upon pollination; pollen grains that fail to germinate are removed with the use of an SDS detergent. The aniline blue stain assay is a qualitative assay that allows the observation of the pollen tube growth behavior in the pistil. These three techniques, described in detail below, have been optimized for *Arabidopsis thaliana*, but can be applied to other species with slight modifications. While not described here, the pollen GUS reporter assay using transgenic pollen carrying the *Lat52p:GUS* transgene can also be used as an alternative to the aniline blue stain assay and has been described in detail by Swanson et al. [32]. Here, we have focused on early post-pollination responses of the stigma during pollen–stigma interactions through testing mutant and/or transgenic stigmas with wild-type pollen, however, these assays are equally applicable for the reverse application of mutant pollen from donor plants on wild-type stigmas of recipient plants [33]. Overall, these tools aid in further elucidating the role of the Brassicaceae stigma in regulating the early stages of sexual reproduction and allow the discovery of novel players regulating complex pollen–stigma interactions.

2 Materials

All solutions should be prepared with deionized water of 18 MΩ cm at room temperature.

2.1 Plant Growth Conditions

1. ½ MS plates—½ Murashige and Skoog (MS) media with 1% sucrose and 0.4% (w/v) phytoagar at pH 5.8.
2. Soil and pots or trays.
3. All Purpose Fertilizer 20-20-20.
4. Growth chamber at 22 °C under a 16 h light/8 h dark cycle.
5. Hygrometer.

2.2 Pollen Hydration Assay

1. Plastic wrap.
2. Fine forceps—Splinter forceps, #5 (Long Points, 4–3/8″).
3. ½ MS plates—½ Murashige and Skoog (MS) media with 1% sucrose and 0.4% (w/v) phytoagar at pH 5.8.
4. *A. thaliana* pollen donor and recipient plants grown under conditions described in Subheading 3.1 (*A. thaliana* Col-0 ecotype for the pollen donor).

5. Stereo microscope with digital camera (here: Nikon SMZ800 stereo zoom microscope, DS-Fi1 color camera, and *NIS Element* software).
6. Stage micrometer for calibration—2 × 0.01 mm slide with 2 mm scale in 0.01 mm subdivisions.

2.3 Pollen Adhesion Assay

1. Plastic wrap.
2. Fine forceps—Splinter forceps, #5 (Long Points, 4–3/8″).
3. *A. thaliana* pollen donor and recipient plants grown under conditions described in Subheading 3.1 (*A. thaliana* Col-0 ecotype or Lat52p:GUS SAIL lines for the pollen donor).
4. 0.1% sodium dodecyl sulfate (SDS) solution.
5. Fixative: 3:1 ethanol (100%): glacial acetic acid.
6. Distilled water.
7. Tissue culture plates 24-well flat bottom.
8. Mounting medium: 50% glycerol or distilled water.
9. Microscope slide, 25 × 75 × 1.0 mm.
10. Micro cover glass, No. 1, 22 × 22 mm, 1 oz./pk.
11. Optional: 1 M sodium hydroxide (NaOH) and nail polish to preserve slide samples.
12. Epifluorescence Microscope (here: Zeiss Axioskop2Plus fluorescence microscope, Lumenera Infinity3S color camera and Lumenera Infinity software).

2.4 Aniline Blue Stain Assay

1. Plastic wrap.
2. Fine forceps—Splinter forceps, #5 (Long Points, 4–3/8″).
3. *A.* thaliana pollen donor and recipient plants grown under conditions described in Subheading 3.1 (*A. thaliana* Col-0 ecotype for the pollen donor).
4. Fixative: 3:1 ethanol (100%): glacial acetic acid.
5. Distilled water.
6. Pistil clearing solution: 1 M sodium hydroxide (NaOH).
7. Aniline blue solution: 0.1% aniline blue dissolved in 0.1 M tripotassium phosphate (K_3PO_4, pH ~ 11, *see* **Note 1**).
8. Tissue culture plates 24-well flat bottom.
9. Mounting medium: 50% glycerol, deionized water or Vectashield.
10. Microscope slide, 25 × 75 × 1.0 mm.
11. Micro cover glass, No. 1, 22 × 22 mm, 1 oz./pk.
12. Optional: nail polish to preserve slide samples.

13. Epifluorescence microscope with digital camera (here: Zeiss Axioskop2Plus fluorescence microscope, Lumenera Infinity3S color camera, and Lumenera Infinity software).

3 Methods

3.1 Plant Growth Conditions

1. Grow *A. thaliana* plants under these conditions prior to conducting pollen hydration, pollen adhesion, and aniline blue assays.
2. Seeds should be sterilized and stratified for at least 2 days at 4 °C, then transferred to ½ MS plates for germination at 22 °C under a 16 h light/8 h dark cycle. One-week old seedlings (4-leaf stage) are transferred to soil with an all-purpose fertilizer 20–20–20 (1.0 g/L) and placed in a growth chamber at 22 °C under a 16 h light/8 h dark cycle. After plants begin to bolt, fertilizer is added on a weekly basis for 4 consecutive weeks to promote flowering. The primary inflorescence emerging from the *Arabidopsis* rosette is excised when it reaches ~10 cm in height. This is to increase the number of flowers produced by relieving the secondary inflorescences from apical dominance [34] and to minimize any variability by avoiding potential developmental differences in the primary inflorescence flowers [35].
3. For conducting the post-pollination assays, use hygrometers to check the ambient humidity and keep the levels optimally at ~30–50% in the growth chamber and lab (may require a dehumidifier). High humidity can supersede the regulation of pollen hydration by the stigma and lead to spontaneous pollen hydration [11, 36]. Pollen donor and recipient plants should always be grown together.

3.2 Pollen Hydration Assay

Day 1

1. Emasculate stage 12 flower buds and lightly wrap with a small amount of plastic wrap. (*see* Fig. 1a and **Note 2**).

Day 2

2. The stereo zoom microscope and camera system should be turned on and ready for use. The image capture software should be set to taking high-resolution images. To ensure accurate and reproducible measurements, use the stage micrometer with the 1.5× objective to calibrate if needed (for the *NIS Element* software, the calibration can be saved and reused).
3. Remove a ½ MS plate from the fridge and place under the microscope with a dark background. A black velvet cloth or a

Fig. 1 Pollen hydration assay examples with wild-type *A. thaliana* Col-0 samples. (**a**) An emasculated stage 12 pistil covered in plastic wrap. (**b**) A top-down view of the stigma on a stage 13 pistil. Note that the stigmatic papillae are fully elongated without any noticeable damages. (**c**) An upright pistil place in a ½ MS plate. (**d**) Microscope setup for the pollen hydration assay. (**e–f**) Col-0 pollen applied to a Col-0 stigma at 0 min (**e**) and 10 min (**f**) post-pollination

sheet of paper with the black background can be used (*see* **Note 3**).

4. Carefully remove the plastic wrap from the emasculated flower bud to uncover the now stage 13 pistil (*see* Fig. 1b and **Note 2**).

At this point, it is important to check the ambient humidity using a hygrometer, and the humidity should be less than 60% (*see* Subheading 3.1).

5. Excise the pistil at the pedicel and place it upright in ½ MS plate with the stigmatic surface facing upwards (*see* Fig. 1c).
6. Use this pistil to focus on to the stigmatic surface, starting from the lowest (1.5×) to the highest magnification (6.3×). Adjust the overhead light and exposure time; auto exposure should be turned off for consistency (*see* Fig. 1d).
7. Remove one mature anther from the wild-type Col-0 flower with forceps and lightly apply monolayer of pollen grains onto the stigmatic surface (*see* **Note 4**). Pollinate *horizontally* by lightly brushing the stigma. Take care not to damage the stigma by pushing down on the surface.
8. Immediately take a high-resolution photo of the pollinated pistil for the time point of 0-min (*see* Fig. 1e) and start the timer for 10 min. Save the image file with a clearly identifiable name.
9. Allow the pollen to hydrate for 10-min, then take another high-resolution photo and save the image file (*see* Fig. 1f and **Note 5**).
10. Repeat **steps 8** and **9** for at least three pistils per wildtype and per mutant line.
11. Using the measurement function on a software program, measure the diameter of at least ten random pollen grains per pistil. Always measure laterally across the smallest section of the oval pollen grain (*see* **Note 6**). At least 30 pollen grains (ten pollen/pistil) should be measured for each time point and each plant line. Calculate means for the 0- and 10-min diameter measurements and use a one-way ANOVA with a Tukey post-hoc test to identify significant differences between the wild-type and mutant lines.

3.3 Pollen Adhesion Assay

Day 1

1. Emasculate stage 12 flower buds and lightly wrap with a small amount of plastic wrap (*see* Fig. 1a and **Note 2**).

Day 2

2. Prepare 1.0 ml of 0.1% SDS solution for each pistil that will be pollinated (*see* **Note 7**).
3. Carefully remove the plastic wrap from the emasculated flower bud to uncover the now stage 13 pistil (*see* Fig. 1b and **Note 2**).

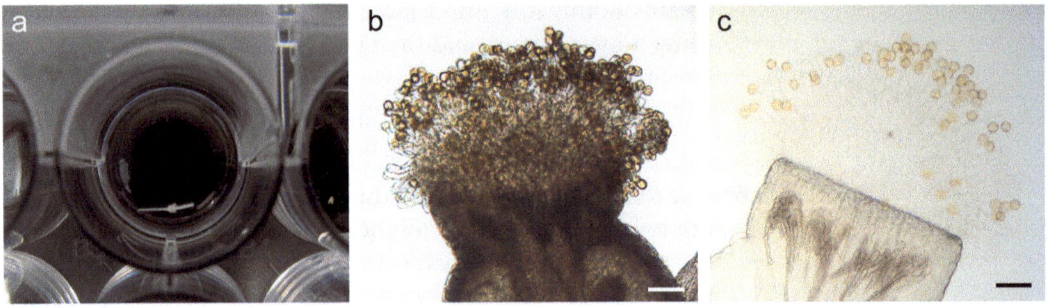

Fig. 2 Pollen adhesion assay examples with wild-type *A. thaliana* Col-0 samples. (**a**) A pollinated pistil with its pedicel has been placed in a well of a 24-well plate containing the 3:1 100% ethanol:glacial acetic acid solution. (**b**, **c**) Representative brightfield images of pollinated Col-0 pistils that were prepared without the 1 M NaOH treatment (**b**) or with the 1 M NaOH treatment (**c**). Scale bar = 100 μM

4. Remove one mature anther from the wild-type Col-0 flower with forceps and lightly apply a monolayer of pollen grains onto the stigmatic surface (*see* **Note 4**).
5. Wait 30 min, then cut the pistil at the pedicel and insert it into the Eppendorf tube with 0.1% SDS solution and vortex for 10 s (*see* **Note 8**). Incubate in 0.1% SDS for 5 min and vortex again for 10 s.
6. Transfer each pistil into an individual well of a 24-well plate containing 300 μl of a 3:1 ethanol (100%): glacial acetic acid solution per well. Fix the pistil for at least 30 min at room temperature (*see* Fig. 2a). *See* **Note 9** for an optional step of counting pollen grains that have washed off [33].
7. Additional pistils can be pollinated during the 30 min in **step 4**, but carefully track the 30-min timepoint for each pollinated pistil as well as the 0.1% SDS wash steps. The pollinated pistils can remain in the 3:1 ethanol (100%): glacial acetic acid solution until all the samples have been processed and the last pistil sample has been fixed for 30 min.
8. Remove the fixative and rinse each pistil three times for 1 min each with distilled water (*see* **Note 10**).
9. Add 500 μl of 1 M NaOH and incubate pistils for 45 min in a 60 °C chamber to clear the pistil tissues (*see* **Note 11**). Remove the NaOH solution and rinse each pistil three times for 1 min each with distilled water (*see* **Note 10**). While this step is optional, it does improve the quality of the pistil images captured in the next step (*see* Fig. 2b, c).
10. Use either 50% glycerol or distilled water to mount the pistil samples on a slide (*see* **Note 12**). Image the slide at 10× magnification under the brightfield setting on the fluorescence microscope.

11. Count the number of pollen grains (*see* Fig. 2c and **Note 13**), calculate the means (at least 10 pistils/line), and use a one-way ANOVA with a Tukey post-hoc test to identify significant differences between the wild-type and mutant lines.
12. To seal the pistil samples for storage, apply nail polish around all the edges of the coverslip on the slide.

3.4 Aniline Blue Stain Assay

Day 1

1. Emasculate stage 12 flower buds and lightly wrap with a small amount of plastic wrap (*see* Fig. 1a and **Note 2**).

Day 2

2. Carefully remove the plastic wrap from the emasculated flower bud to uncover the now stage 13 pistil (*see* Fig. 1b and **Note 2**).
3. Remove one or two mature anthers from the wild-type Col-0 flower with forceps and lightly apply a monolayer of pollen grains onto the stigmatic surface (*see* **Note 4**). Mark each pollinated pistil with tape or marker and wait 2 h (*see* **Notes 2** and **14**).
4. Remove each pistil, including a small section of pedicel for easier handling, and place into an individual well of a 24-well plate containing 300 μl of a 3:1 ethanol (100%): glacial acetic acid solution per well. Fix the pistil for at least 30 min in room temperature (*see* Fig. 2a).
5. Additional pistils can be pollinated during 2 h in **step 2**, but carefully track the 2-h timepoint for each pollinated pistil. The pollinated pistils can remain in the 3:1 ethanol (100%): glacial acetic acid solution until all the samples have been processed and the last pistil sample has been fixed for 30 min (*see* **Note 15**).
6. Remove the fixative and rinse each pistil three times for 1 min each with distilled water (*see* **Note 10**).
7. Add 500 μl of 1 M NaOH and incubate pistils for 45 min in a 60 °C chamber to clear the pistil tissues (*see* **Note 11**). Remove the NaOH solution and rinse each pistil three times for 1 min each with distilled water (*see* **Note 10**).
8. Add 500 μl of the 0.1% aniline blue solution to each well (*see* **Note 1**). Cover the samples with aluminum foil and incubate at room temperature for 30 min or overnight at 4 °C.
9. Prepare slides by placing a drop of distilled water in the center of the slide (*see* **Note 16**).
10. Remove a pistil from the aniline blue solution by grasping the pedicel with forceps, and gently release it into the water drop on the slide. Another pair or forceps may be used to push the pistil into the drop. Place a cover slip over the drop of water.

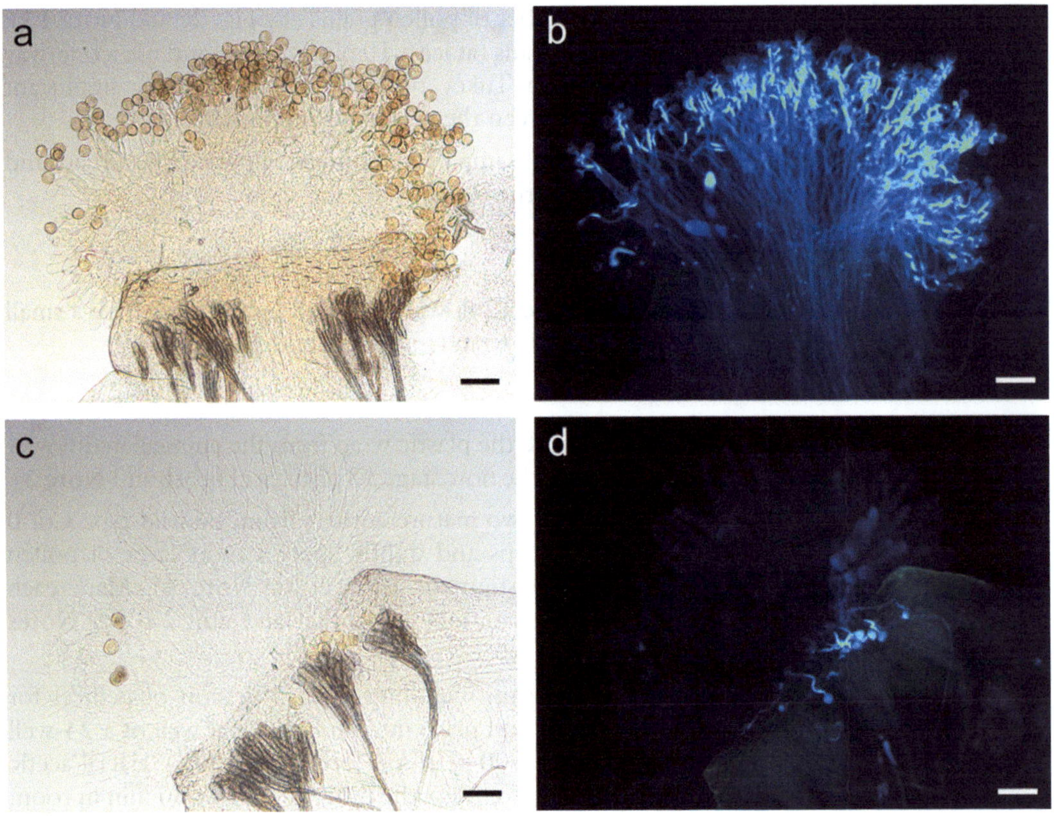

Fig. 3 Aniline blue stain assay for compatible and self-incompatible *A. thaliana* pollinations. (**a–b**) A self-compatible pollination example using wild-type *A. thaliana* Col-0 pistils and self-pollinated for 2 h. Abundant adhered pollen grains are seen under brightfield (**a**) and abundant pollen tubes growing through the stigma are seen under UV fluorescence (**b**). (**c**, **d**) A self-incompatible pollination example using self-incompatible pistils from an *A. thaliana* Col-0 *SCR-SRK-ARC1* transgenic line [26] and self-pollinated for 2 h. Note that very few adhered pollen grains are seen under brightfield (**c**) and resulting in very few pollen tubes in the UV fluorescence images (**d**). Scale bar = 100 μM

11. To seal the pistil samples for storage, apply nail polish around all the edges of the coverslip on the slide.
12. Image the slide at 10× magnification under both brightfield and UV fluorescence settings on the fluorescence microscope. Pollen grains should be visible under brightfield while pollen tube growth through the stigmatic tissues should be visible under UV fluorescence (*see* Fig. 3). At this point, any mistakes that transpired during the protocol (such as improper flower staging or damage to the stigma) should be evident in the brightfield and UV fluorescence images. For some common issues that may come up and solutions to rectify them, *see* Fig. 4.

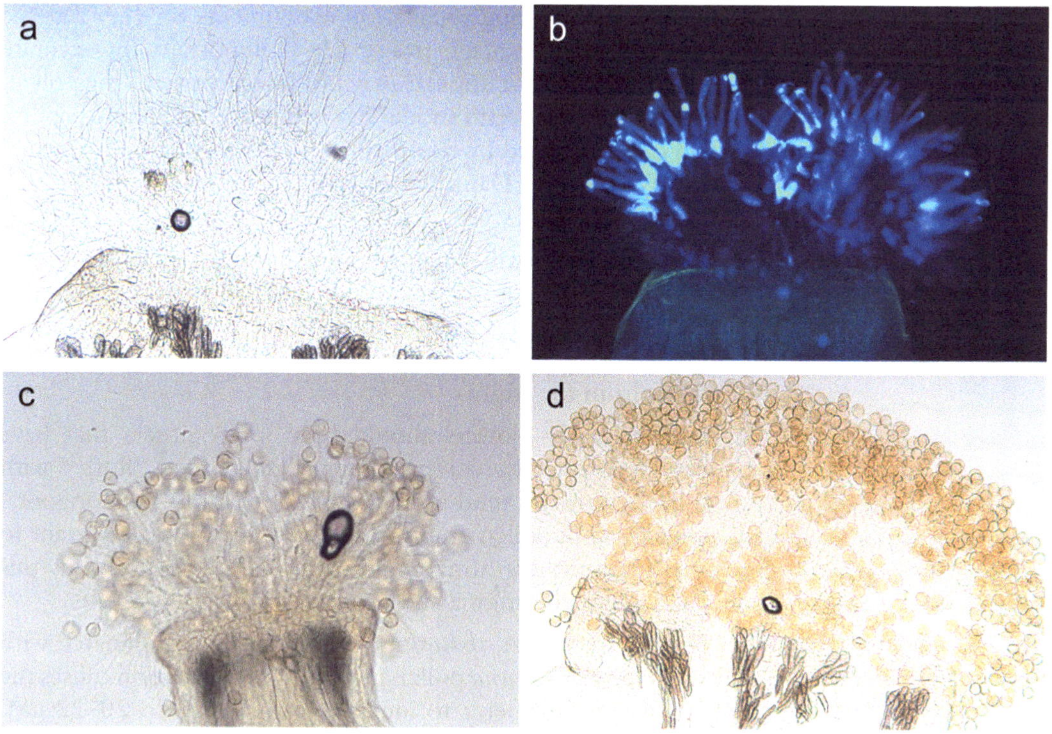

Fig. 4 Troubleshooting guide for the aniline blue stain assay. (**a**) Brightfield image of stigmatic tissue destroyed during mounting. Solution: Either reduce the 1 M NaOH incubation time or reduce the NaOH concentration. (**b**) UV fluorescence image of a stigma that was wounded during emasculation and/or pollination tissue. Wounding leads to bright fluorescence of callose deposits in the stigmatic papillae. Solution: Pollinate and wrap more gently. (**c**) Brightfield image of stigmatic tissue that is not on a single focal plane and as a result pollen tubes will be harder to visualize under UV fluorescence. Solution: Apply more gentle pressure to the coverslip to flatten the stigmatic tissue. (**d**) Brightfield image of an over-pollinated stigma which may mask more subtle post-pollination phenotypes. Solution: Reduce the number of anthers used and gently pollinate

4 Notes

1. When making the 0.1% aniline blue solution [37, 38], leave overnight in the dark at 4 °C to decolorize before use. This process changes the solution's color from blue to yellow-brown. The solution should be stored at 4 °C in the dark, wrapped in aluminum foil.
2. Stage 12 is the optimal stage for emasculation of flower buds on *Day 1*. At this stage, the petals on the flower buds are white, and are at an equal length to sepals and the anthers are yellow, but have not yet dehisced [39, 40]. A stage 12 closed flower bud is typically adjacent to a freshly opened flower. To keep the plastic wrap from falling off the emasculated pistil, a small strip of tape should be used to anchor the plastic wrap. The tape should be applied lightly without damaging the pistil and

pedicel. Conducting pollination experiments on *Day 2* at the correct flowering stage (stage 13) is important for accurate results as immature stigmas may not be receptive to the pollen. The emasculated pistil should have elongated stigmatic papillae and can be verified by comparing to a pistil in a freshly opened wild-type flower. Damaged or crushed stigmas should not be used for any assays.

3. Since the pistil will be positioned upright, backlight is not necessary and only the overhead lights should be used. The black background is used to increase the contrast between the pistil and the background, which helps to distinguish pollen grains from the stigma.
4. Early stage 14 flowers should have long anthers that have extended above the stigma. These anthers look "fluffy" with pollen grains and tend to give the best result in the horizontal application of pollen grains onto the stigma. It is important to brush the anthers horizontally, not tapping vertically, as this will damage the stigma and produce inconsistent results.
5. When wild-type *A. thaliana* Col-0 stigmas are pollinated with wild-type *A. thaliana* pollen grains, pollen hydration causes the pollen grain diameter to increase from 10–12 to 20–22 μM, but this may vary depending on the microscope calibration. At 10 min, hydrated pollen grains will appear translucent and rounder in shape (changing from an oval shape).
6. Avoid measuring in a cluster of pollen grains that are stacked on top of each other; only measure the isolated pollen grains that are clearly in contact with the stigmatic papillae. In the *NIS Element* software, pollen diameter measurements were acquired through the semi-axis measurement option. This measurement option generates a circle around the pollen grain periphery, and five marked points around the pollen grain are used to identify and measure the radius. The measurements are then exported to an Excel file for further analysis (radius values are doubled to give the diameter values and then used to calculate means).
7. The 0.1% SDS treatment removes pollen grains that fail to germinate. For pollination experiments using Lat52p:GUS pollen, SDS should be avoided as it reduces GUS activity in the pollen. As an alternative treatment for GUS activity, 0.5% Triton-X can be used, though it is not as stringent in removing ungerminated pollen grains [17, 32].
8. Different time points can be tested (10 and 20 min) depending on the context of the mutantpistils being assessed. For *A. thaliana* pistils, the rate of pollen germination reaches a plateau at 30 min [32].

9. It is optional to keep the Eppendorf tubes used to expose pistils with 0.1% SDS. The "washed-off" pollen can be spun down at 13,000 × *g* for 10 min and can be fixed for counting purposes. During fixation, apply aceto-orcein stain to create a final concentration of 2% and expose the pollen overnight at room temperature. Spin for 13,000 × *g* for 10 min and resuspend the pollen in 20 μl of 50% glycerol. The resuspended pollen can be mounted on a slide. Once the total number of pollen grains applied is measured, the % of germinated pollen grains can be calculated [33].
10. Pistils can be damaged at this step due to the vigorous pipettor usage. Careful to not aspirate the pistils, placing the 24-well plate on a darker background will allow for greater visibility of the pistils.
11. A 45-min incubation is optimal for *A. thaliana* pistils, while longer incubation times should be used for *Brassica* sp. pistils. Pistils are transparent in color and very delicate after NaOH incubation. Rinsing should be done as gently as possible to maintain pistil integrity. Again, use a darker background to help to increase the visibility of the pistils.
12. To better visualize the pistil and the pollen grains adhered, use the coverslip to slightly crush the pistil tissue. Do not apply a direct force onto the tissue but apply a light force on the edges of coverslip. If NaOH was used to clear the pistil tissue, it should be softened enough and does not require this step.
13. Figure 2c shows a representative example of a Col-0 pistil that has been pollinated with a single Col-0 anther where the pollen grains are easily seen for counting. One can also use 2–3 anthers/stigma to increase the number of pollen grains that are being applied, but it is important to be consistent in the number of anthers applied per sample.
14. The 2-h post-pollination timepoint is sufficient for the Col-0 pollen tube to reach the transmitting tract of the Col-0 pistil (*see* Fig. 3). Earlier and/or later time points can be tested depending on the phenotypes presented in the mutant pistil. To observe pollen tube growth down to the base of the transmitting tract, pollinated samples should be left for at least 12 h. For viewing naturally self-pollinated flowers, simply remove anthers and petals from a stage 14 flower and start with **step 3** in the aniline blue stain assay.
15. The pollinated pistils can be left in the 3:1 ethanol (100%): glacial acetic acid solution for several days given that this fixative does not dry out. Wrap the 24 well plates with parafilm to prevent evaporation and store plates at 4 °C.

16. Fifty percentage glycerol or Vectashield can be used on the slide to mount the pollinated pistil instead of distilled water. The 50% glycerol minimizes pistil movement whereas Vectashield can be used to prevent photobleaching in case of extended imaging.

Acknowledgement

We thank members of the Goring lab for critically reading this manuscript. H.K.L. was supported by an Ontario Graduate Scholarship, and research in DRG's laboratory is supported by a Discovery Grant from the Natural Sciences and Engineering Research Council of Canada.

References

1. Boavida LC, McCormick S (2007) Temperature as a determinant factor for increased and reproducible in vitro pollen germination in *Arabidopsis thaliana*. Plant J 52(3):570–582. https://doi.org/10.1111/j.1365-313X.2007.03248.x
2. Palanivelu R, Preuss D (2006) Distinct short-range ovule signals attract or repel *Arabidopsis thaliana* pollen tubes in vitro. BMC Plant Biol 6:7. https://doi.org/10.1186/1471-2229-6-7
3. Cheung AY, Duan QH, Costa SS, De Graaf BHJ, Di Stilio VS, Feijo J, Wu HM (2008) The dynamic pollen tube cytoskeleton: live cell studies using actin-binding and microtubule-binding reporter proteins. Mol Plant 1(4):686–702. https://doi.org/10.1093/mp/ssn026
4. Wang H, Jiang L (2011) Transient expression and analysis of fluorescent reporter proteins in plant pollen tubes. Nat Protoc 6(4):419–426. https://doi.org/10.1038/nprot.2011.309
5. Chapman LA, Goring DR (2010) Pollen-pistil interactions regulating successful fertilization in the Brassicaceae. J Exp Bot 61 (7):1987–1999. https://doi.org/10.1093/jxb/erq021
6. Zheng Y-Y, Lin X-J, Liang H-M, Wang F-F, Chen L-Y (2018) The long journey of pollen tube in the pistil. Int J Mol Sci 19(11):pii: E3529. https://doi.org/10.3390/ijms19113529
7. Mizuta Y, Higashiyama T (2018) Chemical signaling for pollen tube guidance at a glance. J Cell Sci 131(2):pii: jcs208447. https://doi.org/10.1242/jcs.208447
8. Dickinson H (1995) Dry stigmas, water and self-incompatibility in *Brassica*. Sex Plant Reprod 8(1):1–10. https://doi.org/10.1007/BF00228756
9. Doucet J, Lee HK, Goring DR (2016) Pollen acceptance or rejection: a tale of two pathways. Trends Plant Sci 21(12):1058–1067. https://doi.org/10.1016/j.tplants.2016.09.004
10. Hiroi K, Sone M, Sakazono S, Osaka M, Masuko-Suzuki H, Matsuda T, Suzuki G, Suwabe K, Watanabe M (2013) Time-lapse imaging of self-and cross-pollinations in *Brassica rapa*. Ann Bot 112(1):115–122. https://doi.org/10.1093/aob/mct102
11. Ma J-F, Liu Z-H, Chu C-P, Hu Z-Y, Wang X-L, Zhang XS (2012) Different regulatory processes control pollen hydration and germination in *Arabidopsis*. Sex Plant Reprod 25 (1):77–82. https://doi.org/10.1007/s00497-011-0173-0
12. Zuberi MI, Dickinson HG (1985) Pollen-stigma interaction in *Brassica*. III Hydration of the pollen grains. J Cell Sci 76:321–336
13. Jany E, Nelles H, Goring DR (2019) The molecular and cellular regulation of Brassicaceae self-incompatibility and self-pollen rejection. Int Rev Cell Mol Biol 343:1–35. https://doi.org/10.1016/bs.ircmb.2018.05.011
14. Elleman CJ, Dickinson HG (1986) Pollen-stigma interactions in *Brassica*. IV Structural reorganization in the pollen grains during hydration. J Cell Sci 80:141–157
15. Elleman CJ, Dickinson HG (1990) The role of the exine coating in pollen–stigma interactions in *Brassica oleracea* L. New Phytol 114 (3):511–518. https://doi.org/10.1111/j.1469-8137.1990.tb00419.x
16. Preuss D, Lemieux B, Yen G, Davis RW (1993) A conditional sterile mutation eliminates surface components from *Arabidopsis* pollen and

disrupts cell signaling during fertilization. Genes Dev 7(6):974–985. https://doi.org/10.1101/gad.7.6.974

17. Zinkl GM, Zwiebel BI, Grier DG, Preuss D (1999) Pollen-stigma adhesion in *Arabidopsis*: a species-specific interaction mediated by lipophilic molecules in the pollen exine. Development 126(23):5431–5440
18. Doughty J, Dixon S, Hiscock SJ, Willis AC, Parkin IA, Dickinson HG (1998) PCP-A1, a defensin-like *Brassica* pollen coat protein that binds the S locus glycoprotein, is the product of gametophytic gene expression. Plant Cell 10 (8):1333–1347. https://doi.org/10.1105/tpc.10.8.1333
19. Luu DT, Heizmann P, Dumas C, Trick M, Cappadocia M (1997) Involvement of SLR1 genes in pollen adhesion to the stigmatic surface in Brassicaceae. Sex Plant Reprod 10 (4):227–235. https://doi.org/10.1007/s004970050091
20. Luu DT, Marty-Mazars D, Trick M, Dumas C, Heizmann P (1999) Pollen-stigma adhesion in *Brassica* spp involves SLG and SLR1 glycoproteins. Plant Cell 11(2):251–262. https://doi.org/10.1105/tpc.11.2.251
21. Takayama S, Shiba H, Iwano M, Asano K, Hara M, Che FS, Watanabe M, Hinata K, Isogai A (2000) Isolation and characterization of pollen coat proteins of *Brassica campestris* that interact with S locus-related glycoprotein 1 involved in pollen-stigma adhesion. Proc Natl Acad Sci U S A 97(7):3765–3770. https://doi.org/10.1073/pnas.040580797
22. Goring DR (2017) Exocyst, exosomes, and autophagy in the regulation of Brassicaceae pollen-stigma interactions. J Exp Bot 69 (1):69–78. https://doi.org/10.1093/jxb/erx340
23. Dickinson H, Elleman C, Doughty J (2000) Pollen coatings – chimaeric genetics and new functions. Sex Plant Reprod 12(5):302–309. https://doi.org/10.1007/s004970050199
24. Elleman CJ, Dickinson HG (1994) Pollen-stigma interaction during sporophytic self-incompatibility in *Brassica oleracea*. In: Williams EG, Clarke AE, Knox RB (eds) Genetic control of self-incompatibility and reproductive development in flowering plants, vol 1. Advances in cellular and molecular biology of plants. Springer, Dordrecht
25. Elleman CJ, Dickinson HG (1996) Identification of pollen components regulating pollination-specific responses in the stigmatic papillae of *Brassica oleracea*. New Phytol 133 (2):197–205. https://doi.org/10.1111/j.1469-8137.1996.tb01886.x
26. Indriolo E, Safavian D, Goring DR (2014) The ARC1 E3 ligase promotes two different self-pollen avoidance traits in *Arabidopsis*. Plant Cell 26(4):1525–1543. https://doi.org/10.1105/tpc.114.122879
27. Iwano M, Shiba H, Matoba K, Miwa T, Funato M, Entani T, Nakayama P, Shimosato H, Takaoka A, Isogai A, Takayama S (2007) Actin dynamics in papilla cells of *Brassica rapa* during self- and cross-pollination. Plant Physiol 144(1):72–81. https://doi.org/10.1104/pp.106.095273
28. Safavian D, Goring DR (2013) Secretory activity is rapidly induced in stigmatic papillae by compatible pollen, but inhibited for self-incompatible pollen in the Brassicaceae. PLoS One 8(12):e84286. https://doi.org/10.1371/journal.pone.0084286
29. Safavian D, Zayed Y, Indriolo E, Chapman L, Ahmed A, Goring D (2015) RNA silencing of exocyst genes in the stigma impairs the acceptance of compatible pollen in *Arabidopsis*. Plant Physiol 169(4):2526–2538. https://doi.org/10.1104/pp.15.00635
30. Samuel MA, Chong YT, Haasen KE, Aldea-Brydges MG, Stone SL, Goring DR (2009) Cellular pathways regulating responses to compatible and self-incompatible pollen in *Brassica* and *Arabidopsis* stigmas intersect at Exo70A1, a putative component of the exocyst complex. Plant Cell 21(9):2655–2671. https://doi.org/10.1105/tpc.109.069740
31. Žárský V, Kulich I, Fendrych M, Pečenková T (2013) Exocyst complexes multiple functions in plant cells secretory pathways. Curr Opin Plant Biol 16(6):726–733. https://doi.org/10.1016/j.pbi.2013.10.013
32. Swanson RJ, Hammond AT, Carlson AL, Gong H, Donovan TK (2016) Pollen performance traits reveal prezygotic nonrandom mating and interference competition in *Arabidopsis thaliana*. Am J Bot 103(3):498–513. https://doi.org/10.3732/ajb.1500172
33. Wang LD, Clarke LA, Eason RJ, Parker CC, Qi BX, Scott RJ, Doughty J (2017) PCP-B class pollen coat proteins are key regulators of the hydration checkpoint in *Arabidopsis thaliana* pollen-stigma interactions. New Phytol 213 (2):764–777. https://doi.org/10.1111/nph.14162
34. Clough SJ, Bent AF (1998) Floral dip: a simplified method for *Agrobacterium*-mediated transformation *of Arabidopsis thaliana*. Plant J 16(6):735–743. https://doi.org/10.1046/j.1365-313X.1998.00343.x
35. Plackett ARG, Powers SJ, Phillips AL, Wilson ZA, Hedden P, Thomas SG (2018) The early

inflorescence of *Arabidopsis thaliana* demonstrates positional effects in floral organ growth and meristem patterning. Plant Reprod 31 (2):171–191. https://doi.org/10.1007/s00497-017-0320-3

36. Safavian D, Jamshed M, Sankaranarayanan S, Indriolo E, Samuel MA, Goring DR (2014) High humidity partially rescues the *Arabidopsis thaliana exo70A1* stigmatic defect for accepting compatible pollen. Plant Reprod 27 (3):121–127. https://doi.org/10.1007/s00497-014-0245-z
37. Kho YO, Baer J (1968) Observing pollen tubes by means of fluorescence. Euphytica 17 (2):298–302
38. Mori T, Kuroiwa H, Higashiyama T, Kuroiwa T (2006) GENERATIVE CELL SPECIFIC 1 is essential for angiosperm fertilization. Nat Cell Biol 8(1):64–71. https://doi.org/10.1038/ncb1345
39. Christensen CA, King EJ, Jordan JR, Drews GN (1997) Megagametogenesis in *Arabidopsis* wild type and the *Gf* mutant. Sex Plant Reprod 10(1):49–64. https://doi.org/10.1007/s004970050067
40. Smyth DR, Bowman JL, Meyerowitz EM (2007) Early flower development in *Arabidopsis*. Plant Cell 2:755–767. https://doi.org/10.2307/3869174

Chapter 3

Decapitation Crosses to Test Pollen Fertility Mutations for Defects in Stigma-Style Penetration

Chrystle Weigand and Jeffrey Harper

Abstract

During sexual reproduction in flowering plants, pollen grains germinate on the stigma surface and grow through the stigma-style tissue to reach the ovary and deliver sperm cells for fertilization. Here, we outline a method to test whether a pollen fertility mutation specifically disrupts pollen penetration through the stigma-style barrier. This method surgically removes the stigma-style (stigma decapitation) to test whether transferring pollen directly onto an exposed ovary surface significantly improves the transmission efficiency (TE) of a mutant allele. To illustrate this technique, we applied stigma decapitation to investigate a loss-of-function mutation in *Arabidopsis* OFT1, a gene encoding a putative o-fucosyl transferase functioning in the secretory pathway. *oft1-3* mutant pollen showed a significant decrease in transmission efficiency compared to wild type. Decapitation crosses (described here) indicated that the removal of the stigma-style barrier alleviated the transmission deficiency from 858-fold to a 2.6-fold, providing evidence that most, but not all, *oft1* pollen deficiencies can be attributed to a reduced ability to penetrate through the stigma-style barrier. This method outlines a genetic strategy to quantify a mutation's impact on the ability of pollen to navigate through the stigma-style barrier on its journey to the ovule.

Key words Stigma-style decapitation, Pollen, Male sterile, Penetration, Pollen transmission, Pollen tube growth, Semi in vivo

1 Introduction

Flowering plants evolved to use pollen tubes to deliver sperm cells to ovules for fertilization. Given that nearly half of the *Arabidopsis* genome is expressed in pollen, there is an expectation for a large number of mutations impacting male fertility [1, 2]. Pollen fertilization is a dynamic process that requires rapid cell growth and communication between pollen and the female pistil. In their journey from the stigma to the ovule, pollen tubes engage in compatible interactions with at least seven different cell or tissue types

Electronic supplementary material: The online version of this chapter (https://doi.org/10.1007/978-1-0716-0672-8_3) contains supplementary material, which is available to authorized users.

Anja Geitmann (ed.), *Pollen and Pollen Tube Biology: Methods and Protocols*, Methods in Molecular Biology, vol. 2160, https://doi.org/10.1007/978-1-0716-0672-8_3,

[1]. In studying pollen fertility mutations, an important challenge is determining the biological and biochemical point(s) of defect through which a mutation disrupts successful fertilization.

When diagnosing a pollen defect, an important question is whether a mutation results in a segregation distortion because a mutant pollen is less competitive than wild type in a race to fertilize an ovule. This question of competitive fitness can be addressed by comparing the transmission efficiencies of a hemizygous mutation through two types of pollen outcrosses, competitive and non-competitive (*see* Fig. 1). In a competitive cross, a manual pollination is conducted with a large number of pollen competing for a fixed number of ovules. In a non-competitive cross, transferring only a few pollen grains ensures that every pollen has an opportunity to find and fertilize an ovule. If a pollen transmission defect is severe under competitive cross conditions, yet displays a normal frequency without competition, this suggests that mutant pollen have either a growth rate deficiency or a reduced ability to negotiate the physical barriers and signaling cues required during the journey from the stigma to the ovule.

The stigma-style tissue represents a significant reproductive barrier. Pollen tubes must penetrate through this tissue to reach the ovary. This penetration requires the secretion of hydrolases to help digest a passage way, while still maintaining the pollen's own cell wall integrity [1]. Additionally, pollen penetration requires compatible interactions with the host pistil to avoid being mistaken as a pathogen. Thus, there are numerous cellular processes that are important during this phase of a pollen's journey, including the ability to regulate gene expression, secrete hydrolases and cell wall materials, operate cell growth machinery, sense and respond to guidance clues, and engage in compatible cell–cell interactions.

Here, we present a method to remove the stigma-style barrier (decapitation) to quantitatively assess whether a mutation has a specific impact on pollen growth through the stigma-style barrier. This method involves comparing a mutation's transmission efficiency using three different pollen outcross strategies: (1) competitive, (2) non-competitive, and (3) stigma decapitation. This method provides a complement to Chapter 9 that outlines how to use a semi-in vivo technique to evaluate the growth rate of mutant pollen tubes through an isolated stigma-style that was removed from the pistil and placed on the surface of an agar slide. We suggest using both methods to corroborate a pollen-penetration defect.

Using both semi-in vivo and decapitation methods, a loss-of-function mutation for *oft1*, a gene encoding a putative o-fucosyl transferase in *Arabidopsis*, was identified as being required for pollen tube penetration through the stigma-style barrier [3]. Competitive male outcrosses (ovules limited) provided evidence that an *oft1* loss-of-function mutation resulted in a male fertility defect with a more than 858-fold decrease in pollen transmission

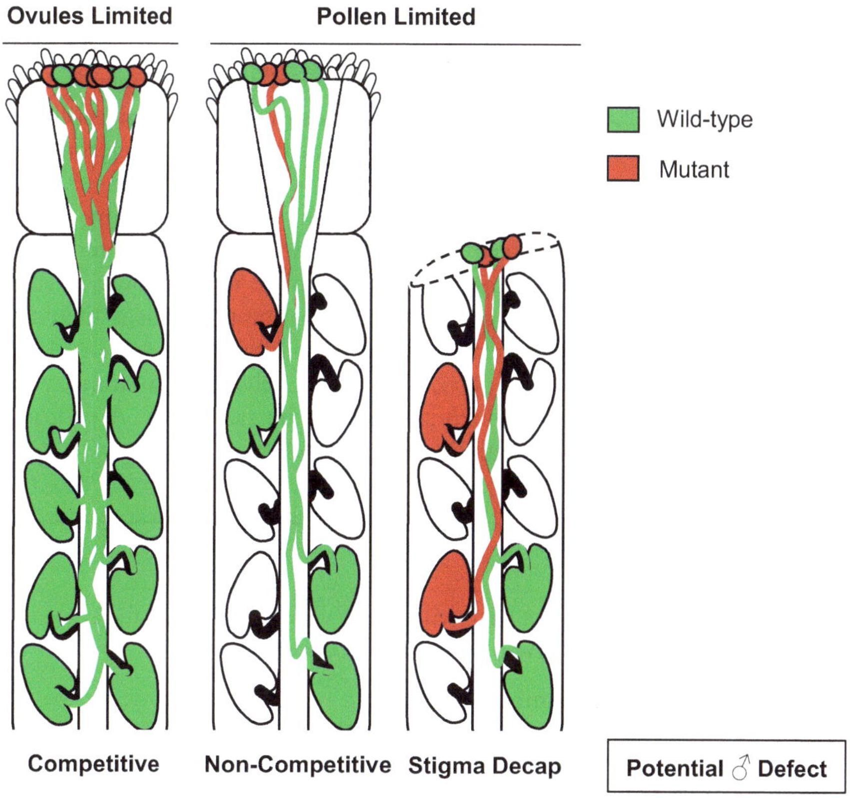

Competitive	Non-Competitive	Stigma Decap		Potential ♂ Defect
0%	50%	50%	←	reduced growth rate
0%	25%	25%	←	shorter pollen tubes or reduced tropism
0%	~0%	50%	←	stigma penetration

Examples of Observed Mutant TE

Fig. 1 Illustration of competitive, non-competitive, and stigma decapitation outcrosses. (*Left*) Competitive pollination (ovules limited). When the number of pollen tubes exceeds a limited number of ovules (approximately 50–60), the transmission frequency of a mutant allele (red) is impacted by competition wild-type pollen (green). In this example, wild-type pollen fertilized all available ovules (indicated by green filling), outcompeting all of the mutant pollen. (*Middle*) Non-competitive pollination (pollen limited). In the absence of wild-type competition, mutant pollen have an equal opportunity to fertilize an ovule because the number of available ovules is greater than the number of pollen. In this example, mutant pollen successfully fertilize ovules (red filling) but still shows a reduced transmission efficiency (TE) compared to wild type. (*Right*) Stigma decapitation (pollen limited). To determine if a transmission deficiency is related to stigma penetration, removing the stigma-style barrier provides an opportunity to compare a TE to a non-competitive condition. In this example, the removal of the stigma barrier restored a mutant TE to be equal to wild type, supporting a hypothesis that a mutation associated with a pollen defect is specifically disrupting a cellular process important for stigma-style penetration (*see* Supplemental Table 1 for a generic Excel worksheet with genetic calculations and an example with *oft1-3*) compared to the wild type. Additional evidence from decapitation crosses (described here) indicated that the removal of the stigma-style barrier alleviated the transmission deficiency from 858-fold to a 2.6-fold, suggesting that most, but not all, *oft1* pollen deficiencies can be attributed to a reduced ability to penetrate through the stigma-style barrier.

These methods are relevant to evaluating any mutation affecting pollen fertility and provide a genetic strategy to quantify a mutation's impact on pollen–pistil interactions. This method can be used with any hemizygous mutation that is linked to an easily scorable assay for the mutant allele (e.g., *T-DNA* with a drug resistance marker), or alternatively, with a homozygous mutant segregating a hemizygous rescue transgene. A protocol for both genetic strategies is outlined below.

2 Materials

2.1 Seed Stocks

Seed stocks harboring the following mutant alleles can be obtained from the ABRC or upon request.

1. Columbia (Col-0).
2. *ms1-1*, (At5g22260) ABRC Germplasm / Stock: CS75.
3. Hemizygous mutant plant (WT/mutant) (e.g., *oft1-3* +/−) (AT3G05320, WiscDsLox489-492M4).

2.2 Female Host Preparation and Pollination

1. Headband Magnifier, 2× Magnification Glass Lens Plate 10″ Focal Length.
2. Forceps—Standard Tip (Fine Scientific Tools Item No. 11251-30).
3. 70% Ethanol.
4. General-purpose laboratory colored tape.
5. Wooden skewers.
6. Micropore tape.
7. Male sterile *ms1* plants to be used as female host.
8. Hemizygous plants harboring a mutant to be analyzed for pollen transmission deficiencies (e.g., *oft1-3* +/− with a linked Basta^{R} marker).

2.3 Stigma Decapitation

1. Suture Cutter 3.5 mm curved cutting edge without lock.
2. Transparent Sun Bag humidity chamber.
3. Supplies from Subheading 2.2.

2.4 Seed Harvest and Scoring Progeny for Mutant Allele Transmission

1. Mature siliques from Subheading 3.2 or 3.3.
2. 1.7 mL Eppendorf tubes.
3. If needed, seed surface sterilization and plating supplies described in **Note 3**.

2.5 Calculating Transmission Efficiency Percentages (TE%) and Transmission Efficiency Ratio (TE_r)

1. Statistical analysis software (Microsoft Excel).
2. Genetics data from Subheading 3.4.

2.6 Using Stigma Decapitation to Generate Homozygous Mutants

1. Supplies from Subheading 2.2 female host preparation and pollination.
2. Hemizygous mutant plant.

3 Methods

3.1 Female Host Preparation and Pollination

1. Cleanse forceps in 90% ethanol and dry with Kimwipe.
2. For the female host, identify 3–4 recently opened *ms1* flowers.
3. Use forceps to gently remove petals and anthers from each flower.
4. Repeat using additional inflorescences as desired (*see* **Note 1**).
5. For the pollen donor, identify several flowers that have been open for 1–3 days, with ample, yellow pollen (e.g., from an *oft1-3 +/− plant*) (*see* Fig. 2a).
6. Gently pluck pollen donor flower at the base of the stem and remove petals to fully expose pollen-bearing anthers.
7. Use forceps to hold and remove 4–6 anthers.
8. Holding anthers by the filament base, dab pollen from the anthers onto the female surface (stigma papillae or exposed ovary from decapitation) (*see* Fig. 2b).
   - For standard competitive male outcross (ovules limited), dab as much pollen as possible onto the stigma surface (*see* Fig. 1). The stigma surface will appear yellow.
   - For a limited pollination outcross (pollen limited), first remove excess pollen from the anthers by flicking or brushing them onto another surface (such as another stigma), and then transfer a single dab to the target stigma chosen for limited pollination. The goal is to transfer fewer pollen than the number of ovules (*see* Fig. 1 and **Note 2**).

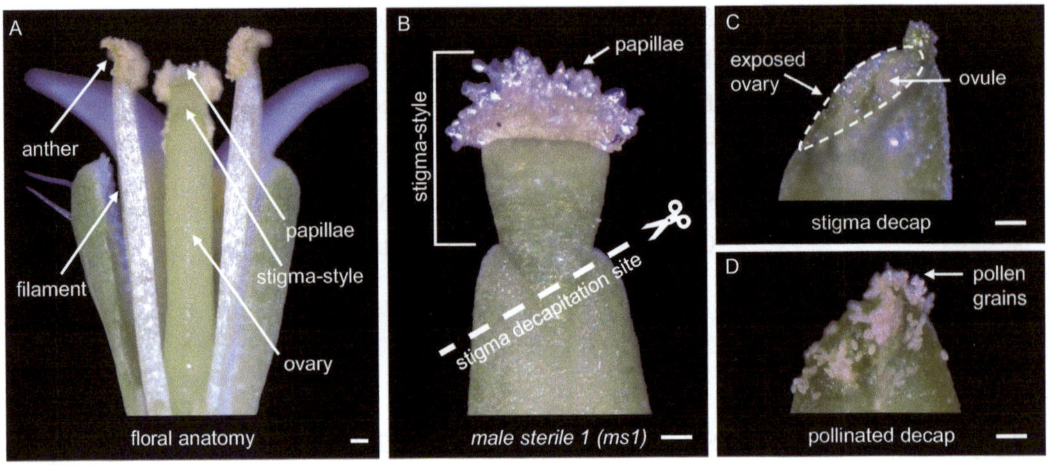

Fig. 2 Illustration of a stigma decapitation cross with *Arabidopsis thaliana*. (**a**) Floral anatomy showing elongated stamen with dehisced anthers depositing mature pollen onto a receptive stigma. (**b**) Unpollinated pistil showing the stigma-style decapitation site. (**c**) Exposed ovary after decapitation. (**d**) Decapitated stigma after manual pollination. Scale bar = 125 μm

9. Wrap a descriptive, colored tape label around the inflorescence of manually pollinated stigmas. A recommended label size is 0.5 × 7 cm with 2–3 mm at each end folded over to facilitate easy removal and label-transfer when seeds are harvested (*see* Fig. 3 inset). See examples of outcross descriptions below.
   - Cross 1: Competitive (ovules limited):
     ♀ *ms1* × ♂ *oft1-3 (+/−)*.
   - Cross 2: Non-competitive (pollen limited):
     ♀ *ms1* × ♂ *oft1-3 (+/−)* pollen limited.
   - Cross 3: Stigma decapitation (pollen limited):
     ♀ *ms1* (decap) x ♂ *oft1-3 (+/−)*.
10. If the tape label weighs the branch down, use micropore tape to attach and support the branch on a wooden skewer.
11. Repeat for each emasculated flower.
12. For pollinating decapitated stigmas (a special case of a pollen-limited cross), proceed to the following section.

3.2 Stigma Decapitation

1. Use suture cutters or razor blade to cut off the stigma-style at an approximate 45° angle, exposing the ovary cavity (*see* Fig. 2b, c). Forceps can also be used to remove the stigma-style but might pinch close the opening to the ovary, blocking subsequent pollen tube access and growth into the ovary.
2. Gently pluck pollen donor flower at the base.
3. Use forceps to hold the flower, peel back the petals, and remove 4–6 anthers.

Fig. 3 Enclosure to maintain humidity after a stigma decapitation cross. Sun bag or alternative enclosure is recommended to maintain high humidity during pollen germination and growth after stigma decapitation

4. Holding anthers by the filament base, dab pollen from the anthers onto the exposed ovary surface. A hint of yellow from the applied pollen should be visible. Because removing the stigma reduces the surface area for pollen adhesion, decapitation crosses are likely to have limited pollination (e.g., pollen numbers less than 30% of available ovules) (*see* Fig. 2d, Fig. 1).
5. Mist decapitated pistils with water and quickly enclose pollinated plants or individual branches in a plastic bag to maintain high humidity for 24–48 h (*see* Fig. 3).
6. Remove bag and let siliques mature under normal growth conditions.

3.3 Seed Harvest and Scoring Progeny for Mutant Allele Transmission

1. Harvest mature, slightly yellow siliques in a 1.7 mL Eppendorf tube.
2. Transfer tape label with cross-description to tube.
3. Leave tube open for 2–3 days to allow siliques to finish drying.
4. Harvest seeds for sterilization and germination on selection plates to test the transmission frequency of the mutant allele (e.g., *oft 1-3* Basta^R) (*see* **Note 3**).
5. Record the number of resistant and sensitive seedlings.

Formulas for TE, TE$_r$, and X^2

Transmission Efficiency TE%	Pearson's Chi-Squared Test	Transmission Efficiency Ratio TE$_r$
$\frac{\# R_{observed}}{n}$	**Step 1)** $X^2 = \sum \frac{(\# R_{obs} - \# R_{exp})^2}{\# R_{exp}} + \frac{(\# S_{obs} - \# S_{exp})^2}{\# S_{exp}}$ **Step 2)** Find *p*-value using critical values table of X^2 distribution. Degrees of freedom = 1	$\frac{\# R_{observed}}{\# S_{observed}}$

Genetic Analyses of *oft1-3*

Male x Female: *oft1-3* (+/- Basta^R) x *ms1*	#Basta^R	#Basta^S	n	OBS TE%	EXP TE%	*p*-value	OBS TE$_r$	EXP TE$_r$	OBS *vs.* EXP TE$_r$ X-fold Diff
1. Competitive Outcross (Ovules Limited)	3	2573	2576	0.0012	0.5	<0.0001	0.0012	1	↓858-fold
2a. Non-competitive Outcross (Pollen Limited)	2	84	86	0.023	0.5	<0.0001	0.024	1	↓ 42-fold
2b. Compare to cross #1 OBS TE (competitive)				0.023	0.0012*	<0.0001	0.024	0.0012*	↑ 20-fold
3a. Stigma Decapitation (Pollen Limited)	103	267	370	0.28	0.5	<0.0001	0.39	1	↓2.6-fold
3b. Compare to cross #2 OBS TE (noncompetitive)				0.28	0.023*	<0.0001	0.39	0.024*	↑ 16-fold
Male x Female: OFT1-GFP (+/- Hyg^R) in *oft1-3* (-/-) x *ms1*	**#Hyg^R**	**#Hyg^S**	**n**	**OBS TE%**	**EXP TE%**	***p*-value**	**OBS TE$_r$**	**EXP TE$_r$**	**OBS *vs.* EXP TE$_r$ X-fold Diff**
4. Competitive Outcross (Ovules Limited)	572	1	573	~1	0.5	<0.0001	572	1	↑572-fold
5a. Non-competitive Outcross (Pollen Limited)	489	98	587	0.83	0.5	<0.0001	4.99	1	↑ 5-fold
5b. Compare to cross #4 OBS TE (competitive				0.83	~1*	<0.0001	4.99	572*	↓117-fold
6a. Stigma Decapitation (Pollen Limited)	80	50	130	0.62	0.5	0.009	1.60	1	↑ 1.6-fold
6b. Compare to cross #5 OBS TE (noncompetitive)				0.62	0.83*	<0.0001	1.60	4.99*	↓ 3-fold

Fig. 4 Statistical analysis of *oft1-3* pollen transmission. (Top, left) Transmission efficiency (TE) of a mutant allele is calculated as a percentage by dividing the number of selection-marker resistant seedlings by the total number of seedlings in an outcross. An expected TE for a typical pollen outcross is 50%. (Top, center) Statistical significance is determined by a chi-squared (X^2) test with degree of freedom = 1. (Top, right) Transmission efficiency ratio (TE$_r$) of a mutant allele is calculated as a ratio by dividing the number of selection-marker resistant seedlings by the number of sensitive seedlings in an outcross. A typical expected TE$_r$ for a pollen outcross is 1 (i.e., equal number of resistant and sensitive F1 progeny). (Outcross #1–3) *oft1-3* (+/– Basta^R) pollen transmission to a female *male-sterile* (*ms1*) host under three outcross conditions: competitive pollination (ovules limited), non-competitive pollination (pollen limited), and stigma decapitation (pollen limited). Removing the stigma penetration barrier, as shown by stigma decapitation, decreased the magnitude of *oft1-3* reduced pollen transmission from 858-fold observed under competitive pollination to 2.6-fold under stigma decapitation. (Outcross #4–6) Pollen transmission of rescue transgene *OFT1-GFP* (+/– Hyg^R) in *oft1-3* (–/–) to a female *male-sterile* (*ms1*) host under three outcross conditions: competitive pollination (ovules limited), non-competitive pollination (pollen limited), and stigma decapitation (pollen limited). Pollen harboring an *OFT1-GFP* (Hyg^R) transgene outcompeted an unrescued *oft1-3* pollen in a competitive cross (572-fold better), consistent with the expectation that the transgene effectively rescues the mutation and mimics a "wild-type pollen." Importantly, pollen harboring a transgene were still more successful in a non-competitive outcross (fivefold better) or stigma decapitation outcross (1.6 fold better). This supports a model in which the removal of the stigma-style barrier can mostly, but not completely, alleviate the deficiencies caused by the *oft1-3* mutation. *Asterisk* previously OBS TE(r) used as EXP TE(r) (*see* **Note 6**)

3.4 Calculating Transmission Efficiency Percentages and Statistical Significance

A reproductive defect, male or female, is determined by quantifying the transmission frequency of a mutant allele. Here, a transmission efficiency % (TE = number mutant (resistant) seedlings divided by total number of seedlings) is measured from pollen outcrossed to a *ms1* female host, where the expected TE of a mutant allele is initially 50% (*see* Fig. 4). A mutation can be classified as resulting in a male deficiency if the mutant allele transmission shows a statically significant change from expected. To determine if a mutation causes a pollen-penetration defect, compare the TE from each of the following types of crosses (described above and *see* Fig. 4, top left, for example with *oft 1-3 with a* Basta^R marker linked to the mutation):

- Cross 1: Competitive (ovules limited).
- Cross 2: Non-competitive (pollen limited).
- Cross 3: Stigma decapitation (pollen limited).

1. For each outcross, conduct a Pearson's Chi-square test (*see* Fig. 4, top center) to determine if the TE is significantly different from 50% transmission of the mutant allele (e.g., *oft 1-3* Basta^R) (*see* Fig. 4, crosses 1–3). The standard null-hypothesis being tested for each outcross is that the mutant allele transmission frequency is equal to the wild-type allele (expected TE = 50%).
2. To determine if mutant pollen transmission is statistically improved under limited pollination, conduct a second Pearson's Chi-square test using an expected TE that corresponds to the actual observed TE of the competitive (ovules limited) results. This tests the null-hypothesis that transmission of the mutant allele is not affected by competition (*see* Fig. 4 cross #2b).
   - If competition does affect the TE of the mutant allele, this might indicate that the mutant pollen are slower to germinate, slower to grow or locate ovules, or less able to penetrate through the stigma (*see* **Note 4**).
3. To determine if a stigma decapitation cross statistically improves mutant pollen transmission, conduct a second Pearson's Chi-square using an expected TE that corresponds to the actual observed TE of the non-competitive (pollen limited) results. This tests the null-hypothesis that transmission of the mutant allele through pollen is not affected by a stigma penetration barrier (*see* Fig. 4 cross #3b and **Note 5**).

3.5 Reporting X-Fold Differences in Pollen Transmission Using a Transmission Efficiency Ratio (TE_r)

In a pollen outcross segregating two male gametes from a hemizygous plant (*oft1-3* +/− Basta^R), the gametes are expected to transmit equally (1 Basta^R mutant (−): 1 wild-type Basta^S (+) allele). A transmission efficiency ratio (TE_r) is determined by dividing the number of mutant (resistant) seedlings by the number of wild-type (sensitive) seedlings (*see* Fig. 4, top right). In a typical pollen outcross, the expected TE_r of a mutant allele is 1. If the observed TE_r is more than 1, then pollen harboring a mutant allele are transmitting more than wild-type pollen. If the observed TE_r is less than 1, then mutant pollen are transmitting less than wild-type pollen. After calculating an observed TE_r, an X-fold change in pollen transmission efficiency can be reported as a change in TE_rs.

1. Calculate the observed (OBS) TE_r by dividing the number of resistant seedlings by the number of sensitive seedlings.
2. To calculate an x-fold difference, compare TE_rs for OBS and EXP by dividing the larger TE_r by the lesser TE_r.
3. Report the quotient as an x-fold increase if OBS > EXP or decrease if EXP > OBS in mutant pollen transmission (*see* Fig. 4 and **Note 6**).

3.6 Alternative Strategy to Compare Pollen TE_rs Using a Rescue Transgene

In the example above, the use of a marker (e.g., $Basta^R$) linked to the *oft1-3* mutation enabled a comparison of pollen transmission deficiencies between a competitive outcross (858-fold), non-competitive outcross (42-fold), and stigma decapitation (2.6-fold) (*see* Fig. 4 crosses 1–3). However, for pollen mutations lacking a selection marker, it is still possible to conduct a decapitation cross using an alternative strategy in which the marker is associated with a rescue transgene. For example, TE_r s are compared within a set of crosses (*see* Fig. 4 crosses 4–6) using a homozygous mutant background (*oft1-3* −/− all $Basta^R$) and scoring for the transmission of a $Hygromycin^R$ selection marker linked to a rescue transgene (*OFT1-GFP*, +/− Hyg^R). In contrast to following the transmission of the mutant allele with a Basta marker, the alternative strategy here for scoring the rescue transgene can be thought of as following the transmission of the "wild-type" allele.

3.7 Using Stigma Decapitation to Generate Homozygous Mutants

For some pollen fertility mutants that appear to be male-sterile, consider using a decapitation cross to generate a homozygous mutant. Once a single homozygous mutant plant is generated, self-fertilization might allow for rare fertilization events that can perpetuate the mutation as a homozygous mutant plant line [4].

1. Perform steps from Subheading 3.2 using a plant hemizygous for the mutant allele as both the pollen donor and decapitated female host (*oft1-3* +/− $Basta^R$).
2. If stigma-style penetration is the only pollen defect, then the expected frequency of homozygous mutant genotype is 25%.
3. Harvest seeds from decapitation crosses, surface sterilize, and germinate on selection plates (*see* Subheading 3.3).

4. Perform genomic DNA extraction and PCR-based genotyping to identify homozygous mutant plants.
5. Observe homozygous mutant plant to test for complete or partial sterility, as well as additional reproductive or vegetative phenotypes.

4 Notes

1. We suggest performing at least 20 stigma decapitation crosses to obtain enough F1 progeny for statistical analyses.
2. Because it is difficult to know if a limiting amount of pollen was transferred, it is important to only harvest siliques that are shorter than expected. Mature siliques derived from limited pollination are expected to be visibly shorter in length, where less than 30% of the available ovules are fertilized.
3. Several seed sterilization methods and plant media recipes are available. A common method is to surface sterilize seeds using chlorine gas. Sterilize no more than 100 μL volume of seed in an open 1.7 mL Eppendorf tube. Obtain a plastic, sealed container that can hold a 100 mL glass beaker and a small tube rack. In a chemical fume hood, place rack with tube of seed and glass beaker into the container. Pour 100 mL generic bleach (6% hypochlorite) into the beaker and add 3 mL hydrochloric acid. Quickly seal. Allow the chlorine gas that is generated to surface sterilize seeds for 3 h. Open the container lid and degas seeds in the chemical hood for 10 min. Plate seeds on desired plant selection plates. Selection plates can be made by adding appropriate drugs or herbicides to a base medium of 0.5× Murashige and Skoog medium (Phytotechnology laboratories) (pH 5.7) with 1% agar, 0.05% MES, and optional 5% sucrose.
4. For example, *oft1-3* Basta^R transmission changed from 0.0012 (competitive TE) to 0.023 under non-competitive pollination. However, non-competitive conditions still failed to restore to near wild-type transmission levels. While this indicates that *oft1-3* pollen is less competitive compared to wild type, it leaves open the question of whether the mutation might simply slow down pollen tube growth, or might attenuate cellular processes involved in stigma-style penetration, tropism, or fertilization.
5. For example, *oft1-3* Basta transmission efficiency significantly increased from 0.023 TE under non-competitive pollination to 0.28 TE after decapitation of the stigma-style barrier. However, the decapitation conditions still failed to restore to wild-type transmission levels, suggesting that there is a still a significant deficiency not completely resolved by removing the

stigma-style barrier. This raises the question of whether the mutation might have additional impacts on cellular processes involved in tropism or fertilization.

6. When comparing TE_rs between competitive and limited pollination assays, use OBS TE_rs from previous assays as EXP TE_rs in the current assay (*see* Fig. 4 Cross 2b, 3b).

References

1. Johnson MA, Harper JF, Palanivelu R (2019) A fruitful journey: pollen tube navigation from germination to fertilization. Annu Rev Plant Biol 70:809–837. https://doi.org/10.1146/annurev-arplant-050718-100133
2. Rahmati Ishka M, Brown E, Weigand C et al (2018) A comparison of heat-stress transcriptome changes between wild-type Arabidopsis pollen and a heat-sensitive mutant harboring a knockout of cyclic nucleotide-gated cation channel 16 (cngc16). BMC Genomics 19:1–19. https://doi.org/10.1186/s12864-018-4930-4
3. Smith DK, Jones DM, Lau JBR et al (2018) A putative protein O-fucosyltransferase facilitates pollen tube penetration through the stigma – style interface. Plant Physiol 176:2804–2818. https://doi.org/10.1104/pp.17.01577
4. Myers C, Romanowsky SM, Barron YD et al (2009) Calcium-dependent protein kinases regulate polarized tip growth in pollen tubes. Plant J 59:528–539. https://doi.org/10.1111/j.1365-313X.2009.03894.x

Chapter 4

Isolation of the Pistil-Stimulated Pollen Tube Secretome

Said Hafidh and David Honys

Abstract

Detection of secreted proteins and peptides during pollen tube guidance has been impeded due to lack of techniques to capture the pollen tube secretome without contamination from the female secreted proteins. Here we present a protocol to detect tobacco pollen tube secreted proteins, semi-in vivo pollen tube secretome assay (SIV-PS), following pollen tube crosstalk with the female reproductive tissues. This method combines the advantages of in vivo pollen tube–pistil interaction and filter-aided sample preparation (FASP) techniques to obtain an in-depth proteome coverage. The SIV-PS method is rapid, efficient, inexpensive, does not require specialized equipment or expertise, and provides a snapshot of the ongoing molecular interplay. We show that the secretome obtained is of greater purity (<1.4% ADH activities) and that pollen tubes are physiologically and cytologically unaffected. A compendium of quality controls is described and a rough guide on downstream bioinformatics analysis is outlined. The SIV-PS method is applicable to all studies of protein secretion using pollen tube as a model and can be easily adapted to other flowering species with modification. The overall duration for this protocol is approximately 8 hours spanning 4 days (an average of 2 h/day per two workers) excluding microscopy and LC-MS/MS analysis.

Key words Sexual plant reproduction, Cell–cell communication, Pollen tube guidance, Protein secretion, LC-MS/MS, Glycosylation

1 Introduction

At the interface between two adjacent cells, the extracellular matrix provides a hub for cell–cell communication. Small peptides secreted to the extracellular space are a unique signature of the cellular response to the external signals perceived. Therefore, studying the secretome or more accurately the "peptidome" of the extracellular matrix, provides an excellent system to obtain a detailed account of the short and long-distance signalling events and more importantly the molecular interplay of the cells involved. The extracellular matrix constitutes of cleaved mature peptides (peptidome), hormones, modified polysaccharides as well as secreted nanovesicles that serve as long-distance signalling vesicles in almost all forms of life [1].

Anja Geitmann (ed.), *Pollen and Pollen Tube Biology: Methods and Protocols*, Methods in Molecular Biology, vol. 2160, https://doi.org/10.1007/978-1-0716-0672-8_4,

In flowering plants, precise guidance of the pollen tube (the male gametophyte) toward the embryo sac of the ovule (the female gametophyte) involves persistence response by the pollen tube to pre-laid secreted peptide signals from the female gametophyte in the transmitting tract tissues [2]. Several of the female secreted peptides have been identified by cell-specific transcriptomic studies [3–6] and modified carbohydrate oligomers by genetic approaches using female mutants defective in pollen tube attraction [7]. This was feasible owing to the accessibility of the female gametophyte in several plant species in particular *Torenia fournieri* where the embryo sac is markedly protruded [5, 8]. Therefore, the crosstalk between the two gametophytes, involving a single polar cell (the pollen tube) and a multiplex of female reproductive tissues (the transmitting tract and female gametophyte) provides an excellent system to study short- and long-distance signalling as well as single cell versus complex tissues mechanisms of communications. For this a network of crosstalk events involving secreted ligands and perceiving receptors needs to be established. Unfortunately, development of methods for isolating the pollen tube secretome responsive to the female guidance signals has proven difficult due to the inaccessibility of the pollen tubes within the female reproductive tissues. Given this limitation, we developed a straightforward and easily adaptable method for studying the pollen tube secretome following penetration through the pistil tissues.

1.1 Purification of Pollen Tube Secretome Using SIV-PS Method

Here we demonstrate an efficient capture of the pollen tube secretome (peptidome) in a seminative environment after the pollen tube penetration through the stigma and style (Figs. 1, 2, 3, and 4). We named this improvised method from the original technique [9], semi-in vivo pollen tube secretome assay (SIV-PS). The procedure involves preparation of in-planta pollinated pistils and control unpollinated pistils in an improvised pollen tube germination chamber, incubation for 24 h, concentration of the secretome with 3/10 kDa protein filters followed by FASP processing and protein identification by liquid chromatography mass spectrometry. The apparatus used in a step-by-step set up and the anticipated results are shown in Figs. 1, 2, 3, and 4. SIV-PS has several advantages over in vitro pollen tube secretome; (1) it first allows interaction of the pollen tubes with the female reproductive tissues, thus rendering the pollen tubes competent to perceive female secreted signals that considerably influence the pollen tube secretome as well as transcriptome [10–13] (2) the pollen tube secretome captured is most likely directly influenced by the perceived female emitted cues, and therefore offers the possibility to identify candidate molecules of the male–female interaction (3) the setup of the protocol does not require specialized equipment and hence remains inexpensive (4) when coupled with FASP processing [14] and gel-free tandem

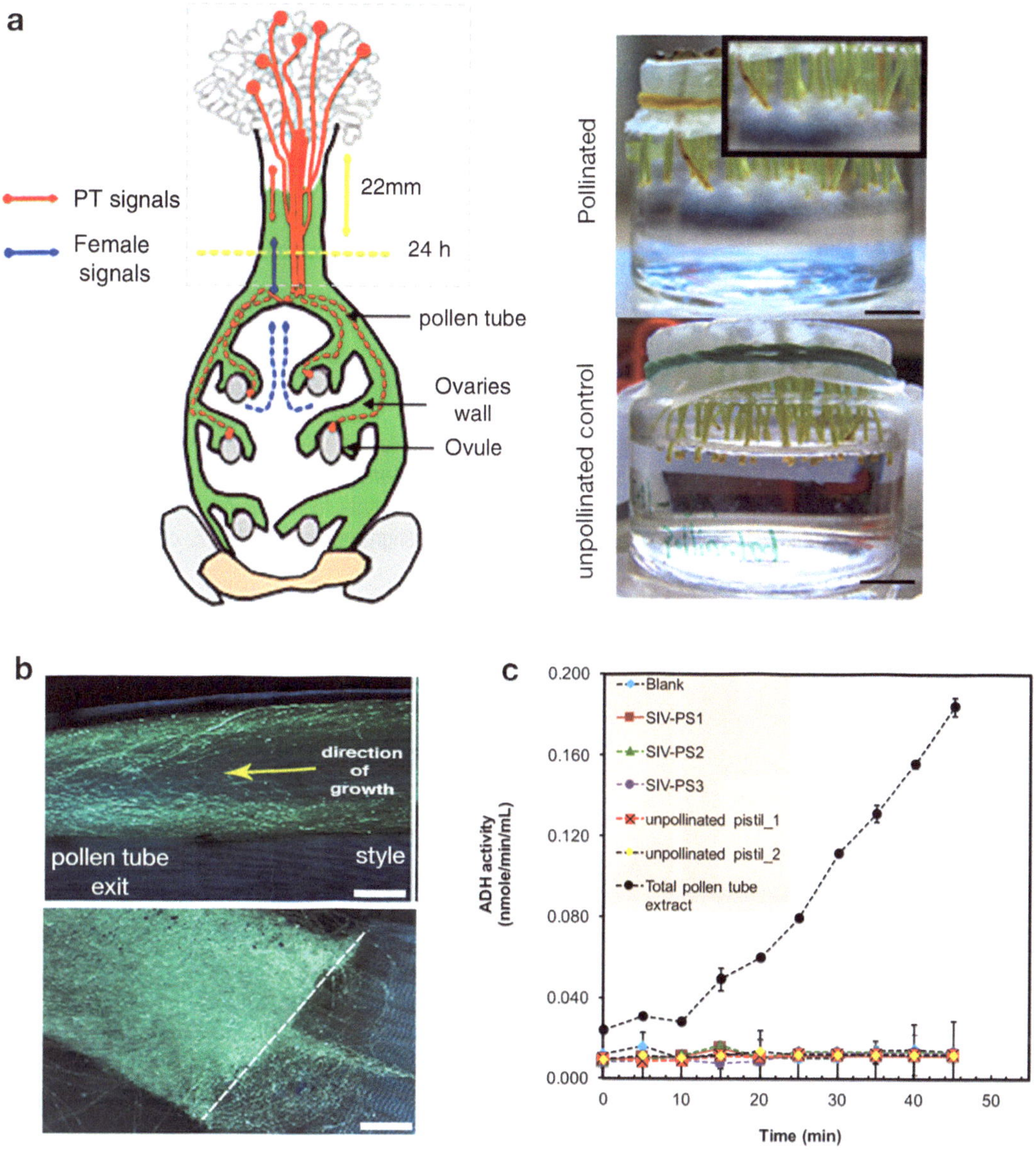

Fig. 1 The SIV-PS system for capturing semi-in vivo pollen tube secretome. (**a**) Schematic of Nicotiana tabacum pollen tube growth through the transmitting tissue of the style and entry into an ovule of the female gametophyte. After 24 h of in-planta germination, an excision is made on the pistil 22 mm (yellow dotted line) from the stigma shoulders and the pistil is transferred to the secretome chamber for in vitro pollen tube germination and secretome collection. Right panel shows an actual secretome chamber of the pollinated pistil (upper panel) and the unpollinated pistil control (lower panel) with successful pollen tube growth through the style tissues into the secretome chamber highlighted in the inset. (**b**) Aniline blue stain of the pollen tube bundle direction and exit from the cut pistil. Scale bars, ~5 μm. (**c**) Exemplified expected results of the secretome purity test from the SIV-PS derived secretome samples conducted using the ADH assay. Purity of the secretome is assessed by comparing maximum ADH activities of the secretome samples after 45 min incubation at 37 °C to the ADH activity of the total pollen tube extract

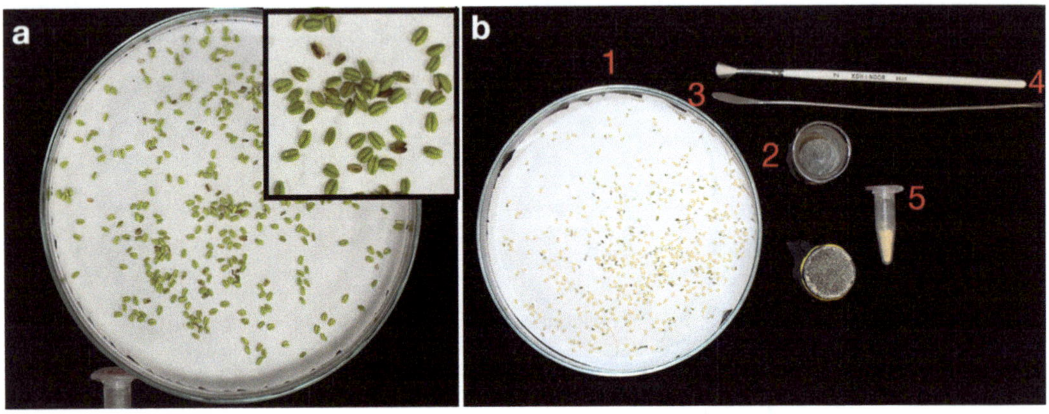

Fig. 2 Collection of dry pollen. (**a**) Anthers at pre-anthesis are collected into a petri dish covered with filter paper and incubated at room temperature overnight (**b**) 1; fully dry anthers 24 h after incubation, 2; chamber for sieving pollen grains into a freshly arranged filter paper, 3; spatula for scooping filtered pollen into an Eppendorf tube, 4; exemplified brush size used for pollination of stigma (*see* **step 18**), 5; collected dry pollen grains for immediate use or storage at −20 °C

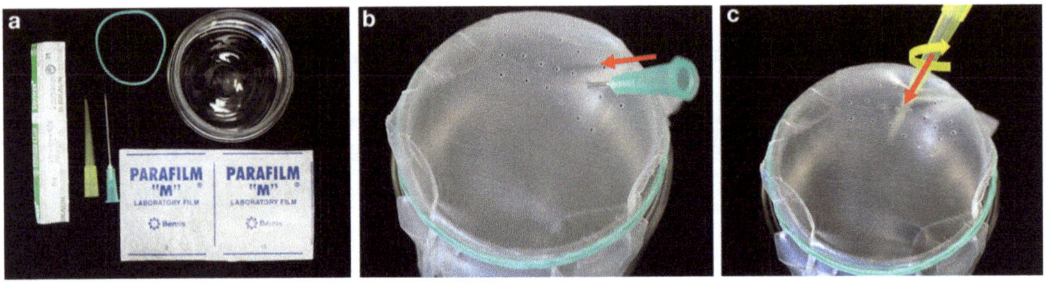

Fig. 3 Preparation of the SIV-PS secretome chamber. (**a**) A 50 mL capacity glass beaker is sterilized by autoclaving. Sterile 21G1 needle, parafilm (ethanol sterilized), yellow tip, and a rubber band are prepared. (**b**) In the laminar flow cabinet, decant tobacco pollen tube germination media into the chamber and seal with parafilm by slight stretching and secure with a rubber band. Perforate holes with the 21G1 needle ~2 mm apart in a sequential order. (**c**) Enlarge the holes for easy pistil penetration using a yellow tip by applying a vertical pressure (red arrow) and a twist (yellow arrow). Secretome chamber ready to use. Prepare just before pistil excision

LC-MS/MS, peptides of low molecular weight (<10 kDa) and of low abundance (as low as 0.45 parts per million) can be detected [10] (5) and unlike in vitro assay, the SIV-PS method has extremely high pollen tube growth reproducibility and allows set up of negative controls to quantitatively assess true secreted proteins (Fig. 1 [10]). The samples obtained by the SIV-PS method are also useful to study posttranslational modifications of secreted proteins as well as pollen tube metabolome to identify secreted modified carbohydrates and other potential signalling macromolecules. It is important to test the purity of the secretome samples. To estimate contamination by cytosol, one of the most abundant cytosolic enzymes, alcohol dehydrogenase (ADH), is used as an indicator

Fig. 4 Purification of the pollen tube secretome with the SIV-PS system. (**a**) The capture of the secretome start with the flower emasculation and in-planta pollination. (**b**) 24 h post-pollination, pistils are collected for processing. (**c**) Pollinated and unpollinated pistils are excised ~22 mm from the stigma shoulders. (**d**) All pistils are then gently inserted into prearranged secretome chamber using a sterilized forceps. (**e–f**) A humid chamber is constructed and the complete SIV-PS system with prearranged pistils is incubated at 28 °C for 24 h. On the next day, the secretome is ready for collection

[10]. Because of the noninvasive nature of the protocol, the purity of the sample is high, nearly 15-fold below detection of the sensitive ADH activities. The benefits of the SIV-PS method are also applicable to several other research areas. In particular, studies of endomembrane dynamics and secretion mechanisms including endocytosis, exocytosis, and unconventional protein secretion in a single cell system, influence of signal transductions on transcription and translation in pollen tubes, cytoskeleton dynamics, architecture of plant cell wall composition as well as in pathogen–plant interaction; e.g., by comparative genome-omics and secretome dynamics when pollen tube penetrates the pistil tissues as to a fungal hyphae penetrating leaf tissues. Lately, we have applied the SIV-PS method and coupled it with phospholipase C cleavage to enrich potential pollen tube secreted glycosylphosphatidylinositol-anchored proteins (GPI-anchor) as first in-line transient receptor proteins that could perceive female guidance signals (Hafidh et al., unpublished). The SIV-PS method can be easily adapted to other flowering species with modification.

1.2 Maximizing Protein Detection by FASP and Label-Free Quantitative LC-MS/MS

Since secreted peptides are predominantly small in size and of low abundance, their detection can be challenging. Efficient ionization of small tryptic peptides for MS/MS detection is highly influenced by downstream processing of the protein samples. Two major strategies are used in preparation of peptides suitable for mass spectrometry. The first method involves solubilization of proteins with detergents, separation of proteins by sodium dodecyl sulfate polyacrylamide gel electrophoresis (SDS-PAGE), and in-gel digestion [15]. The in-gel processing is free from interfering impurities but suffers from lower peptide recovery and hence protein detection. The second method is detergent-free, comprising protein extraction with strong chaotropic reagents such as urea and thiourea, protein precipitation, and in-solution digestion of proteins under denaturing conditions [16]. The disadvantage of the gel-free approach is that the proteome may be incompletely solubilized and the digestion may be impeded by interfering substances. Therefore, we opt to follow a recently described method, filter-aided sample preparation (FASP), which combines the advantages of both methods above [16]. The essential steps of the FASP method include depletion of detrimental components in urea-containing buffer, carboamidomethylation of thiols, on-filter digestion of proteins, and elution of peptides [16]. With the FASP method, the number of unique peptides identified can double the number of peptides identified through in-gel processing [14]. By combining FASP method with label-free LC-MS/MS, we have identified over 800 candidate protein accessions from tobacco SIV-PS using <2 μg of the total secretome sample with an average of 62% of peptides that are less than 20 kDa and identified by >3 total peptides reproducible in at least 3/4 replicates [10]. The majority of the identified proteins were detected by 2–10 total peptides suggesting great sensitivity even of low abundant proteins [10].

1.3 Comparisons with Other Secretome Methods

Prior to our studies [10, 11], there have not been any other studies of pollen tube secretome. We presume the reason was simply due to lack of a technique that can allow capture of pure pollen tube secreted proteins after a crosstalk with the female reproductive tissues. Nevertheless, prevailing methods exist for studying the secretome of the apoplast (extracellular matrix) in plants. They use either vacuum-assisted infiltration collection method (VIC) [17] or a gentle gravity-assisted extraction method (GEM) [18]. Both of these methods have been commonly used in the studies of plant–pathogen interaction with an effort to identify plant/pathogen secreted proteins to the apoplast during plant defense against the infection [19]. Although these methods suffer from cytosolic contaminations due to cell breakage and insensitive detection of the apoplastic proteins because of the collection procedures and amount of materials recovered, they have spearheaded discovery of elicitor secreted peptides, pathogen-associated molecular patterns (PAMPs), as well as PAMP-triggered immunity response proteins of the host cells [20]. Application of these methods is feasible because of the type of tissues used, rosette leaves which constitute of cells with rigid cell wall structures. The only gametophytic secretome study conducted was that of isolated ovule secretome which used a modified GEM method, tissue free (tfGEM) [21]. We therefore set out to develop a rapid, efficient, and an informative method for capturing pollen tube secreted proteins following pollen tube penetration through the female reproductive tissues. To our knowledge, this is the only method that has been tested and published so far [10].

1.4 Limitations of the SIV-PS Method

Our improvised SIV-PS protocol enables efficient and representative identification of male factors that participate in the male–female communication during fertilization without a need for higher level of expertise [10]. As any other technique, the SIV-PS method has its limits. Compared to the in vitro pollen tube secretome method (duration of 1–2 days), the SIV-PS method is time consuming. It requires up to 4 days (with an average of 2 h/day/2 coworkers) to obtain the secretome samples with completed purity and microscopic viability test. This step is independent of the number of workers and therefore its duration cannot be optimized. However, the limited hours spent per day means the SIV-PS method permits running of parallel experiments. Another disadvantage of our SIV-PS method is that it requires more starting materials: approximately 50 emasculated pistils per sample to obtain ~300 μL of 1–2 μg/μL of the pollen tube secretome per sample. However, with two or more coworkers, this might not be a limiting factor, instead more materials can be obtained by more coworkers, though this can increase the chance of sample variability. The main constraint of the SIV-PS method is that it cannot be paused at any time until the secretome samples have been collected.

2 Materials

2.1 Starting Material

1. It is strongly recommended to prepare a minimum of four biological replicates to obtain meaningful reproducibility. Approximately 20–30 flowering tobacco plants are required to obtain four replicates of ~1–2 μg/μL (total of 200–300 μL) protein concentration per sample per single run.
2. Multiple runs can be performed during the summer period if the plants are grown in outdoor green houses. At the first bolting, the primary inflorescence should be cut to encourage formation of secondary inflorescences. Approximately fifty emasculated flowers are required per sample per run.

2.2 Preparation of the Plant Material

1. *Critical*: Even though pistils are not considered fully sterile, all beakers used as germination chambers as well as flasks and falcon tubes must be sterile throughout the procedure to reduce potential fungal contamination. It is important that sample visualization be done immediately and purity tests should be conducted ideally before freezing the secretome samples at −80 °C.

2.3 General Reagents

1. Prepare all solutions using deionized (dH_2O) ultrapure water. Use of RNase free water is not essential unless for a specific application. Prepare and store all reagents at room temperature (unless indicated otherwise). For the N-glycosylation test, the detection buffer (4-chloro-1-naftol, methylalcohol in 10 mM Tris–HCl pH 6.8) should preferably be prepared fresh. Carefully follow all waste disposal regulations when disposing waste materials. The pollen tube germination media can be used for months if stored in cold (<8 °C) and only opened in flow box.
2. EDTA–disodium salt ($C_{10}H_{14}N_2O_8Na_2{\cdot}2H_2O$, Mw 372.24).
3. DAPI for nucleic acid staining (C16H15N5·2HCl, Mw 350.25). *Caution*: DAPI is a suspected mutagen; wear gloves and work in a flow hood while preparing the stock solution.
4. *N,N*-dimethylformamide (DMF, $HCON(CH_3)2$, Mw 73.09). *Caution*: DMF is potentially toxic; wear protective gloves under a chemical hood while handling the solution.
5. Propidium iodide (PI, $C_27H_{34}I_2N_4$, Mw 668.39). *Caution*: PI is suspected as a mutagen; wear gloves while preparing the stock solution.
6. Liquid nitrogen (N_2, Mw 28.01). *Caution*: Wear safety glasses or a face shield when handling liquid nitrogen; wear gloves when you are touching any object cooled by liquid nitrogen.
7. Ultrapure ethanol (C_2H_6O, Mw 46.07).
8. Acetic acid, glacial ($C_2H_4O_2$, Mw 60.05). *Caution*: Always pour acid slowly into the solution.
9. Glycerol ($C_3H_8O_3$, Mw 92.09).
10. Triton X-100 ($(C_2H_4O)nC_{14}H_{22}O$).
11. Tris–HCl.
12. Boric acid (H_3BO_3, Mw 61.83). *Caution*: Boric acid can cause respiratory sensitilization and a range of health hazards, wear gloves and mask while preparing the stock solution.
13. Calcium nitrate hydrous ($Ca(NO_3)_2.4H_2O$), Mw 236.19). *Caution*: Calcium nitrate is corrosive and toxic, wear safety glasses or a face shield, gloves and mask while preparing the stock.

14. Magnesium sulfate heptahydrous ($MgSO_4 \cdot 7H_2O$, Mw 246.47).
15. Potassium nitrate (KNO_3), Mw 101.1.
16. 2-(N-Morpholino)ethanesulfonic acid sodium salt (MES), Mw 209.26.
17. Sucrose.
18. Murashige-Skoog (MS) medium.
19. SAVO bleach.
20. Agar.
21. Acrylamide/Bis solution.
22. Sodium dodecylsulfate (SDS), Mw 288.37.
23. Ammonium persulfate (APS), Mw 228.2.
24. Glycine.
25. Dithiothreitol (DTT).
26. Coomassie Bromophenol Blue (CBB).
27. Isopropanol (C_2H_6O, Mw 46.07).
28. Methanol.
29. NaCl.
30. $MgCl_2 \cdot 6H_2O$, Mw 203.3.
31. Milk powder.
32. Bovine serum albumin.
33. Nitro blue tetrazolium chloride (NBT).
34. 5-Bromo-4-chloro-3-indolyl phosphate (BCIP).
35. 1-StepTM NBT/BCIP ready-made mix.
36. Concavalin A.
37. Horseradish peroxidase.
38. 4-Chloro-1-naftol.
39. Hydrogen peroxide.
40. Primary antibodies.
41. Secondary antibodies.
42. Prestained protein marker.
43. Alcohol dehydrogenase activity assay kit.
44. ProteoSilver silver stain kit.

2.4 General Equipment

1. Nitrocellulose membranes.
2. Blotting papers Whatman, thickness 5 (alternatively four pieces of Whatman papers, thickness 3), size cca 7 × 9.5 cm per one blot.

3. Mini PROTEAN® 3 System, glass plates, filter pads, gel holder cassettes, buffer tank, Plastic spatula, and lid.
4. Ice blocks (MiniPROTEAN) for cooling.
5. Plastic trays for processing the gels and assembly of blot sandwiches.
6. Polypropylene boxes 10 × 10 cm square.
7. Disposable cuvettes.
8. Spectrophotometer capable of measuring absorbance at 480 nm.
9. Soil composition—Jiffy-7 tablets (seed pellets, peat pellets; Jiffy Products International; http://www.Jiffygroup.com), diameter 41 mm.
10. Square plastic plant pot (3.2 × 3.2 cm).
11. Tweezers.
12. Scissors.
13. 2× 1 L plastic beakers.
14. 50 mL polypropylene conical centrifuge tubes.
15. Eppendorf 5810R centrifuge with swing-bucket rotor or similar alternative, max. RCF 4500 × *g*.
16. Pipettes.
17. 0.80 × 120 mm needle for protein loading.
18. 2-ml syringe to aspirate the gel holes before loading.
19. 1.5-ml transparent cap-locked microtube.
20. Microscope slides.
21. Coverslips of 0.13–0.16 mm thickness; 18 × 18 mm.
22. Laminar flow box.
23. Magnetic stirrer.
24. Glass mortar and pestle (inner diameter 8 cm). *Critical*: Store these in a refrigerator or in a cold room (4 °C) 1 day before homogenization.
25. Outdoor green house or a growth room.
26. Epifluorescence or, ideally, a confocal laser scanning microscope equipped with a spinning disk module.
27. Knife (blade).
28. 28 °C incubator.
29. Millipore 3 and 10 kDa filters.
30. −20 and −80 °C freezers.
31. Garden tying strings.
32. Optional; eBlot TM semidry transfer system.

2.5 Pollination, Dissection Tools, Germination Chambers, and Pollen Tube Germination Media

1. Bright brush bristle (#2/4) (KOH-I-NOOR Hardtmuth a.s; http://eshop.koh-i-noor.eu/shop/bright-brush-9936-bristle-4, cat. no. 9936006011BL).
2. 21G1–1.5 needles.
3. A sterile 3.5 cm × 4.5 cm round beaker (40 mL capacity).
4. A 500 mL glass beaker to create humidity chamber.
5. Sterile PARAFILM M.
6. Elastic rubber bands.
7. Tobacco pollen tube germination media (SMM-MES): dissolve 60 g of sucrose in 500 mL deionized water, 0.1 g H_3BO_3, 0.708 g $Ca(NO_3)_2{\cdot}4H_2O$, 0.2 g $MgSO_4{\cdot}7H_2O$, 0.1 g KNO_3, and 4.88 g 2-(N-Morpholino)ethanesulfonic acid sodium salt (MES). Make up the volume to 900 mL and adjust the pH with KOH to 5.9. Adjust the final volume to 1 L. Sterilize the media in water bath >100 °C for 40 min. Repeat the sterilization after 24 h. Store the media in a cold room (<8 °C).

2.6 In-Gel N-Glycosylation Test

1. Concavalin A.
2. Horseradish peroxidase.
3. Hydrogen peroxide.
4. Detection buffer: 45 mg 4-chloro-1-naftol, 15 ml methylalcohol, and 60 ml 10 mM Tris–HCl pH 6.8.

2.7 Alcohol Dehydrogenase Assay (ADH)

1. ADH kit.
2. Disposable cuvettes.
3. Spectrophotometer capable of measuring absorbance at 480 nm.

2.8 SDS Polyacrylamide Gel

1. Resolving gel buffer: 1 M Tris–HCl, pH 8.8. Weigh 36.33 g Tris–HCl and transfer to a beaker containing 900 mL of water. Mix and adjust the pH with HCl (*see* **Note 1**). Make up the volume to 1 L with water. Store at 4 °C.
2. Stacking gel buffer: 0.5 M Tris–HCl, pH 6.8. Weigh 18.24 g Tris–HCl, mix with 900 mL of water, and adjust the pH as above. Adjust the volume to 1 L solution and store at 4 °C.
3. 30% acrylamide/Bis solution (29.2:0.8) acrylamide:Bis: Weigh 29.2 g of acrylamide monomer and 0.8 g Bis (cross-linker) and transfer to a 100 mL graduated cylinder containing about 40 mL of water. Add a spatula of AG 501-X8 (D) mixed-resin beads and mix for about 30 min. Make up to 100 mL with water and filter through a 0.45 μm Corning filter (*see* **Note 2**). Store at 4 °C, in a bottle wrapped with aluminum foil (*see* **Note 3**).

4. 10% (w/v) Sodium dodecylsulfate (SDS): Dissolve 30 g SDS in 200 mL water. Adjust the volume to 300 mL and store at room temperature.
5. 40% (w/v) Ammonium persulfate (APS): Dissolve 20 mg in 50 μL of water (*see* **Note 4**).
6. Electrode buffer (10× stock solution without SDS): Weigh 30.3 g Tris (250 mM) and 144 g of glycine (1.92 M) and mix with 800 mL of water. Make up the volume to 1 L.
7. SDS-PAGE running buffer: 25 mM Tris, 192 mM glycine, 0.1% SDS. Dilute 100 mL of 10× electrode buffer to 990 mL with water and add 10 mL of 10% SDS. Care should be taken not to mix the solution vigorously since SDS generates bubbles.
8. 3× 1D sample buffer; Mix 15 mL of 0.5 M Tris–Cl pH 6.8 (final 0.15 M), 20 mL of 75% (v/v) glycerol (final 30%), 3 g SDS (final 6% w/v), 1.157 g DTT (final 0.15 M), 1.5 mL of 1% bromophenol blue in 50% isopropanol (final 0.03%) (*see* **Note 2**). Adjust the volume to 50 mL, aliquot in to 2 mL tubes, and store at −20 °C. To prepare a working 1D sample buffer, simply dilute the 3× 1D sample buffer by one-third with water or protein sample.

2.9 Immunoblotting

For wet transfer system (Bio-Rad MiniProtean Blotter, Catalogue 170-3930).

1. Nitrocellulose membranes.
2. Blotting buffer: 0.025 M Tris–HCl, 0.192 M glycine, 20% methanol (*see* **Note 3**).
3. Tris-buffered saline (TBS; 10×): 1.5 M NaCl, 0.1 M Tris–HCl, pH 8.0.
4. TBS containing 0.05% Tween-20 (TBST 1×): Mix 100 mL 10× TBS with 0.5 mL Tween 20. Adjust to 1 L final volume.
5. Blocking solution: 5% milk in TBS (*see* **Note 5**). Store at 4 °C.
6. 1% Bovine serum albumin (BSA): Dissolve 3 g of bovine serum albumin V (Sigma, cat. A4503) IN 300 mL of 1× TBST
7. Primary antibodies: Dilute 500–5000× with 1× TBST with 0.5% BSA in a 25 mL solution per blot (*see* **Note 6**).
8. Secondary antibodies: Dilute 5000× with 1× TBST with 0.5% BSA in a 25 mL solution per blot (*see* **Note 6**).
9. Mini PROTEAN® 3 System glass plates, filter pads, gel holder cassettes, buffer tank, and lid.
10. Plastic trays for processing the gels and assembly of blot sandwiches.

11. Blotting papers Whatman, thickness 5 (alternatively four pieces of papers Whatman, thickness 3), size cca 7 × 9.5 cm per one blot.
12. Ice blocks (MiniPROTEAN) for cooling.
13. 1 M Tris–Cl (pH 9.5): Dissolve 12.12 g to 80 mL of water. Adjust the pH to 9.5 with HCl. Make up the solution to 100 mL.
14. 2 M NaCl: Dissolve 11.7 g to 100 mL of water.
15. 1 M $MgCl_2 \cdot 6H_2O$: Dissolve 20.3 g of $MgCl_2 \cdot 6H_2O$ to 100 mL water. Autoclave at 120 °C for 30 min and store at 4 °C.
16. Alkaline phosphate buffer (AP buffer): Mix 10 mL of 1 M Tris–Cl (pH 9.5), 5 mL of 2 M NaCl and 0.5 mL of 1 M $MgCl_2 \cdot 6 \cdot H_2O$. Adjust to a total volume of 100 mL.
17. 70% Dimethylformamide (DMF). Mix 1.4 mL dimethylformamide with 0.6 mL water.
18. Nitro blue tetrazolium chloride (NBT). Dissolve 50 mg in 1 mL of 70% dimethylformamide. Store at 4 °C.
19. 5-Bromo-4-chloro-3-indolyl phosphate (BCIP): Dissolve 25 mg BCIP in 1 mL of 70% DMF. Protect from light, store at −20 °C.
20. Ready to use AP buffer: Add 132 μL of BCIP and 132 μL of NBT to 20 mL of alkaline phosphatase buffer just before adding to membrane. Alternatively, use 1-StepTM NBT/BCIP ready-made mix.

3 Methods

3.1 Overview of the Method

1. The first step of the protocol involves sowing of tobacco seeds (*Nicotiana tabacum*, cv. Samsun) and growing tobacco plants.
2. Day 1: Flowers at 1 day before anthesis are emasculated.
3. Day 2: The pistils are pollinated and pollen tube germination chambers are then prepared for the incubation of pollinated as well as unpollinated control pistils.
4. Day 3: Pollinated and unpollinated flowers from day 2 are harvested, excised 22 mm below the stigma shoulder, arranged in germination chambers, and incubated for 24 h at 28 °C under full humidity.
5. Day 4: The pollen tube germination media containing secreted proteins are collected and concentrated using protein size exclusion filters and samples are split for purity check, SDS-PAGE separation, LC-MS/MS peptide identification, and for long-term storage. Pollen tubes emerged from the cut

pistil can also be excised and stored if desired for total protein extraction or for RNA sequencing.

6. The duration allocated to some steps can be reduced if more coworkers are involved, however, high care must be taken as this might result in bigger variabilities between samples as a result of individual handling.

3.2 Seed Germination and Plant Growth

1. For plants that will be grown to the outdoor green house, seeds should be sown 3 months before the start of the experiment.
2. Compost mixture preparation: Rehydrate Jiffy tablets with water in an appropriate container (e.g., a beaker). After rehydration, transfer the soaked tablets to pots and/or trays for subsequent sowing and/or planting (*see* **Note 5**).
3. Seed germination: Use the compost mixture to fill each pot (8 cm in diameter) and resurface the top soil. Sow approximately 100 seeds per pot and slightly cover the seeds with soil using a sieve. Label the pots and place them on a large tray to facilitate easy watering from the bottom of the tray.
4. Place the pots in a cold room (7 °C) for 2 days to stratify the seeds. Next, transfer the pots to the growth room for seed germination (*see* Subheading 2. Materials for plant material).
5. Alternative seed germination for plant growth in growth chamber: sterilize the seeds in 1% (w/v) sodium hypochlorite (90% bleach SAVO, Biochemie group a.s.) for 10 min, wash with sterilized water four times, and plate on half strength MS agar. After stratification at 4 °C in the dark for 3 days, move the plates to the growth chamber set at 25 °C under a 16-h-light/ 8-h-dark photoperiod for plant growth. After 8–10 days, transplant the seedlings to soil.
6. *Pause point*: Seedlings should be grown for 2 weeks.
7. Seedling planting: Fill 20–40 18-cm-diameter pots with a compost mixture. Make a small indentation into the soil with an index finger. Transfer the seedlings into the holes and let them grow under standard greenhouse conditions, 22–25 °C under short day for 6 weeks.
8. Growing plants in an outdoor green house under natural day–night photoperiods in spring and summer: Cut the stems of adult plants with fully developed roots to leave four bottom leaves. Let the new inflorescence branches to emerge (~2 weeks) and transfer into soil spacing the plants approximately one meter apart.
9. *Pause point*: Plantlets should be grown for 4–6 weeks.

3.3 Collection of Mature Pollen Grains for Pollination (Timing 90 min)

1. Approximately one week prior to the start of SIV-PS experiment, mature pollen grains to be used for pollination need to be collected on a daily basis and stored at −20 °C . Most effective method is to collect the immature flowers from 20–30 plants at stage 6 of development (52–55 mm in length) containing immature pollen grains 1 day before anthesis (Fig. 2) [22]. These anthers are dried overnight at room temperature and pollen grains are easily filtered out using prearranged glass cylinder and sieve.
2. Prepare one or two 1 L plastic beakers layered with wet paper towels. First remove and dispose all open flowers containing dehisced mature pollen grains. Collect all stage 6 flowers that are 1 day before anthesis (52–55 mm) into the beakers (*see* **Note 6**).
3. Bring the flowers to the laboratory, prepare two-third 16 cm glass Petri dishes layered with 0.1 cm Whatmann paper or equivalent. Manually open the flower petals longitudinally to expose the anthers and pull-off the anthers into the Petri dish (Fig. 2). Continue until the bottom of the Petri dish is covered with anthers with enough space between anthers, approximately $5/cm^2$ density. If more flowers are left, start a new Petri dish (*see* **Note** 7).
4. Leave the anthers on a laboratory bench to dry at room temperature overnight.
5. On the next day, install a legging net (or equivalent) slightly stretched on an open ends glass cylinder 3 cm × 3 cm and secure it with a rubber band (Fig. 2). Gather the now dehisced anthers into the prearranged glass cylinder and filter through the pollen grains by gently milling the anthers with a spatula.
6. Transfer the obtained pollen grains powder into a sterile 1.5 ml microcentrifuge tube.
7. Weight the pollen grains collected and store at −20 °C (*see* **Note 8**).

3.4 Flower Emasculation [Day 1] (Timing 120 min)

1. Counted as SIV-PS Day 1, the next step is the emasculation of N. tabacum flowers (Fig. 4).
2. Prepare 30–40 netting bags (approximately 10 cm × 10 cm) and equivalent number of garden tying strings.
3. Remove all open flowers containing dehisced mature pollen and discard them.
4. Manually open stage 6 flowers (52–55 mm closed buds containing undehisced anthers [22]) and remove all the anthers. Continue to do so until all flowers available from a single plant have been emasculated. (*see* **Notes 9** and **10**).

5. Using the netting bags, cover the inflorescences holding the emasculated flowers and secure them with tying strings (Fig. 4).
6. Leave the flowers overnight for stigma to fully mature.

3.5 Pollination of Emasculated Flowers [Day 2] (Timing 60 min)

1. On day 2, using a bright brush bristle, load the brush with pollen and pollinate each stigma sufficiently without over saturating the stigma. Pollinated flowers are covered with the net and left in planta for another 24 h.
2. Remove, from −20 °C, an aliquot of frozen tobacco mature pollen grain powder and let it acclimatize to room temperature for 5 min (*see* **Note 11**).
3. Remove the netting bags with care not to damage the flowers or the inflorescences.
4. Using a bright brush bristle, pollinate each stigma (*see* **Note 12**).
5. After completing the pollination of all emasculated flowers, cover back all the inflorescences with the net bag and leave for pollen tubes to germinate overnight in planta.

3.6 Harvesting of Pollinated Flowers [Day 3] (Timing 60 min)

1. After the 24 h of pollen tube growth in vivo, pollinated pistils are collected for the final step of in vitro germination. Flowers are collected into a plastic beaker layered with wet paper towels to avoid pistil dehydration especially during hot summers. The collection should not last more than 30 min and ideally should be done in the early morning. Collection of bigger number of flowers at once is not recommended. If many flowers have been pollinated to obtain several replicates, the collection can be done in batches of four replicates per round of collection by two efficient coworkers.
2. If more than one worker is involved, prepare an equivalent number of 1 L plastic beakers layered with 2–3 wet paper towels.
3. Remove all the netting bags and manually detach each emasculated flower (pollinated and unpollinated flowers) and place them in to the 1 L beaker.
4. Cover the collected flowers with one wet paper towel for transportation to the lab.
5. Leave the plants for 1–2 days to recover before the second experiment can be repeated.

3.7 Preparation of Germination Chamber (Timing 20 min)

1. Unpack the germination chambers from aluminum foil in the flow box.
2. Titrate 40 mL of tobacco pollen tube germination media into each germination chamber (Fig. 3).
3. Seal the top of the germination chambers with sterile parafilm and secure with a rubber band (Fig. 3, *see* **Note 13**).
4. Puncture holes in a clockwise manner with a 21 G × 1–1.5″ needle leaving 2–3 mm gaps between holes (Fig. 3). Continue until the center of the parafilm has been perforated. This normally equates to approximately ×50 holes. Enlarge the hole using a sterile yellow tip for easy penetration of the pistils (Fig. 3).
5. Label the side of each germination chamber and transfer them to the bench.

3.8 Arrangement of Pistils on Pollen Tube Germination Chamber (Timing 120 min)

1. Prepare new 21 G × 1–1.5″ needle, ruler, and clean forceps.
2. Take a flower and manually detach the pistil at the attachment with the ovary.
3. Align the pistil with a ruler on a paper towel. Using the 21 G × 1–1.5″ needle, excise the pistil at 22 mm below the stigma shoulders (Fig. 4, *see* **Note 14**, *see Troubleshooting*—Table 1).
4. Using forceps, place the excised pistil into one of the holes of the germination chamber containing fresh pollen tube germination media (SMM-MES). Arrange the pistils spirally (Fig. 4). The pistil cut end should be fully submerged in to the germination media (Fig. 1, *see* **Note 15**).
5. Follow similar steps for pollinated and unpollinated pistils.
6. Continue until all flowers have been processed.

3.9 Semi-In Vivo Pollen Tube Germination (Timing 20 min)

1. Prepare an equivalent number of Petri dish bottoms to pollen germination chambers layered with wet paper towels.
2. Transfer the pistil-loaded germination chambers on to a Petri dish bottoms,
3. Cover the germination chambers with 500 mL beakers to create a humid chamber (*see* **Note 16**).
4. Incubate the complete germination chambers for 24 h at 28 °C without shaking.

3.10 Secretome Collection [Day 4] (Timing 30 min)

1. The PT secretome is collected after 24 h of SIV pollen tube growth; suspended pistils with protruded pollen tubes are removed and pollen tubes can be excised and collected or discarded if not required. The media containing pollen tube secreted proteins is transferred into falcon tubes and kept on

Table 1
Troubleshooting

Step	Problem	Possible reason	Solution
3.8	No pollen tube growth from pistil explant	Old pollen or bad media	Check the stored pollen and the media by in vitro pollen tube germination
3.10	No pollen tube emergence from the pistil	Mutilated excision of the pistil	When excising the pistil, use a sharp 21G1 needle to minimize tissue damage
3.12	Very low secretome protein concentration	Poor pollen tube germination	Make sure you have pollen germination >90% of the pollinated pistils
		Not enough pistils arranged	Arrange enough pistils (~50) per secretome chamber
		The secretome sample was not concentrated enough	Concentrate the samples further if necessary to <500 µl or use less starting pollen germination media depending on the type of chamber used
		High amount of proteases in the media	Addition of proteases inhibitors is not recommended as it might influence pollen tube secretome
3.13	Excessive cytosolic contamination	Bad pollen tube germination media or not well buffered	Make fresh media and check the pH regularly. Check pollen tubes integrity after incubation
3.18	High rate of nonviable pollen tubes	pH of the media is off	Check the pH
		Pollen tubes were left too long	Assess pollen tube viability immediately after collection
		Staining solution induced pollen tube burst	Check and quantify the pollen tubes by light microscopy prior to any staining

ice. Pistil materials are used for cytological tests including pollen tube viability and pharmacological treatments to assess pollen tube endomembrane secretion activities. For pollen tube excision (Subheading 3.11) from the cut pistils, pistils are placed on a sterile pre-wet glass slide under a dissection microscope and pollen tubes are excised along the cross-section with a sterile 21 G × 1–1.5″ needle. The pollen tube pellet is immediately transferred into a cool 1.5 mL microcentrifuge tube on ice containing buffer with RNase inhibitor. Once the collection of pollen tube is completed, the samples are flash frozen in liquid nitrogen and stored at −80 °C. The media containing secreted proteins is then processed (Subheading 3.12).

2. Check if the pollen tube bundles have emerged from the pistil explant before proceeding (Fig. 1, *see Troubleshooting*—Table 1).

3. If pollen tube bundles need to be collected, first perform pollen tube excision prior to collecting the secretome (*see* Subheading 3.11).
4. Remove gently the rubber band and the parafilm from the germination chamber.
5. Decant all the germination media containing secreted proteins into sterile 50 mL Falcon tubes. Leave the tubes on ice.
6. Proceed with secretome concentration (Subheading 3.12).
7. *Pause point*: Collected secretome can be stored at −20 °C for shorter time or at −80 °C for longer time storage prior to protein concentration.

3.11 Excision of Pollen Tube Bundles (Timing 30 min)

1. Sterilize a microscope glass slide by wiping with 100% pure ethanol.
2. Place the slide under dissection microscope. Wet the slide with approximately 20 μL of the pollen tube germination media.
3. Using forceps, lift off a pistil explant with pollen tube bundle and lay it on a slide so that the pollen tubes are extended horizontally along the slide without too much tangling.
4. Use a sterile 21 G × 1–1.5″ 1 needle and excise the pollen tube bundle longitudinally.
5. Transfer the pollen tube bundles using the needle into a prearranged cooled 1.5 mL microcentrifuge tube.
6. Once all pistil explants have been processed, resuspend the pollen tube bundles with RNA extraction buffer containing RNases inhibitors or a protein extraction buffer containing protease inhibitors for later RNA or protein extraction. Freeze the samples immediately in liquid nitrogen and store at −80 °C (*see* **Note 17**).

3.12 Concentration of the Secretome (Timing 60–90 min)

1. Obtained secretome samples are generally in low ng/μL levels in 50 mL pollen tube germination medium. For more reliable and accurate downstream applications, secretome samples should be concentrated using 3 kDa or a 10 kDa ultracentrifugation filters. This also allows partial dialysis of the samples from metabolized amino acids, salts, and sugars present in pollen tube germination medium that could interfere with LC-MS/MS. The duration for sample concentration is approximately 30–60 min to reduce the volume to ~300 μL per sample. This might vary depending on the protein concentration of the secretome. To avoid freeze-thaw cycles, concentrated samples are aliquoted for multiple applications and stored.
2. Arrange a required number of Amicon Ultra 2 mL 10K Centrifugal filters (Merk Millipore) (*see* **Note 18**).

3. Load each filter with 2 mL of the secretome supernatant.
4. Gently arrange the filters in a fixed angle rotor and centrifuge at 7500 × *g* for 10 min at 15 °C (*see* **Note 19**).
5. Discard the flow through and reverse spin to collect the retained secretome.
6. Continue until the whole secretome has been concentrated from 50 mL down to 0.5–2 mL (*see* **Note 19**).
7. Load the filter with the remaining 2 mL of the secretome and centrifuge at 1000 × *g* for 2 min at 15 °C. Continue until 300–500 μL is achieved (*see* **Note 20**).
8. Aliquot approximately 150–200 μL to be used for protein concentration measurements (e.g., 2-D Quanti kit, GE-healthcare), purity test by enzymatic assay and SDS-PAGE analysis (*see Troubleshooting*—Table 1).
9. Store the remaining samples at −80 °C for gel-free LC-MS/MS protein identification.
10. *Pause point*: Store the SIV-PS secretome at −80 °C until further use.

3.13 Purity Test Using ADH Calorimetric Assay (Timing 90 min)

1. All secretome samples are subjected to contamination by proteins from the cytoplasm and to a laser extent from the cell wall depending on the method of secretome collection used. Therefore, purity test must be conducted to estimate the levels of contamination and evaluate whether the obtained samples are of reliable quality for downstream analysis. Enzymatic assays provide sensitive, reliable, and quantitative estimation of secretome purity due to their abundance. Commonly used enzymes as biomarkers of cytosolic contamination include alcohol dehydrogenase (ADH), glucose-6-phosphate dehydrogenase (G6PDH), glyceraldehyde-3-phosphate dehydrogenase (GAPDH), and malate dehydrogenase (MDH). Secretome purity is conducted by measuring ADH activities in secretome samples and compared with ADH activities in a total cell extract. Ideally, the same materials used to obtain the secretome should be used to obtain the total cell extract. Instead of ADH activities, western blot can be performed to compare ADH levels between the samples (Subheadings 3.14–3.16). The advantage of the ADH assay is that it provides a direct quantitative measure with reliable estimation of contamination, whereas western blot does not unless it is done very carefully with specialized imaging equipment and appropriate controls [23]. ADH activity is measured over time and secretome samples purity is established (Fig. 1).
2. Extract total proteome from SIV pollen tubes and quantify amount of proteins for each sample.

3. Switch on a spectrophotometer capable of measuring absorbance at 450 nm (*see* **Note 21**).
4. Acclimatize the ADH assay buffer to room temperature.
5. Reconstitute the NADH standard and the developer solution with ultrapure water.
6. Prepare NADH standard by diluting 10 μL of the 10 mM NADH stock solution with 90 μl of the ADH assay to generate a 1 mM standard solution. Pipette 0, 2, 4, 6, 8, and 10 μL of the1 mM standard solution into a 96 well plate in duplicates. Add ADH Assay Buffer to each well to bring the total volume to 50 μL per well (*see* **Note 22**).
7. Aliquot 3, 6, and 9 μg (or at least two different concentrations) of the secretome samples and their controls in duplicate into the same 96 well plate. Bring the final volume to 50 μL per well with the ADH buffer.
8. Prepare 100 μL of reaction mix per sample by mixing 82 μL ADH assay buffer, 8 μL developer solution, and 10 μL of 2 M UV pure ethanol (*see* **Note 23**).
9. Pipette 100 μL of the appropriate reaction mix to each well. Mix well using a horizontal shaker.
10. Incubate the plate at 37 °C in a dark without shaking.
11. After 2 min of incubation (T_{initial}), record the initial absorbance at 450 nm ($A450_{\text{initial}}$).
12. Return the plate to the incubator and continue recording the absorbance at 5 min intervals.
13. When the absorbance value of the most active sample is greater than the value of the highest standard (10 nmol/well), stop with the measurements to avoid values exceeding the end of the linear range of the standard curve.
14. For calculating ADH activities, use the final measurement ($A450_{\text{final}}$) and time (T_{final}) of the most active sample prior to exceeding the end of the linear range of the standard curve.
15. Subtract the final measurement $A450_{\text{final}}$ of the 0 (blank) NADH standard from the $A450_{\text{final}}$ of the standards and the samples.
16. Plot the NADH standard curve of the absorbance $A450_{\text{final}}$ (nm) against time (min).
17. Obtain the change in absorbance ($\Delta A450$) between T_{initial} and T_{final} by subtracting $A450_{\text{final}} - A450_{\text{initial}}$, i.e., $\Delta A450 = A450_{\text{final}} - A450_{\text{initial}}$.
18. Use this absorbance value to obtain the amount of NADH generated between T_{initial} and T_{final} (B) from the standard

curve using a Y-intersection formula (Microsoft Excel or similar).

19. Compute the ADH activities (nmol/min/mL, mU/mL) for each sample as follows:

$$\text{ADH activity (nmol/ min /mL)} = \frac{B \times \text{dilution factor of the sample}}{T_{\text{final}} - T_{\text{initial}} \times V(\text{sample volume added})}.$$

$B =$ amount (nmol) of NADH between T_{initial} and T_{final}.

20. Plot a scatter graph of ADH activities against time for all secretome samples and their respective positive controls for comparison and estimation of contamination levels (*see* **Note 24**, *see Troubleshooting*—Table 1).

3.14 SDS-PAGE Analysis for Western Blotting (Timing 60 min)

1. SDS-PAGE of the secretome samples is not essential in the SIV-PS method since protein detection is done by gel-free LC-MS/MS. However, SDS-PAGE needs to be performed when western blot is a desired method for the secretome purity test or when performing a glycosylation test to establish the fraction of glycosylated secreted proteins. Secretome samples are mixed with SDS-PAGE running buffer in a 1:1 (v/v) ratio, sonicated for 5 min, and standard SDS-PAGE procedures are followed using *12.5% Mini PROTEAN® 3*. To obtain higher sensitivity of protein bands on the gel, silver stain should be used instead of Coomassie Brilliant Blue G250 stain. Generally, ~40 μg per well is run on an SDS-PAGE to obtain reliable sensitivity and visualization of differential protein profiles.
2. Protein transfer to nitrocellulose membrane for downstream application such as immunoblotting or N-glycosylation test can be done using wet transfer or a semidry transfer system. The wet transfer system is lengthier (at least 20 min of preparation and approx. 45 min running) but provides an overall efficient transfer of all protein size range and abundance. Conversely, the semidry transfer system is much more convenient in handling (10 min of preparation) and has shorter running time (10 min) but it is less efficient in overall protein transfer. For routine usage and in longer runs, the semidry system is also more expensive. We recommend usage of the semidry system only during immunoblotting of known secreted protein or for a trial test of any downstream application. Wet transfer should be performed for any unknown secretome sample at least for the first attempt as it improves transfer of low MW proteins.
3. Mix 2.5 mL of resolving buffer, 3.33 mL of acrylamide mixture, and 4 mL water in a 50 mL conical flask. Degas with helium for 10 min. Add 100 μL of SDS, 80 μL of ammonium persulfate, and 10 μL of TEMED and cast gel within a 7.25 cm × 10 cm × 1.5 mm gel cassette. Allow space for stacking gel and gently overlay with isobutanol or water.

4. Prepare the stacking gel by mixing 1.25 mL of resolving buffer, 0.67 mL of acrylamide mixture, and 3 mL water in a 50 mL conical flask. Degas with helium for 10 min. Add 100 μL of SDS, 40 μL of ammonium persulfate, and 5 μL of TEMED. Insert a 10-well gel comb immediately without introducing air bubbles.
5. Heat aliquots of bovine Ro 60, RBC membranes, and human spectrin antigens at 95 °C for 5 min. Do not add lysis buffer to the prestained protein standard or subject it to heat. Centrifuge the heated samples at 3000 × *g* for 30 s to bring down the condensate. Load increasing amounts of Ro antigen (1–4 μg) on one gel and the same amounts of spectrin (3 μg/lane) or RBC membrane antigens on two other gels along with protein standards (10 μL/well–2 μg/marker/lane). Add protein standards in every other lane (alternating with spectrin) in the gel with spectrin. Electrophorese at 15 mA until the sample has entered the gel and then continue at 25 mA till the dye front (from the BPB dye in the samples) reaches the bottom of the gel.
6. Following electrophoresis, pry the gel plates open with the use of a spatula. The gel remains on one of the glass plates. Rinse the gel with water and transfer carefully to a container with western blot transfer buffer.
7. Cut a nitrocellulose membrane to the size of the gel and immerse in methanol. Rinse twice in distilled water and once with transfer buffer.
8. Protein transfer to nitrocellulose membrane can be performed in a wet system (Subheading 3.15) using Mini PROTEAN® 3 protein transfer system (Bio-Rad, Subheading 3.15) or by a semidry method (Subheading 3.16) using eBlot protein transfer system.

3.15 Western Blot Using Mini PROTEAN® 3 Protein Transfer System (Timing 60 min)

1. After gel electrophoresis (Subheading 3.14), carefully wrench the gel plates with the use of a plastic spatula. The gel remains attached on one of the glass plates. Using a glass spatula, cut off the stacking gel and carefully transfer the resolving gel into a plastic container containing blotting buffer.
2. Submerge the gel into blotting buffer and incubate for 10 min.
3. Change to a fresh blotting buffer and incubate for another 10 min.
4. Repeat **step 3** and incubate for additional 10 min (*see* **Note 25**).
5. Cut a nitrocellulose membrane to the size of the gel and immerse in blotting buffer. Incubate for 30 min.

6. Assemble the blotting sandwich in a plastic tray containing blotting buffer according to MiniProtean manufacture instructions (*see* **Note 26**).
7. Remove any air bubbles by rolling a glass rod on the blotting sandwich.
8. Load the blotting sandwich into the electrode module with the black side of the sandwich facing the black side of the electrode module. Place the complete module on top of magnetic table stirrer.
9. Slot the ice container to the electrode module to allow cooling during blotting.
10. Place and start the magnetic stirrer.
11. Fill the electrode tank with blotting buffer and plug the electrodes to the powerpack.
12. Blot the gel for 1 h at 350 mA.
13. Remove the blots out of the sandwich and proceed with immunoblot protein detection.

3.16 Western Blot Using GenScript eBlot Semidry Transfer System (Timing 20 min)

1. After electrophoresis (Subheading 3.14) rinse the gel with deionized water for 10 min on a horizontal shaker.
2. Decant the water and repeat **step 1** two more times.
3. Set up the eBlot apparatus according to manufacture instructions.
4. Equilibrate a precut nitrocellulose membrane for 1 min in equilibration buffer.
5. Place the equilibrated membrane on the eBlot Anode pad and aspirate any air bubbles using a plastic shovel.
6. Carefully place the pre-run SDS-PAGE gel on top of the membrane and remove any air bubbles.
7. Place an appropriate size gel window (88 × 78 mm) ensuring the gel window fully covers the margins of the membrane.
8. Place the eBlot Cathode pad and close the lid.
9. Set running time to 10 min and start the eBlot run.
10. Remove the blots and proceed with immunoblot protein detection and glycosylation test (*see* **Note 27**).

3.17 Pollen Tube Viability, Intactness, and Secretory Activities

1. Viability and intactness is assessed by visualizing cytoplasmic streaming of the pollen tubes using differential interference contrast microscopy (DIC) (Subheading 3.18) as well as using Alexander staining (Subheading 3.19) and propidium iodide (PI) (Subheading 3.20) to estimate proportions of pollen tube burst. Pollen tubes from the SIV-PS assay are

reproducibly intact with active cytoplasmic streaming. The secretion activities of the SIV-PS pollen tubes can be tested using a combination of pharmacological with endocytic markers.

2. Secreted proteins that follow the canonical secretory pathway are glycosylated while translocated into the lumen of endoplasmic reticulum. To detect the N-glycoproteome of the SIV-PS secretome, secreted proteins are separated on SDS-PAGE, and thereafter transferred onto nitrocellulose filter by western blot. Obtained blots are incubated with Concavalin A (ConA) as primary antibody followed by the incubation with horseradish peroxidase (HR) as a secondary antibody. Glycosylated fraction of the secretome is then visualized by monitoring bound peroxidase activities in the presence of 4-chloro 1-naftol and 30% hydrogen peroxide as a substrate. The chemiluminescent reaction results in the development of purple colored bands as sites of HR activity. The reaction is stopped by the addition of water. Stained protein bands can be excised and identified by LC-MS/MS.

3.18 Cell Viability: Pollen Tubes Cytoplasmic Streaming (Timing 20 min)

1. Place a single pollen tubes bundle onto a microscope glass slide containing 20 μL of fresh pollen tube germination media. Cover the slide with a cover slip.
2. Using a confocal microscope or any DIC equipped microscope, set the microscope to DIC mode, and observe streaming of the pollen tube cytoplasm. A viable pollen tube should show a fast inwards fountain like flow of the cytoplasm originating from the pollen tube base toward the tip. Simultaneously, quantify integrity of the pollen tubes (*see* **Note 28**, *see Troubleshooting*—Table 1).

3.19 Cell Viability: Alexander Staining (Timing 30 min)

1. Transfer a single pollen tube bundle onto a microscope glass slide containing 10 μL of Alexander staining solution. Cover the slide with a cover slip.
2. Incubate the pollen tube samples in the staining solution at room temperature for few minutes.
3. Visualize by bright-field microscopy. Viable pollen tubes are stained pink-to-purple, nonviable cells are stained green.

3.20 Cell Viability: PI Staining (Timing 30 min)

1. Transfer a single pollen bundle onto a microscope glass slide containing 10 μL of 20 μg/mL PI solution. Cover the slide with a cover slip and incubate in the dark for 10 min.
2. Visualize the staining with appropriate filter setting, e.g., TRITC HYQ filter set (590–650 nm, Nikon) in the Nikon TE2000 fluorescence microscope or the RFP channel configuration using the 560 nm laser on the confocal microscope.

Viable cells should be stained only in the pollen tube cell wall and the dye should not have penetrated the pollen tube. Viability can be estimated by scoring the ratio of viable:nonviable pollen tubes.

3.21 N-Glycosylation Test

1. Secreted proteins that follow the ER-Golgi pathway are post-translationally modified by addition of glucan moieties termed N-glycosylation. Detection of protein glycosylation provides independent verification of the pathway of protein secretion.
2. Resolve the secretome, ~40 μg per sample, on SDS-PAGE (Subheadings 3.14–3.16) and perform western blot.
3. Stain the nitrocellulose membrane with Ponceau S staining to verify western blot protein transfer.
4. Incubate the membrane in TTBS buffer and shake at 25 rpm for 30 min.
5. Exchange with fresh TTBS buffer, incubate for another 30 min.
6. Incubate the membrane with Concavalin A (2.5 mg/100 mL) in TTBS buffer for 1 h.
7. Wash the membrane with TTBS buffer for 30 min.
8. Incubate the membrane with horseradish peroxidase (5 mg/100 mL) in TTBS buffer for 45 min.
9. Transfer the membrane to a container containing detection buffer, 45 mg 4-chloro 1-naftol in 15 mL methyl alcohol and 60 ml 10 mM Tris–HCl, pH 6.8.
10. To visualize the staining, slowly add 25 μL of 30% hydrogen peroxide at one time to the detection while constantly mixing.
11. Continue to add 25 μL of 30% hydrogen peroxide (maximum 75 μL) until desired intensity of the bands has been achieved.
12. Stop the reaction by replacing the detection buffer with deionized water (*see* **Note 29**).

3.22 Tentative Bioinformatics Workflow

Following label-free quantitative LC-MS/MS, evaluate secreted proteins as follows;

1. Consider protein groups as secreted based on Top3 protein algorithms stipulating that (a) the ratio of the calculated median of a protein group in a sample to the average median of the control unpollinated samples is >3 and (b) the number of total peptides is, at a minimum 3, and that (c) the protein group must be upregulated in at least half of the replicates. For sample comparison, individual protein accessions from replicates of all sample types are combined into supergroups (SGs).

2. Filter out protein groups according to the algorithms above. Alternative (less stringent) approaches can be adopted after thorough evaluations based on custom LC-MS/MS output.
3. *Using the obtained putative "secreted list" perform a batch analysis as follows:*
   (a) Assess presence of N-terminal signal peptide using SignalP v5.0 (http://www.cbs.dtu.dk/services/SignalP/) and iPSORT (http://ipsort.hgc.jp) algorithms.
   (b) For the remaining non-signal peptide containing proteins, assess their secretion probabilities as unconventionally secreted proteins using SecretomeP v2.0 server (http://www.cbs.dtu.dk/services/SecretomeP/) with animal sequence trained algorithms.
   (c) In parallel, use TargetP v2.0 (http://www.cbs.dtu.dk/services/TargetP-2.0/index.php) and WoLF PSORT (https://wolfpsort.hgc.jp) servers to predict subcellular localization of all accessions from the identified putative secreted list.
   (d) For posttranslational modification, use NetNGlyc v1.0 server (http://www.cbs.dtu.dk/services/NetNGlyc/) to predict protein glycosylation and complement the conventional or unconventional secretion pathway predicted from the above analysis.
   (e) To assess membrane affinity, use PredGPI (http://gpcr.biocomp.unibo.it/predgpi/), FragAnchor and big-PI (http://www.embl-heidelberg.de/beisenha/gpi_p_prediction.html) and CSS-Palm v4.0 (http://csspalm.biocuckoo.org) to predict if a protein contains a glycosylphosphatidylinositol (GPI)-anchor regions for plasma membrane anchoring or sites for posttranslational palmitoylation.
   (f) Use TMHMM v2.0 (http://www.cbs.dtu.dk/services/TMHMM/), DAS-TMFilter (http://www.enzim.hu/DAS/DAS.html), HMMTOP v2.0 (http://www.enzim.hu/hmmtop/) and PROSITE (https://prosite.expasy.org) servers to eliminate plasma membrane destined proteins that could have contaminated the extracellular secretome.
   (g) Above workflow is a guideline only; consider using additional and latest servers for more confident prediction.
   (h) From the verified in silico list, select representative candidate accessions to verify their secretion by immunolocalization and fluorophore tagging by live cell imaging.

3.23 Anticipated Results

Capturing and purification of pollen tube secreted proteins using the SIV-PS workflow yields typically 4 μg/μL of protein, but varies between 1 and 6 μg/μL depending on the efficiency of pollen tube germination from the pistil explants. This amount of protein is sufficient for performing LC-MS/MS mass spectrometry for sequencing of secreted peptides, ADH purity test, glycosylation test, SDS-PAGE protein profiling, and immunodetection. For detection of externally anchored proteins, such as GPI-anchored proteins, an enrichment step is necessary to increase detection based on our empirical results. A treatment of the pollen tubes after germination prior to secretome collection with phospholipase C should allow enrichment of the GPI-anchored proteins in the SIV-PS secretome. The estimated cytosolic contamination of the SIV-PS procedure is expected to be <1.5% ADH activities, whereas pollen tube burst consistently remained under 13%. The number of detected proteins (peptides) is variable between samples (likely reflecting secretome dynamics). In our hands we observed an average of 800 protein groups that are reliably detected. The number of groups obtained combined with quantitative LC-MS/MS allows critical evaluation of secretome dynamic between replicates and in different treatments (such as with/without ovule stimulation/BSA/Wortmanin pharmacological treatment) as well as authenticity of the pollen tube secreted proteins. Application of stringent algorithms prior to bioinformatics analysis is necessary to construct a pollen tube specific pistil induced secretome.

4 Notes

1. Preferably should be <5 months old to achieve better protein band resolution.
2. To prepare 1% (w/v) bromophenol blue, mix 0.5 g bromophenol blue with 50 mL of 50% (v/v) isopropanol.
3. To prepare blotting buffer, mix 200 mL of methanol with 100 mL of 10× electrode buffer. Adjust the volume to 1 L with water.
4. Preferably prepare the working stock fresh. However, a pre-prepared 40% stock can be stored at −20 °C for a week.
5. When the tablets are rehydrated, the volume increases fivefold. This fact should be taken into consideration when planning the container size. The diameter of a fully swollen tablet is ca 3.5 cm.
6. Stage 6 flowers can be easily identified by the pink coloration of the petals tip ends in the closed state. Care must be taken not to collect younger flowers. If younger flowers are collected, the anthers will not mature and dehisce to release mature pollen

grains. It is better to collect flowers at early mornings, at 7–8 a.m., as greenhouses get warmer in later times during the summer period.

7. If too many anthers are collected in one Petri dish, drying of the anthers will not be efficient and fewer pollen grains will be recovered.
8. Pollen grains should be collected for at least two weeks consecutively (or can be randomized over the summer to achieve an equivalent of two weeks collection) to ensure that enough materials are available for pollination. Collected tobacco pollen grains are viable for at least 2 years if stored properly at −20 °C. Pollen grain viability can be tested by in vitro germination at an interval of three months following storage. One summer collection provides enough pollen material for 2–3 years worth of SIV-PS experiment or for a general pollen grain usage.
9. It is highly recommended to set up a negative control for the secretome study. Therefore, when emasculating flowers, it should be taken into consideration that half of those flowers will be left unpollinated to establish the background set of proteins present in the media that would originate from the unpollinated pistil itself.
10. If desired, gloves can be used to avoid tanning stain on fingers. Although rarely happens, processed flowers should be netted overnight to avoid potential cross pollination and allow pistil maturity (Fig. 4). Emasculation of higher proportion of immature flowers, <48 mm, increases the chance of lack of pollen tube germination through the pistil, and thus substantially reducing the amount of protein recovered in the secretome.
11. We estimate that approximately 300 μL (~200 mg) equivalent of pollen grains are sufficient for a limited pollination of 50 emasculated flowers. This amount is adjustable depending on the individual worker.
12. Do not over pollinate the stigmas as excess pollen grains will germinate not only through the pistil but also sideways and likely to encourage fungal growth on top of the germination chamber. Fungal secreted proteins might then contaminate the secretome of the pollen tube.
13. Parafilm can be sterilized simply with 100% ethanol for 1 min and let to dry in the flow box before use.
14. For 50 pistils (equivalent of one replicate), a single 21 G × 1–1.5″ needle is sufficient for the excision of the style ends before the needle loses its sharpness.
15. If the germination medium is not sufficient to cover the pistil bottoms, additional medium can be added by pipetting through one of the pre-made holes.

16. Ensure that the interface between the beaker and the petri dish bottom is fully sealed with a wet paper towel to maintain humidity. More coworkers can be recruited to shorten the time of flower collection.
17. It is important to proceed promptly with the dissection of the pollen tube bundles to avoid potential loss of pollen tube viability and RNA degradation.
18. For faster secretome concentration, more than one filter can be used per secretome sample. At later stages of concentration, these can be pooled again to represent a single sample.
19. See Merck Millipore ultracentrifugation guide lines for your specific centrifuge.
20. Do not centrifuge at higher RCF as this will permeabilize the membrane leading to protein loss, particularly low MW proteins.
21. Where available, a plate reader provides a much more convenient way as well as accuracy of recording absorbance at the time intervals required and should be used instead of a standard single sample spectrophotometer.
22. If performing the ADH assay for the first time, include a positive control supplied. The standard curve generated does not need to be repeated if all measurements will be performed within 1 week. Where possible, use the black bottom 96 well plate to minimize light interference during incubation.
23. Free NADPH and NADH in the samples will generate background signal, therefore for each sample prepare a reaction mix without ethanol. The ethanol can be substituted with ADH assay buffer. The obtained background values will be subtracted from each sample reading prior to calculating ADH activities of the samples.
24. Always use ultrapure water to reconstitute the reagents. When reconstituting the developer solution, mix well by pipetting but do not vortex the mixture. Aliquot the developer solution in 100 μL into nontransparent 1.5 ml microcentrifuge tubes or wrap the tubes with aluminum foil to protect from direct light. Store at −20 °C and use within 2 months. After reconstituting the NADH standard, also use within 2 months.
25. The blotting buffer must be prepared fresh and should not be reused after gel equilibration.
26. The order of the sandwich layers must be kept to avoid protein transfer to the wrong layer. The orientation of the gel will be reversed (mirrored) after blotting, this should be taken into consideration with the orientation of the protein marker.

27. After the eBlot run, do not leave the membrane in the assembly unit for too long as this will result in protein diffusion and loss of protein band resolution. To ensure efficient transfer, make sure the titanium plates of the eBlot system are clean and free from salts and the nitrocellulose membrane is sufficiently equilibrated.
28. The cytoplasmic streaming events can be recorded in a time series as a proof of viability.
29. The 30% hydrogen peroxide has to be added gradually to avoid oversaturation of the blot. Once oversaturated, the staining is not reversible.

Troubleshooting advice can be found in Table 1.

Acknowledgments

We thank Jana Feciková for her help with flowers emasculation and secretome concentration and Anna J. Wiese for help with imaging. This work was supported by the Czech Science Foundation grants 17-23203S, 18-07027S and 18-02448S, and by Ministry of Education, Youth and Sport CR from European Regional Development Fund-Project "Centre for Experimental Plant Biology" (No. CZ.02.1.01/0.0/0.0/16_019/0000738). We acknowledge the Imaging Facility of the Institute of Experimental Botany, Czech Acad. Sci. supported by the MEYS CR (LM2018129 Czech-BioImaging and OPPK CZ.2.16/3.1.00/21519 Projects). *Author contributions*: S.H. designed and wrote the protocol. D.H. edited the manuscript.

References

1. Frantz C, Stewart KM, Weaver VM (2010) The extracellular matrix at a glance. J Cell Sci 123 (Pt 24):4195–4200
2. Johnson MA, Harper J, Palanivelu R (2019) A fruitful journey: pollen tube navigation from germination to fertilization. Annu Rev Plant Biol 70:809–837
3. Dresselhaus T, Franklin-Tong N (2013) Male-female crosstalk during pollen germination, tube growth and guidance, and double fertilization. Mol Plant 6(4):1018–1036
4. Kessler SA, Shimosato-Asano H, Keinath NF, Wuest SE, Ingram G, Panstruga R, Grossniklaus U (2010) Conserved molecular components for pollen tube reception and fungal invasion. Science 330(6006):968–971
5. Okuda S, Tsutsui H, Shiina K, Sprunck S, Takeuchi H, Yui R, Kasahara RD, Hamamura Y, Mizukami A, Susaki D et al (2009) Defensin-like polypeptide LUREs are pollen tube attractants secreted from synergid cells. Nature 458(7236):357–361
6. Sprunck S, Rademacher S, Vogler F, Gheyselinck J, Grossniklaus U, Dresselhaus T (2012) Egg cell-secreted EC1 triggers sperm cell activation during double fertilization. Science 338(6110):1093–1097
7. Higashiyama T (2015) The mechanism and key molecules involved in pollen tube guidance. Annu Rev Plant Biol 66:393–413
8. Higashiyama T, Kuroiwa H, Kawano S, Kuroiwa T (1998) Guidance in vitro of the pollen tube to the naked embryo sac of *Torenia fournieri*. Plant Cell 10(12):2019–2032
9. Palanivelu R, Preuss D (2006) Distinct short-range ovule signals attract or repel *Arabidopsis*

thaliana pollen tubes in vitro. BMC Plant Biol 6:7

10. Hafidh S, Potesil D, Fila J, Capkova V, Zdrahal Z, Honys D (2016) Quantitative proteomics of the tobacco pollen tube secretome identifies novel pollen tube guidance proteins important for fertilization. Genome Biol 17 (1):81
11. Hafidh S, Potesil D, Fila J, Fecikova J, Capkova V, Zdrahal Z, Honys D (2014) In search of ligands and receptors of the pollen tube: the missing link in pollen tube perception. Biochem Soc Trans 42(2):388–394
12. Leydon AR, Beale KM, Woroniecka K, Castner E, Chen J, Horgan C, Palanivelu R, Johnson MA (2013) Three MYB transcription factors control pollen tube differentiation required for sperm release. Curr Biol 23 (13):1209–1214
13. Qin Y, Leydon AR, Manziello A, Pandey R, Mount D, Denic S, Vasic B, Johnson MA, Palanivelu R (2009) Penetration of the stigma and style elicits a novel transcriptome in pollen tubes, pointing to genes critical for growth in a pistil. PLoS Genet 5(8):e1000621
14. Wiśniewski JR, Ostasiewicz P, Mann M (2011) High recovery FASP applied to the proteomic analysis of microdissected formalin fixed paraffin embedded cancer tissues retrieves known colon cancer markers. J Proteome Res 10:3040–3049
15. Shevchenko A, Tomas H, Havli J, Olsen JV, Mann M (2006) In-gel digestion for mass spectrometric characterization of proteins and proteomes. Nat Protoc 1(6):2856–2860
16. Wiśniewski JR, Zougman A, Nagaraj N, Mann M (2009) Universal sample preparation method for proteome analysis. Nat Methods 6 (5):359–362
17. Terry ME, Bonner BA (1980) An examination of centrifugation as a method of extracting an extracellular solution from peas, and Its use for the study of indoleacetic acid-induced growth. Plant Physiol 66(2):321
18. Jung Y-H, Jeong S-H, Kim SH, Singh R, Lee J-E, Cho Y-S, Agrawal GK, Rakwal R, Jwa N-S (2008) Systematic secretome analyses of rice leaf and seed callus suspension-cultured cells: workflow development and establishment of high-density two-dimensional gel reference maps. J Proteome Res 7(12):5187–5210
19. Agrawal GK, Jwa N-S, Lebrun M-H, Job D, Rakwal R (2010) Plant secretome: unlocking secrets of the secreted proteins. Proteomics 10 (4):799–827
20. Alexandersson E, Ali A, Resjo S, Andreasson E (2013) Plant secretome proteomics. Front Plant Sci 4:9
21. Liu Y, Joly V, Dorion S, Rivoal J, Matton DP (2015) The plant ovule secretome: a different view toward pollen-pistil interactions. J Proteome Res 14(11):4763–4775
22. Tupý J, Süss J, Hrabětová E, Řihová L (1983) Developmental changes in gene expression during pollen differentiation and maturation in *Nicotiana tabacum* L. Biol Plant 25 (3):231–237
23. Taylor SC, Posch A (2014) The design of a quantitative western blot experiment. Biomed Res Int 2014:361590

Chapter 5

Restricted Pollination for Tracing Individual Pollen Tubes in a Pistil

Taro Takahashi, Toshiyuki Mori, and Tomoko Igawa

Abstract

As one of the essential steps to complete sexual reproduction, a pollen tube is precisely guided to an embryo sac to deliver the sperm cells. This ovule targeting by a pollen tube is one of the essential steps in pollen tube guidance. To assess the ovule targeting ability of the pollen tube from a certain mutant line, comparative analysis of pollen tube behaviors between wild-type and mutant genotypes is important. Here, we provide a protocol that traces all pollen tubes germinated from the *quartet* tetrad in a pistil by restricted pollination and aniline blue staining. By this analysis, statistical comparison between wild-type and the mutant pollen tube functions under the same in vivo condition is possible.

Key words Pollen tube, Ovule targeting, Micropylar guidance, Restricted pollination, *Arabidopsis thaliana*, *quartet* tetrad, Aniline blue staining

1 Introduction

Through evolution, plants have developed appropriate reproductive mechanisms, adapting to the shift in habitats from water to land. Plant species such as algae, mosses, and ferns that live in environments with high humidity release spermatozoa, which swim in the water toward eggs. In further evolved spermatophytes that live in relatively dry land, pollen has emerged as the male gametophyte containing male gametes [1]. The pollen of gymnosperms is mostly delivered by the wind and pollinates directly to the ovule. Male and female gametes mature during the pollen tube elongation within the ovule for several months [1]. In several gymnosperm species, the pollen tube ruptures and releases the sperms into the fluid surrounding the ovule, and the sperm swims toward the egg cell to fertilize [2]. In angiosperm pollination, pollen attaches to the stigma located at the tip of the pistil, which is far from the ovule. Since the sperm cells of angiosperms are immotile [3], the pollen tube elongates to reach the synergid cell in an ovule. Via rupture of the pollen tube, the sperm cells locate

Anja Geitmann (ed.), *Pollen and Pollen Tube Biology: Methods and Protocols*, Methods in Molecular Biology, vol. 2160,
https://doi.org/10.1007/978-1-0716-0672-8_5,

between the female gametes. Thus the mission to let male and female gametes meet is largely entrusted to a pollen tube. The merit of this strategy is that male gametes can be delivered accurately to female gametes. Most of the angiosperm pollen tubes reach the ovule within several hours after pollination, showing faster elongation than that of the gymnosperm [4]. Thus, angiosperms have acquired a mechanism to achieve rapid sexual reproduction under suitable environmental conditions [4]; that is to say, accurate pollen tube elongation is extremely important for the sexual reproduction of angiosperms.

After pollination, the pollen vegetative cell protrudes a tube, which grows by extension of the tip of the pollen tube [5]. To deliver the sperm cells to the embryo sac, angiosperms have a pollen tube guidance mechanism. Germinated pollen tubes elongate through the stigma, style, and transmitting tract toward the base of the pistil. During this process, pollen tubes acquire the ability to respond to the attractants from female gametophytes [6]. Subsequently, a pollen tube emerges from the transmitting tract into the surface of the septum, where an ovule is located (pollen tube emergence) [7, 8]. It further grows along the surface of the funiculus (funicular guidance) and eventually invades the micropyle, being attracted by the synergid cells in an embryo sac (micropylar guidance) [8, 9]. As the molecule involved in micropylar guidance, LUREs, secreted defensin-like cysteine-rich peptides that are expressed in the synergid cells, have been reported [10, 11]. In addition, receptor-like kinases MALE DISCOVERER 1 (MDIS1), MDIS1-INTERACTING RECEPTOR-LIKE KINASE (MIK1), MIK2 [12], and POLLEN RECEPTOR-LIKE KINASE 6 (PRK6) [13] localized to the plasma membrane of the pollen tube have been identified as the receptors of LUREs.

Although recent studies have discovered several micropylar guidance molecules, many factors are still unknown [8]. To evaluate the function of a candidate molecule regarding pollen tube guidance, accurate analysis that enables the comparison of the behavior of pollen tubes between the wild-type and the mutant lines for the molecule under in vivo condition is required. However, under conditions of a general pollination test in which the number of pollen grains applied to a stigma exceeds the number of ovules in a pistil, several influences affect the assessment of the phenotype of the pollen tubes. For instance, male–male repulsion between the pollen tubes [14], and a fertilization recovering system [15] may interfere with the behavior of pollen tube. To avoid these problems, restricted artificial pollination, in which the number of pollen grains is reduced to fewer than four, can be performed. This method allowed the observation of the individual pollen tubes in a pistil so that ovule targeting by each pollen tube is accurately assessed [16].

In this chapter, we provide a protocol utilizing the *quartet* (*qrt*) mutant as the pollen parent to take advantage of its physical and genetical character. The *qrt*mutant produces a pollen tetrad due to the failure of pollen grains to separate after meiosis [17], making it easy to handle and count the pollen grains. Moreover, a pollen tetrad derived from the heterozygous mutant is comprised of wild-type and mutantgenotypes at a 2:2 ratio. When pollen tubes from more than half of the pollen of the tetrad are observed in a pistil, at least one must surely be a mutant pollen tube. Therefore, the behavior of a mutant pollen tube can be statistically compared to that of wild type in the same in vivo condition. Restricted artificial pollination combined with the staining by aniline blue allows tracing individual pollen tubes in a pistil and provides evidence for the ovule targeting ability of each pollen tube.

2 Materials

2.1 Plants

Arabidopsis thaliana seeds with *qrt* background [17].

2.2 Aseptic Seeding

1. Surface-sterilizing solution: Adjust sodium hypochlorite solution (antiformin) to 1% of effective chlorine concentration with distilled water. Add one drop of Tween-20 per 50 ml of the solution.
2. Germination medium: Murashige and Skoog (MS) medium (pH 5.8) supplemented with 1% sucrose, 0.5 g/L 2-morpholinoethanesulfonic acid (MES), 0.8% agarose, and a suitable kind and concentration of antibiotics for the particular seed. After autoclaving, dispense the medium into a 9-cm plate.

2.3 Aniline Blue Staining

1. Fixation solution: mix acetic acid and ethanol at 1:3 (v/v) ratio.
2. Ethanol series for rehydration: 70%, 50%, and 30% ethanol.
3. Alkaline treatment solution: 8 M NaOH.
4. Aniline blue solution: 0.1% (w/v) aniline blue in 108 mM K_3PO_4. After dissolving, store at 4 °C overnight. Then filtrate the solution through one spoon of active carbon powder placed on filter paper set on a funnel for decolorization. After the filtration, add glycerol at the final concentration of 2% (v/v). Store at 4 °C.

2.4 Specific Laboratory Equipment and Supplies

1. Temperature- and light-controlled growth chambers.
2. Stereo microscope.
3. Tapered bristle of a toothbrush.
4. Head magnifying glasses.
5. Forceps (No. 5 is preferred).

6. 1.5-mL microtube.
7. 18-G injection needle.
8. 20-mL plastic syringe.
9. 24-Well plate.
10. Epifluorescence microscope, equipped with objective lenses (4×, 10×, and 20×), a digital camera, and an appropriate excitation and emission filter for UV irradiation.

3 Methods

3.1 Growing the Plant Material

1. Put the seeds into a 1.5-mL tube.
2. Sterilize the surface of the seeds. Add 1 mL of surface-sterilizing solution to the tube then invert for 5–10 min. Replace the surface-sterilizing solution with sterilized distilled water (DW) and invert several times. Repeat this step 3–5 times to wash out the remaining sodium chloride completely.
3. Place the seeds separate from each other on the germination medium using a micropipette.
4. Cover the medium plate with aluminum foil, then leave it at 4 °C at least one night (or up to 7 days) for vernalization.
5. Transfer the plate to a growth chamber where the environment is set at 22 °C and a 16-h light/8-h dark cycle.
6. Acclimatize the seedlings grown on the plate medium for about 2 weeks (*see* **Note 1**). Transfer the seedlings to the pots filled with soil (*see* **Note 2**). Cover the pots with plastic wrap to keep high humidity for a few days. In a week, make several holes in the wrap, gradually broaden them, and finally remove the wrap.
7. Cut the first bolted flower stalk to enhance the growth of lateral buds (*see* **Note 3**).

3.2 Restricted Pollination

1. Half a day or 1 day before the artificial pollination (*see* **Note 4**), emasculate the female parent. Remove sepals, petals, and stamens from the bud at stage 11 [18] of which only the tips of petals appear at the bud top, using forceps (*see* Fig. 1).
2. Under a stereo microscope, dissect the *qrt*-background flower at stage 13 [18] of which the anther is dehiscent.
3. Pick up one tetrad from the anther using a tapered bristle of a toothbrush (*see* Fig. 2a), confirming under a stereo microscope.
4. Pollinate one *qrt* tetrad to the emasculated stigma (*see* Fig. 2b) (*see* **Note 5**). Using the head magnifying glasses is helpful to handle a tapered bristle.

Fig. 1 Emasculation of the flower bud. (**a**) Flower bud at stage 11 [18] showing only the tips of the petals at the top. (**b–d**) Flower bud with removed sepals (**b**) and petals (**c**) and that was emasculated completely (**d**). Anther dehiscence has not occurred yet at this stage. Scale bar = 0.5 mm

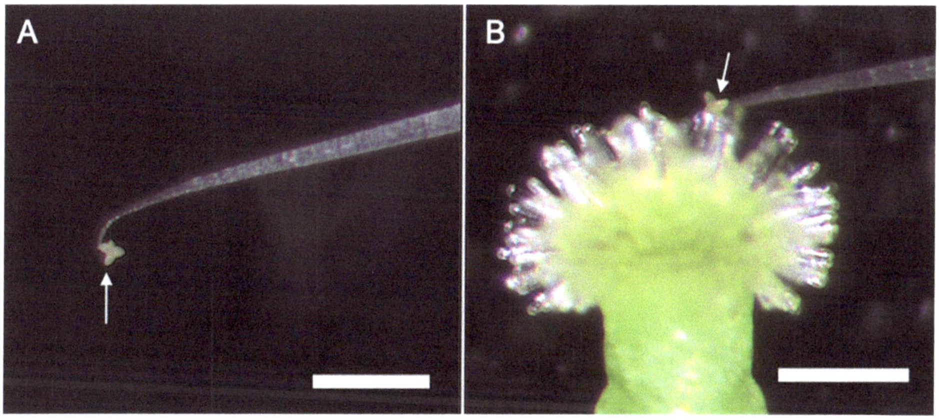

Fig. 2 Restricted pollination using a *qrt* tetrad. (**a**) A *qrt* tetrad (arrow) picked up with a tapered bristle of a toothbrush. (**b**) Restricted pollination of a single *qrt* tetrad (arrow). Although the image was captured with a stereo microscope, actual handling should be done with the unaided eye or through head magnifying glasses. Scale bar = 200 μm

3.3 Aniline Blue Staining (See Note 6)

1. At least 12 h after pollination, collect the pollinated pistils into a 1.5-mL microtube using forceps (*see* **Notes 7** and **8**).
2. Add 1 mL of fixation solution to the microtube and seal the closed lid with parafilm.
3. Aspirate the air in the tube using an 18-gauge injection needle attached to a 20-mL syringe until no bubbles come out from the pistils.
4. Leave the tube for at least 2 h after sealing the hole on the lid with parafilm or cellophane tape.

5. Remove the fixation solution, then add 1 mL of 70% ethanol. Leave for 10 min.
6. Replace 70% ethanol with 50% ethanol, then leave for 10 min.
7. Do same treatment with 30% ethanol and distilled water, as described in **step 5**.
8. Transfer the pistils to one well of the 24-well plate filled with 1–2 mL of alkaline treatment solution (*see* **Notes 9** and **10**). Leave the plate overnight.
9. Carefully replace the alkaline treatment solution with an equal volume of distilled water. Leave the plate for 10 min.
10. Remove the water and add 1 mL of aniline blue solution. Then cover the plate with aluminum foil and leave it for at least 2 h.

3.4 Microscopy

1. Transfer the pistils to a drop of 25 μL of aniline blue solution on a slide glass (*see* **Note 11**).
2. Put a cover glass gently on the specimen from the side of the flower stem (*see* **Note 12**).
3. Observe the stained pollen tubes using an appropriate objective lens under a UV irradiation condition. Acquire images using the equipped digital camera.
4. Confirm whether the number of pollen tubes in the pistil is more than half of the pollinated pollen from the tetrad (*see* **Notes 13** and **14**).

4 Notes

1. A seedling with 2–4 true leaves that are well expanded is tolerant for acclimatization.
2. Use of autoclaved soil is recommended to reduce the risk of biological contamination. Add some water and mix with the soil before autoclaving if the soil is extremely dry. Up to three seedlings per 6-cm-diameter pot are suitable for the growth of the plants. Supplying liquid fertilizer enhances the vigorous growth.
3. The first flower stalk can be cut as soon as it appears (about 1 week after acclimatization). Two to three weeks after cutting, the lateral buds should grow into suitable size (about 25 cm in height) with many vigorous flower buds available for artificial pollination.
4. Use of flower buds with non-dehiscent anthers is strongly recommended to avoid natural self-pollination. However, the pistil at stage 11 has not reached maturation as shown by short papilla cells (*see* Fig. 1b–d). Papilla cells that are too short seem

unready for pollen germination. Therefore, it is better to use pistils after half a day to 2 days from emasculation for artificial pollination. In addition, emasculation of flower buds at a younger stage may attenuate the pistil that is still not tolerant to the outer environment.

5. Water absorption of pollen grains from papilla cells is essential for the pollen tube germination. Due to the tetrapod-like arrangement of a *qrt* pollen tetrad, not all of the pollen grains can attach to the papilla cell if the tetrad is only placed on the stigma. The tetrad should be placed into the papilla cells using the tapered bristle of a toothbrush.
6. All procedures are performed at room temperature unless specified otherwise.
7. In the case of excess pollination, almost all the ovules will accept a pollen tube in 10 h from pollination [19]. Since the time point may alter depending on each experimental condition, it is recommended to collect the pollinated pistil at least after 12 h from pollination. Pollen tubes can be visualized by aniline blue until 48 h after pollination.
8. Pistils collected together with the flower stem (ca. 1 cm) are easy to handle during aniline blue staining.
9. Before the transfer, wipe off the water from the pistils to avoid diluting the alkaline treatment solution.
10. If the pistils float on the alkaline treatment solution, steep them gently using forceps.
11. After the alkaline treatment, pistils become extremely softened. Handle the pistils gently by pinching the flower stalk region.
12. Avoid pushing the cover glass. The weight of the cover glass pushes the pistil to expand the ovary wall.
13. Not always that all the pollen of the tetrad will germinate (*see* Fig. 3a).
14. The full length of each pollen tube (from the papilla to micropyle) cannot be observed because of the high autofluorescence in the stigma tissue (*see* Fig. 3b). However, each pollen tube from the transmitting tract to the ovule is observed when Subheadings 3.4, **steps 1** and **2** are performed correctly (*see* Fig. 3c).

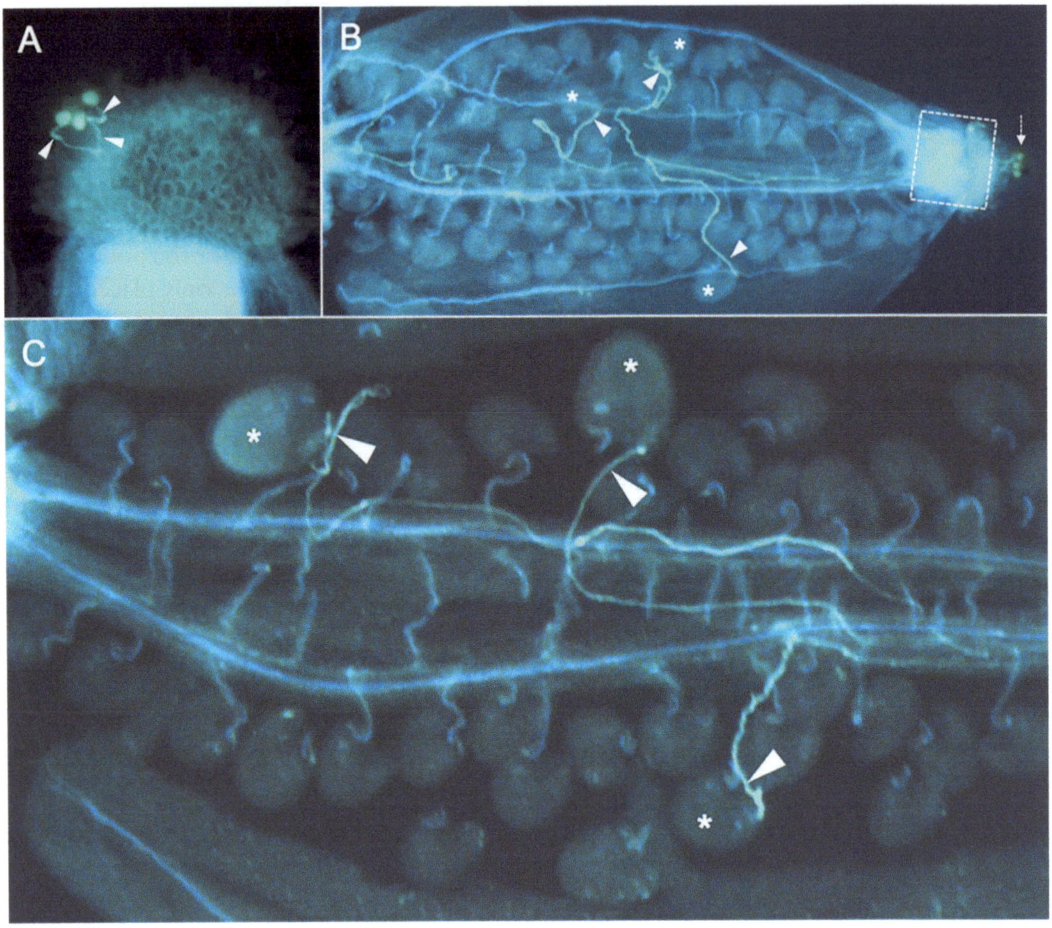

Fig. 3 Pollen tube images in restricted pollination test. (**a**) Germinated pollen tubes (triangles) from a *qrt* tetrad on the papilla cells. (**b**) Image of a whole pistil pollinated with a single *qrt* tetrad (dashed arrow). The area marked with a dashed box represents the stigma region showing the intense autofluorescence. Pollen tubes and the ovules accepting a pollen tube are labeled with triangles and asterisks, respectively. In this pistil, three pollen germinated. (**c**) A good example image. All the ovules that accepted a pollen tube (asterisks) are in the same focus plane. The trace of each pollen tube (triangles) from the transmitting tract to each ovule is well visualized. In this pistil, three pollen of a *qrt* tetrad germinated

Acknowledgments

This work was supported by JSPS KAKENHI Grant Number JP17H05832 (to T. I.), in addition to the NAGASE Science Technology Foundation (to T.I.) and the Strategic Priority Research Promotion Program on Phytochemical Plant Molecular Sciences, Chiba University.

References

1. Fernando DD, Quinn CR, Brenner ED et al (2010) Male gametophyte development and evolution in extant gymnosperms. Int J Plant Dev Biol 4:47–63
2. von Aderkas P, Prior NA, Little SA (2018) The evolution of sexual fluids in gymnosperms from pollination drops to nectar. Front Plant Sci 9 (1844). https://doi.org/10.3389/fpls.2018.01844
3. Hodges ME, Wickstead B, Gull K et al (2011) Conservation of ciliary proteins in plants with no cilia. BMC Plant Biol 11(185). https://doi.org/10.1186/1471-2229-11-185
4. Higashiyama T (2017) Pollen tube navigation can inspire microrobot design. Sci Robot 2(8). https://doi.org/10.1126/scirobotics.aao1891
5. Zonia L (2010) Spatial and temporal integration of signalling networks regulating pollen tube growth. J Exp Bot 61(7):1939–1957. https://doi.org/10.1093/jxb/erq073
6. Johnson MA, Harper JF, Palanivelu R (2019) A fruitful journey: pollen tube navigation from germination to fertilization. Annu Rev Plant Biol 70:809–837. https://doi.org/10.1146/annurev-arplant-050718-100133
7. Hulskamp M, Schneitz K, Pruitt RE (1995) Genetic evidence for a long-range activity that directs pollen tube guidance in Arabidopsis. Plant Cell 7(1):57–64. https://doi.org/10.1105/tpc.7.1.57
8. Mizuta Y, Higashiyama T (2018) Chemical signaling for pollen tube guidance at a glance. J Cell Sci 131(2). https://doi.org/10.1242/jcs.208447
9. Palanivelu R, Tsukamoto T (2012) Pathfinding in angiosperm reproduction: pollen tube guidance by pistils ensures successful double fertilization. Wiley Interdiscip Rev Dev Biol 1 (1):96–113. https://doi.org/10.1002/wdev.6
10. Okuda S, Tsutsui H, Shiina K et al (2009) Defensin-like polypeptide LUREs are pollen tube attractants secreted from synergid cells. Nature 458(7236):357–361. https://doi.org/10.1038/nature07882
11. Takeuchi H, Higashiyama T (2012) A species-specific cluster of defensin-like genes encodes diffusible pollen tube attractants in Arabidopsis. PLoS Biol 10(12). https://doi.org/10.1371/journal.pbio.1001449
12. Wang T, Liang L, Xue Y et al (2016) A receptor heteromer mediates the male perception of female attractants in plants. Nature 531 (7593):241–244. https://doi.org/10.1038/nature16975
13. Takeuchi H, Higashiyama T (2016) Tip-localized receptors control pollen tube growth and LURE sensing in Arabidopsis. Nature 531(7593):245–248. https://doi.org/10.1038/nature17413
14. Shimizu KK, Okada K (2000) Attractive and repulsive interactions between female and male gametophytes in Arabidopsis pollen tube guidance. Development 127:4511–4518
15. Kasahara RD, Maruyama D, Hamamura Y et al (2012) Fertilization recovery after defective sperm cell release in Arabidopsis. Curr Biol 22:1084–1089. https://doi.org/10.1016/j.cub.2012.03.069
16. Takahashi T, Honda K, Mori T et al (2017) Loss of GCS1/HAP2 does not affect the ovule-targeting behavior of pollen tubes. Plant Reprod 30(3):147–152. https://doi.org/10.1007/s00497-017-0305-2
17. Preuss D, Rhee SY, Davis RW (1994) Tetrad analysis possible in Arabidopsis with mutation of the QUARTET (QRT) genes. Science 264:1458–1460. https://doi.org/10.1126/science.8197459
18. Smyth DR, Bowman JL, Meyerowitz EM (1990) Early flower development in Arabidopsis. Plant Cell 2:755–767. https://doi.org/10.1105/tpc.2.8.755
19. Kasahara RD, Maruyama D, Higashiyama T (2013) Fertilization recovery system is dependent on the number of pollen grains for efficient reproduction in plants. Plant Signal Behav 8(4):e23690. https://doi.org/10.4161/psb.23690

Chapter 6

Semi-In Vivo Assay for Pollen Tube Attraction

Sheng Zhong, Zhijuan Wang, and Li-Jia Qu

Abstract

In flowering plants, each pollen tube delivers two sperm cells into the ovule to complete double fertilization. During the process, pollen tubes need to be navigated into the ovule, where accurate and complex pre-ovule guidance and ovule guidance are required. In recent years, different methods have been established to study those genes involved in the regulation of pollen tube guidance. Semi-in vivo ovule targeting mimics in vivo pollen tube micropylar guidance, and the semi-in vivo ovule targeting assay has been used to investigate function of genes involved in micropylar guidance. Moreover, the ovule targeting assay is the best way to do live cell imaging, which facilitates observation of pollen tube reception, synergid cell degeneration, and semi-in vivo gamete fusion. Meanwhile, semi-in vivo pollen tube attraction assay is another useful method to directly determine whether a certain molecule has pollen tube attraction activity.

Key words Pollen tube guidance, Ovule targeting, Pollen tube attraction, Pollen tube competition, Live cell imaging

1 Introduction

Pollen tubes are one of the most important structures that are evolved in angiosperm plants, because, leaving the water environment, the two sperm cells are passive cargos and can only be transported by pollen tubes into the ovule to complete double fertilization [1]. Successful double fertilization requires the fine-tuned but complex regulation of pollen tube guidance which navigates pollen tubes to accurately grow into the micropyle of a receptive ovule [2, 3]. Pollen tube guidance is divided into two major processes, pre-ovular guidance and ovular guidance [4]. Pre-ovular guidance refers to the guidance from the stigma to the transmitting tract as well as in the transmitting tract. Several factors were identified to be involved in the pre-ovular guidance because these factors either possess the pollen tube attraction activity in vitro or loss of their function resulted in obvious pollen tube guidance defects in the style or in the transmitting tract [5–8]. Compared with pre-ovular guidance, ovular guidance is much more intensively studied. Ovular guidance seems to start from

Anja Geitmann (ed.), *Pollen and Pollen Tube Biology: Methods and Protocols*, Methods in Molecular Biology, vol. 2160,
https://doi.org/10.1007/978-1-0716-0672-8_6,

pollen tube emerge from the transmitting tract and is further divided into two subprocesses, funicular guidance and micropylar guidance. Peptide/receptor signaling pathways have been reported to play essential roles in these two subprocesses [9–13].

Several different strategies have been established and used in the previous pollen tube guidance studies [10, 14–17]. Among them, semi-in vivo ovule targeting assay and pollen tube attraction assay are the two most popular methods. Semi-in vivo ovule targeting assay is a useful system to investigate whether a protein-of-interest is involved in micropylar guidance [17, 18]. Previous studies also show that the funiculus affects pollen tube targeting efficiency in the semi-in vivo ovule targeting assay, suggesting an important role of the funiculus in pollen tube guidance [17]. In addition, the semi-in vivo ovule targeting assay is also a very powerful method to do live cell imaging, which has been successfully applied to examine degeneration of synergid cells, dynamic changes of Ca^{2+} in pollen tubes during the pollen tube reception, and gamete fusion [19–22]. The semi-in vivo pollen tube attraction assay is also a very useful assay. It directly determines whether a molecule has pollen tube attraction activity or not. In addition, it is also used to detect the response of a pollen tube to the identified pollen tube attractants [10–12].

Here we provide a detailed protocol for both assays. We first introduce the protocol for semi-in vivo pollen germination, since perfectly geminated pollen tubes are the prerequisite for both assays. For the ovule targeting assay, the mature ovules are placed in front of the stigma-style. Use of fluorescence-labeled pollen tubes are strongly recommended to better observe the behavior of pollen tubes in the vicinity of the micropyle. For the pollen tube attraction assay, making gelatin beads properly mixed with pollen tube attractants is the most difficult step, and requires practicing.

2 Materials

2.1 Plants Material and Equipment

1. Plants: wild-type plants, *ms1* (CS301367, Col-0 background) mutant [23], LAT52pro:GUS transgenic plants; all the plants are in bolting condition.
2. Microscope: spinning disc confocal microscope (Zeiss Cell Observer SD, Zeiss, Germany).

2.2 Semi-In Vivo Pollen Tube Growth

Semi-in vivo pollen tube growth medium: for both semi-in vivo ovule targeting assay and pollen tube attraction assay.

1. Semi-in vivo pollen germination medium for *Arabidopsis thaliana* pollen [13]: 0.01% H_3BO_3, 5 mM $CaCl_2$, 5 mM KCl, 1 mM $MgSO_4$, 15% sucrose, adjusted pH to 7.5 with KOH. The stock can be stored at 4 °C for no longer than 1 month.

Add 1.5% low gelling temperature agarose II™ into the stock when use.

2. Glass bottom culture dish.
3. Growth chamber.

2.3 Isolation of Mature Ovules (for Ovule Targeting Assay)

1. Glass slide.
2. Double-sided tape.
3. 1.0 mL syringe needle.
4. Straight tweezer.
5. Dissecting microscope.

2.4 Pollen Tube Attraction Assay

2.4.1 Preparation of Glass Needles

1. Borosilicate glass.
2. Horizontal puller.

2.4.2 Preparation of Beads

1. Silicone oil.
2. Gelatin.
3. Thermomixer comfort.
4. Ultrasonic cleaning.

2.4.3 Micromanipulation

1. Micromanipulator (e.g., TransferMan NK2, Eppendorf).
2. Adapter.
3. Grip head.

3 Methods

3.1 Semi-In Vivo Pollen Tube Growth

1. Add 0.012 g agarose into 1 mL pollen tube growth medium in a 1.5 mL microcentrifuge tube and melt the agarose in hot water or microwave (*see* **Note 1**).
2. Add pollen germination medium (50–80 μL) on glass bottom of the culture dish (*see* **Note 2**, Fig. 1a, b).
3. Once when the medium is completely solidified, cover the culture dish with a lid, and keep the medium at room temperature until use (*see* **Note 3**).
4. Cut the stigma-style from *ms1* or emasculated plant by 1.0 mL syringe needle under a dissecting microscope (*see* **Note 4** and Fig. 2a, b).
5. Transfer the excised stigma-style to the solid medium as quickly as possible (Fig. 2c).
6. Pollinate appropriate number of pollen grains on the excised stigma (*see* **Notes 5–7**).

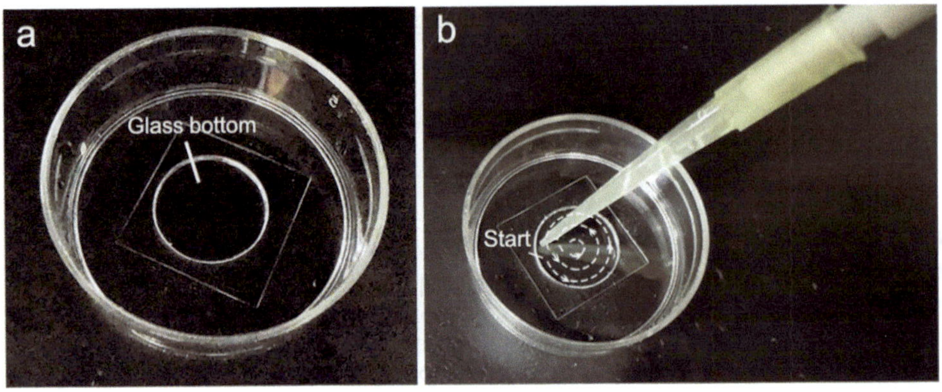

Fig. 1 Preparation of pollen tube growth medium. (**a**) The glass bottom culture dish. (**b**) Add pollen tube growth medium on the glass bottom. The white dotted line indicates the way and direction to add the pollen germination medium

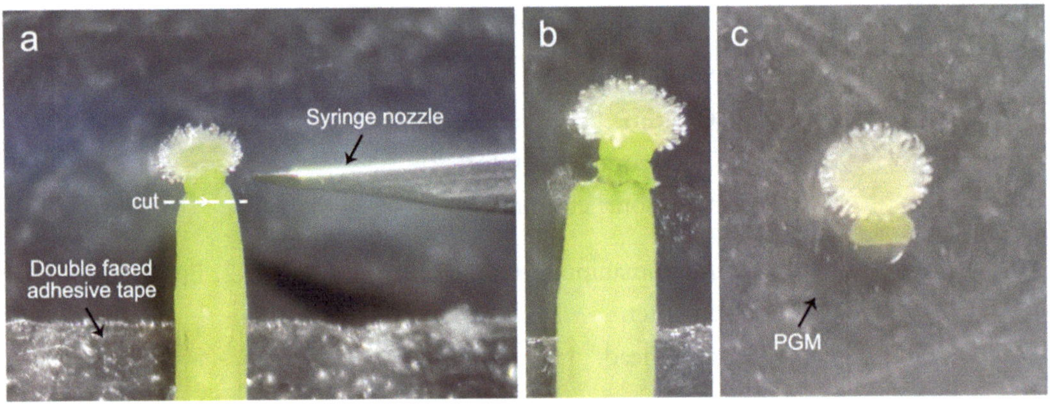

Fig. 2 Cut a stigma-style from a pistil. (**a**) 1.0 mL syringe needle is used to cut the stigma-style. The white dotted line indicates the cutting position. (**b**) The pistil-style is cut by the syringe needle. (**c**) The excised pistil-style has been quickly transferred to the medium. *PGM* pollen germination medium

7. Seal the culture dish with parafilm, and place the culture dish in a growth chamber.
8. Incubate at 22 °C for 2.5–4 h.

3.2 Ovule Targeting Assay

After **step 4** in Subheading 3.1, the following steps are proceeded for ovule targeting assays.

1. Put a pistil of *ms1* or emasculated plants on the double-faced adhesive tape on the glass slide.
2. Cut the carpel open along the ventral suture with a 1.0 mL syringe needle or straight tweezer (*see* **Note 8**).
3. Remove the mature ovules (with funiculus attached) from the septum (*see* **Note 9**).

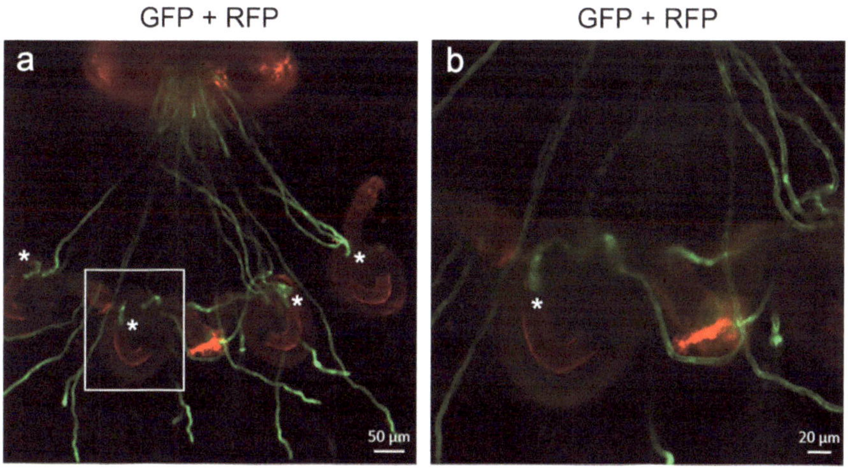

Fig. 3 Ovule targeting assays. (**a**) Pollen tubes enter the ovule 2.5–4 h after incubation, 10× objective. Scale bar: 50 μm; (**b**) Enlarged view of white box in (**a**), 20× objective. *LAT52pro:GFP* pollen donor is used. The asterisk indicates the pollen tube that has entered the micropyle. Scale bars, 20 μm

4. Put the excised ovules on the medium as quickly as possible (*see* **Notes 10–12**).
5. Adjust the stigma-style to the appropriate position on the medium (*see* **Note 13**).
6. Perform **steps 6–8** in Subheading 3.1.
7. Observe pollen tube targeting under a spinning disc confocal microscope (Fig. 3a, b).

3.3 Pollen Tube Attraction Assay

After **step 4** in Subheading 3.1, the following protocols in Subheadings 3.3.1–3.3.3 are executed for pollen tube attraction assays.

3.3.1 Making Glass Needles

1. Switch on horizontal puller and set the parameters (Fig. 4a, b).
2. Put the borosilicate glass into the needle puller symmetrically (Fig. 4c).
3. Start the needle puller (Fig. 4d).
4. Take out the borosilicate glass needle (Fig. 4e) and store them in a clean box.
5. Cut the sharp tip of the borosilicate glass needle with a mini scissor (Fig. 4f).

3.3.2 Preparation of Beads

1. Add 0.012–0.014 g of gelatin in 100 μL pollen growth medium (without agarose) in a microcentrifuge tube.
2. Melt it at 50 °C for about 25–30 min (*see* **Note 14**).
3. Ultra-sonicate it for 1 min (*see* **Note 15**).
4. Mix 1.5 μL of molecule-of-interest together with 1.5 μL of melting gelatin in a 1.5 mL microcentrifuge tube (*see* **Note 16**).

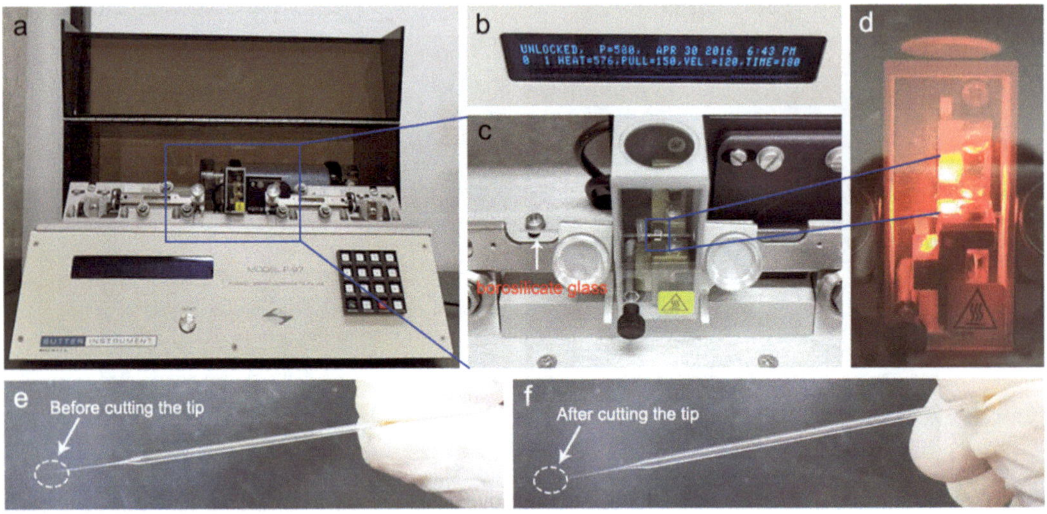

Fig. 4 Preparation of borosilicate glass needle for beads picking. (**a**) The horizontal puller. (**b**) Parameters for pulling borosilicate glass needle. (**c**) The enlarged view of the pulling section. (**d**) Borosilicate glass in melting process. (**e**) The borosilicate glass needle right after pulling process. (**f**) The borosilicate glass needle right after the sharp tip is cut

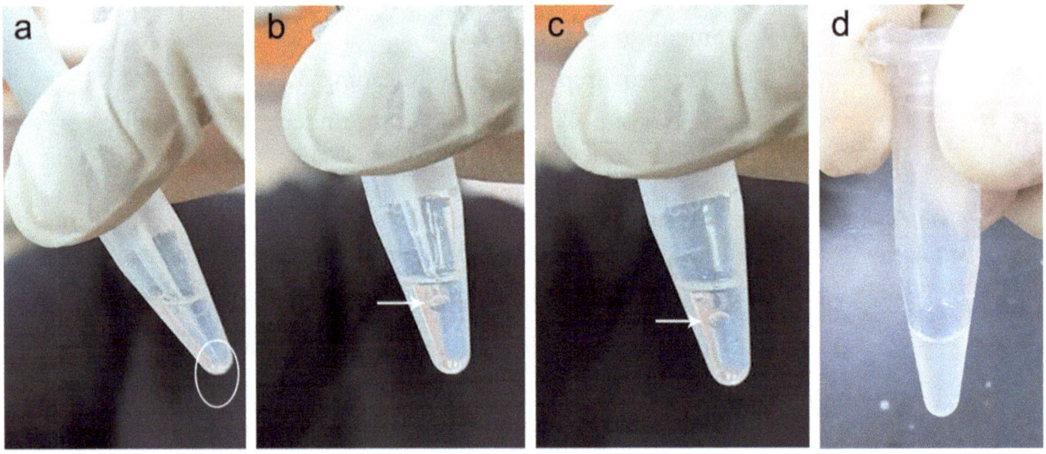

Fig. 5 Preparation of beads. (**a**) Aspirate the liquid mixture at the bottom with a pipette. (**b**) Release the liquid drop at the center of the silicone oil. (**c**) The liquid drop (i.e., gelatin and molecule/protein mixture) at the center of the silicone oil. (**d**) The liquid drop has been dispersed into tiny beads by vortex. The white circle indicates the mixture at the bottom of the microcentrifuge tube, the white arrow indicates the liquid drop

5. Add 150 μL silicone oil into the 1.5 mL microcentrifuge tube from **step 4**.
6. Aspirate the 3 μL liquid mixture from the bottom of the tube by a 10 μL pipette, and release the liquid drop in the center of the silicone oil in the tube (*see* **Note 17**, Fig. 5a–c).
7. Vortex vigorously for more than 15 s (*see* **Note 18**, Fig. 5d).

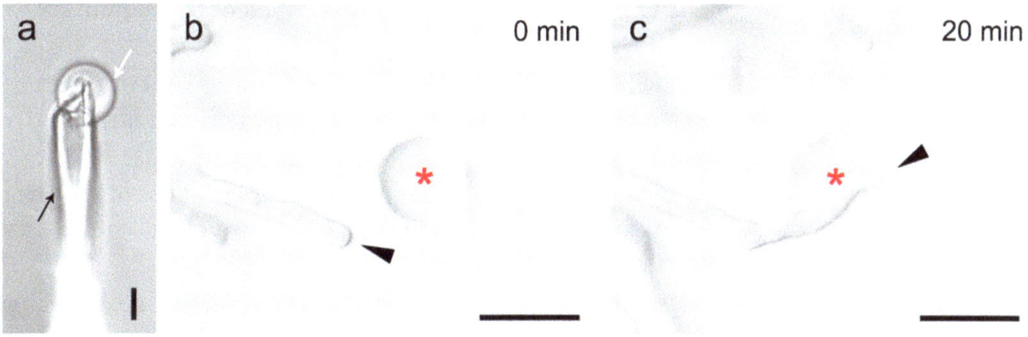

Fig. 6 Micromanipulation. (**a**) One gelatin bead is picked with a glass needle. The white arrow indicates the gelatin bead, the black arrow indicates the glass needle. Scale bars, 10 μm. (**b**) The bead is placed besides a pollen tube. (**c**) The pollen tube grows and turns toward the bead. AtLURE1.2 peptide is embedded in the gelatin beads. Arrowheads indicate the direction of pollen tube growth. Red asterisks indicate the center of the beads. Scale bars, 25 μm

8. Put the microcentrifuge tube on the ice immediately (*see* **Note 19**).

3.3.3 Micromanipulation

1. Fix the prepared glass needle in the grip head of the micromanipulator (*see* **Note 20**).
2. Add trace amounts (less than 0.5 μL) of the beads on the pollen tube growth medium.
3. Pick up one of the beads with the glass needle tip (Fig. 6a).
4. Transfer the beads and place it at the front of a pollen tube tip (*see* **Note 21** and Fig. 6b).
5. Take a picture immediately after the bead is placed to record the original position of the pollen tubes (Fig. 6b).
6. Repeat **steps 3–5** with other pollen tubes (*see* **Note 22**).
7. Seal the culture dish with parafilm.
8. Observe and record the pollen tube behavior 20–30 min after the beads are placed (*see* **Note 23**, Fig. 6c).

4 Notes

1. Use hot water (above 90 °C) or microwave to melt low-melting agarose in the pollen tube growth medium. Vortex the medium to make sure that all the agarose dissolved.
2. Add medium from the edge gradually to the center for better spreading.
3. In order to avoid moisture loss, do not leave the medium exposed in the air too long. Less than 8–12 min of air exposure is recommended.

4. A 1.0 mL syringe needle is more convenient than a mini scissor to cut the stigma-style.
5. Pollination is recommended to proceed at 10 a.m., because pollen of the opening flowers are in the best condition at that time of a day.
6. Use of fluorescence-labeled pollen tubes facilitates observation and judging whether or not a pollen tube has entered the micropyle.
7. Do not pollinate too many pollen grains on one stigma, since excess germinated pollen tubes would jeopardize microscopic observation.
8. When cutting the ovary open with a syringe needle or a straight tweezer, do not cut the ovary too deep to avoid scratching the ovules.
9. Make sure that the funiculus is still attached to the mature ovule, since the funiculus affects pollen tube targeting efficiency.
10. The excised ovules will dry very quickly.
11. Four to six ovules are recommended to place in front of an excised stigma.
12. Two ovules can be placed symmetrically in front of an excised stigma to do a pollen tube competition assay.
13. Place the ovule in front of an excised stigma at the distance of 2–3 ovule length. It is difficult for the pollen tubes to grow beyond that distance to reach the ovules.
14. At 10 min and 20 min, stir the gelatin for 10 s each with a pipette tip.
15. After ultra-sonication, do not expose the microcentrifuge tube to cold temperature for long. Holding the tube in your hand would help to keep the temperature.
16. The molecule-of-interest should be mixed with the melting gelatin uniformly as quickly as possible.
17. This step should be executed as quickly as possible. The mixture will be visible as a liquid drop in the center of the silicone oil. In addition, a good liquid drop can stay stably in the center of the silicone oil.
18. The liquid drop can be dispersed by vortex into tiny beads in the silicone oil, making the solution like an emulsion.
19. Well-prepared beads with high quality maintain an emulsion-like state on ice for a long time (longer than 3 h).
20. Adjust the altitude of the glass needle, adapting to the culture dish on the microscope stage.

21. Place the beads at one side of the pollen tube tip rather than in the front center.
22. Usually, three or four semi-in vivo pollen tube growth assays, each with 3–4 pollen tubes, are recommended in one culture dish.
23. Pollen tubes are considered to be attracted when the deflection angle is greater than 20° [10].

Acknowledgements

The research in the Qu laboratory is supported by grants from National Natural Science Foundation of China (31830004, 31620103903 and 31621001) and by the Peking-Tsinghua Joint Center for Life Sciences.

References

1. Zhang J, Huang Q, Zhong S, Bleckmann A, Huang J, Guo X, Lin Q, Gu H, Dong J, Dresselhaus T, Qu L-J (2017) Sperm cells are passive cargo of the pollen tube in plant fertilization. Nat Plants 3:17079. https://doi.org/10.1038/nplants.2017.79
2. Zhong S, Qu L-J (2019) Peptide/receptor-like kinase-mediated signaling involved in male-female interactions. Curr Opin Plant Biol 51:7–14. https://doi.org/10.1016/j.pbi.2019.03.004
3. Johnson MA, Harper JF, Palanivelu R (2019) A fruitful journey: pollen tube navigation from germination to fertilization. Annu Rev Plant Biol 70:809–837. https://doi.org/10.1146/annurev-arplant-050718-100133
4. Higashiyama T, Takeuchi H (2015) The mechanism and key molecules involved in pollen tube guidance. Annu Rev Plant Biol 66:393–413. https://doi.org/10.1146/annurev-arplant-043014-115635
5. Cheung AY, Wu HM (1995) A floral transmitting tissue-specific glycoprotein attracts pollen tubes and stimulates their growth. Cell 82:383–393. https://doi.org/10.1016/0092-8674(95)90427-1
6. Wu HM, Cheung AY (1995) A pollen tube growth stimulatory glycoprotein is deglycosylated by pollen tubes and displays. Cell 82:395–403. https://doi.org/10.1016/0092-8674(95)90428-X
7. Kim S, Mollet JC, Dong J, Zhang K, Park SY, Lord EM (2003) Chemocyanin, a small basic protein from the lily stigma, induces pollen tube chemotropism. Proc Natl Acad Sci U S A 100:16125–16130. https://doi.org/10.1073/pnas.2533800100
8. Lu Y, Chanroj S, Zulkifli L, Johnson MA, Uozumi N, Cheung AY, Sze H (2011) Pollen tubes lacking a pair of K^+ transporters fail to target ovules in Arabidopsis. Plant Cell 23:81–93. https://doi.org/10.1105/tpc.110.080499
9. Márton ML, Broadhvest J, Dresselhaus T (2005) Micropylar pollen tube guidance by egg apparatus 1 of maize. Science 307:573–576. https://doi.org/10.1126/science.1104954
10. Okuda S, Tsutsui H, Shiina K, Sprunck S, Takeuchi H, Yui R, Kasahara RD, Hamamura Y, Mizukami A, Susaki D, Kawano N, Sakakibara T, Namiki S, Itoh K, Otsuka K, Matsuzaki M, Nozaki H, Kuroiwa T, Nakano A, Kanaoka MM, Dresselhaus T, Sasaki N, Higashiyama T (2009) Defensin-like polypeptide LUREs are pollen tube attractants secreted from synergid cells. Nature 458:357–361. https://doi.org/10.1038/nature07882
11. Takeuchi H, Higashiyama T (2012) A species-specific cluster of defensin-like genes encodes diffusible pollen tube attractants in Arabidopsis. PLoS Biol 10:e1001449. https://doi.org/10.1371/journal.pbio.1001449
12. Takeuchi H, Higashiyama T (2016) Tip-localized receptors control pollen tube growth and LURE sensing in Arabidopsis. Nature 531:245–248. https://doi.org/10.1038/nature17413

13. Zhong S, Liu M, Wang Z, Huang Q, Hou S, Xu YC, Ge Z, Song Z, Huang J, Qiu X, Shi Y, Xiao J, Liu P, Guo YL, Dong J, Dresselhaus T, Gu H, Qu L-J (2019) Cysteine-rich peptides promote interspecific genetic isolation in Arabidopsis. Science 364:eaau9564. https://doi.org/10.1126/science.aau9564
14. Higashiyama TKH, Kawano S, Kuroiwa T (1998) Guidance *in vitro* of the pollen tube to the naked embryo sac of *Torenia fournieri*. Plant Cell 10:2019–2032. https://doi.org/10.1105/tpc.10.12.2019
15. Willemse MTM, Plyushch TA, Reinders MC (1995) In vitro micropylar penetration of the pollen tube in the ovule of *Gasteria verrucosa* (Mill.) H. Duval and *Lilium longiflorum* Thunb.: conditions attraction and application. Plant Sci 108:201–208. https://doi.org/10.1016/0168-9452(95)04133-F
16. Prado AM, Colaco R, Moreno N, Silva AC, Feijo JA (2008) Targeting of pollen tubes to ovules is dependent on nitric oxide (NO) signaling. Mol Plant 1:703–714. https://doi.org/10.1093/mp/ssn034
17. Palanivelu R, Preuss D (2006) Distinct short-range ovule signals attract or repel *Arabidopsis thaliana* pollen tubes *in vitro*. BMC Plant Biol 6:7. https://doi.org/10.1186/1471-2229-6-7
18. Guan Y, Lu J, Xu J, McClure B, Zhang S (2014) Two mitogen-activated protein kinases, MPK3 and MPK6, are required for funicular guidance of pollen tubes in Arabidopsis. Plant Physiol 165:528–533. https://doi.org/10.1104/pp.113.231274
19. Iwano M, Ngo QA, Entani T, Shiba H, Nagai T, Miyawaki A, Isogai A, Grossniklaus U, Takayama S (2012) Cytoplasmic Ca^{2+} changes dynamically during the interaction of the pollen tube with synergid cells. Development 139:4202–4209. https://doi.org/10.1242/dev.081208
20. Duan Q, Kita D, Johnson EA, Aggarwal M, Gates L, Wu HM, Cheung AY (2014) Reactive oxygen species mediate pollen tube rupture to release sperm for fertilization in Arabidopsis. Nat Commun 5:3129. https://doi.org/10.1038/ncomms4129
21. Hamamura Y, Nishimaki M, Takeuchi H, Geitmann A, Kurihara D, Higashiyama T (2014) Live imaging of calcium spikes during double fertilization in Arabidopsis. Nat Commun 5:4722. https://doi.org/10.1038/ncomms5722
22. Denninger P, Bleckmann A, Lausser A, Vogler F, Ott T, Ehrhardt DW, Frommer WB, Sprunck S, Dresselhaus T, Grossmann G (2014) Male-female communication triggers calcium signatures during fertilization in Arabidopsis. Nat Commun 5:4645. https://doi.org/10.1038/ncomms5645
23. Wilson ZA, Morroll SM, Dawson J, Swarup R, Tighe PJ (2001) The *Arabidopsis MALE STERILITY1* (*MS1*) gene is a transcriptional regulator of male gametogenesis, with homology to the PHD-finger family of transcription factors. Plant J 28:27–39. https://doi.org/10.1046/j.1365-313X.2001.01125.x

Chapter 7

Isolation and Cloning of Suppressor Mutants with Improved Pollen Fertility

Steven Beuder and Cora A. MacAlister

Abstract

Mutant screens remain among the most powerful unbiased methods for identifying key genes in a pathway or process of interest. However, mutants impacting pollen function pose special challenges due to their genetic behavior. Here we describe an approach for isolating pollen mutants based on screening for suppressors of a low pollen fertility starting genotype. By identifying suppressor mutants with improved pollen fertility, we are able to identify new genes which are functionally relevant to pathway(s) causing low seed set in the original background. With this method, the low fertility of the genetic background may be due to one or more mutations or transgenes disrupting any aspect of pollen development or function. Furthermore, screening for improved pollen fertility biases toward recovery of the desired mutants due to their enhanced male transmission. The causative mutation is cloned using next-generation sequencing. The procedure uses both genetic and bioinformatics approaches to ultimately yield a very small list of candidate causative mutations speeding the transition from mutant phenotype to underlying gene.

Key words Mutagenesis, Suppressor screen, High-throughput sequencing, Cloning, Fertility, Pollen, Seed set

1 Introduction

Mutagenesis screens have proven to be a highly powerful method for identifying genes involved in a phenotype or biological process of interest. However, most approaches for isolating and cloning plant mutants focus on those affecting the diploid sporophyte generation. Mutants affecting the haploid gametophytes (the pollen and embryo sac) do not have the same genetic behavior as sporophytic mutants, complicating their initial isolation and cloning. Pollen mutants are especially problematic. The female gametophytes occupy fixed positions within an ovary and there is a direct correlation between the number of viable female gametophytes and ultimate seed yield. A plant that is heterozygous for a female gametophytic lethal mutant will produce seed only from 50% of ovules containing the wild-type allele. This strong reduction in seed

Anja Geitmann (ed.), *Pollen and Pollen Tube Biology: Methods and Protocols*, Methods in Molecular Biology, vol. 2160, https://doi.org/10.1007/978-1-0716-0672-8_7, © Springer Science+Business Media, LLC, part of Springer Nature 2020

set leads to an easily detected fertility phenotype. Pollen however, is produced in significant excess relative to the number of available ovules, leading to competition between pollen grains. Plants that are heterozygous for a pollen lethal mutation generally maintain high seed set since the 50% wild-type pollen they produce is sufficient to fertilize all ovules. Though a homozygous pollen-defective mutant would have reduced seed set, such homozygotes are unlikely to appear in segregating populations due to strong bias against male transmission of the mutant allele. Therefore, a mutagenesis screen for reduced seed set is unlikely to recover pollen defective mutants. However, gametophytic mutants have been successfully identified based on screening for distorted transmission ratios. In such screens, the primary screening criterion is generally deviation from the expected Mendelian inheritance pattern of a mutagenic reporter construct (e.g., T-DNA or transposon insertion; [1–4]). Such screens identify mutants which are defective in male and/or female transmission and at any stage of gametophyte development or function. If a particular stage of development or molecular pathway is of interest, secondary screening criteria are required to identify appropriate mutants following the initial screen.

Here we describe a strategy for isolating novel Arabidopsis pollen mutants by screening for increased seed set in a low-fertility genetic background. If a reduced pollen fertility mutant in a pathway of interest has already been identified, for example through reverse genetics, this low-fertility background serves as a convenient starting point for a suppressor screen. Screening for suppressors with improved fertility has several advantages. Firstly, the low-fertility starting genotype serves as a sensitized background allowing identification of suppressors of the original phenotype. The genetic interaction between the suppressors and the background genotype allows the direct identification of genes acting in the pathway(s) causing the low starting seed set. Secondly, both heterozygous and homozygous suppressor mutants will likely display increased seed set since increasing fertility of just half of the pollen will still allow more ovules to be fertilized. Thirdly, suppressed pollen will have a transmission advantage over non-suppressed pollen biasing toward recovery of suppressors in a segregating population (*see* **Note 1**).

The choice of starting genetic background is crucial for the success of a suppressor screen. The starting background must be sufficiently fertile to reliably propagate; but to efficiently screen for increased fertility, it must also have low enough seed set that an increase will be apparent. Here, we use mature silique length as the primary screening criterion since it correlates well with seed number and is an easy and rapid screening method, requiring no sample preparation and minimal handling (Fig. 1). Unlike traditional map-based cloning, the cloning strategy described here does not

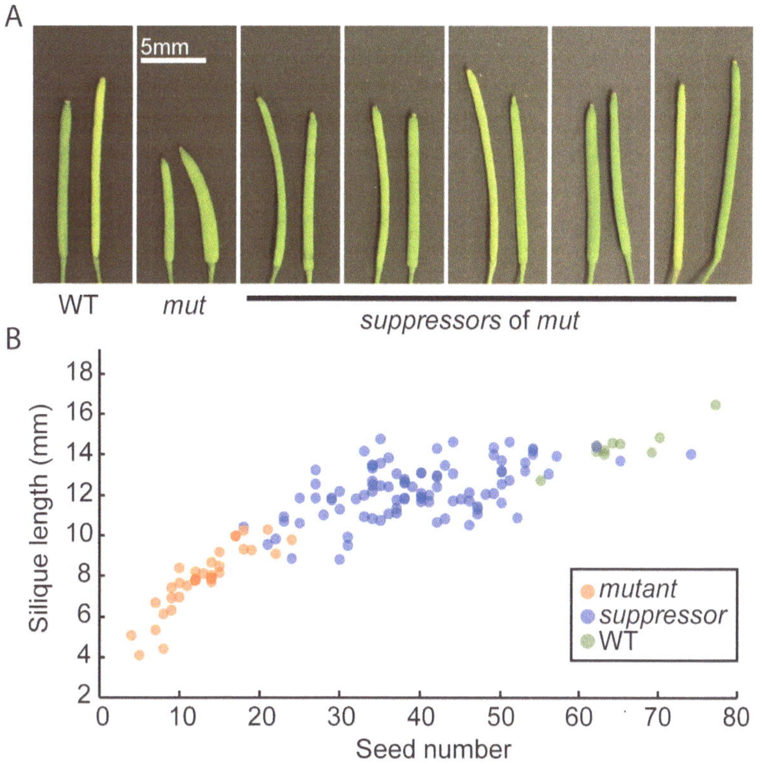

Fig. 1 Silique length approximates seed set. (**a**) Two representative nearly mature siliques from wild type, the low-fertility starting background mutant (*mut*) and five different suppressors of *mut*. Note that the mutant background siliques are shorter than wild type while the suppressors are longer than *mut*, though they may not reach wild-type lengths. (**b**) Plot of silique length vs. seed number for siliques from the background mutant (orange), wild type (green) and a collection of suppressors (blue; correlation coefficient = 0.86)

require any outcrossing from the initial genetic background. Therefore, the starting phenotype can be due to a single mutation, or a more complex genotype including multiple mutations and/or transgenes regardless of their genomic locations. But, no matter how complex the background, it should be fully homozygous and phenotypically stable (i.e., the phenotype is consistent between individuals and from one generation to the next). The background mutation(s) causing low fertility also require a reliable genotyping assay. Any wild-type contamination either from cross pollination or from stray seeds will behave like a strong suppressor and must be excluded based on genotype.

Once suppressor mutants are identified, our strategy for cloning the causative mutation relies on a combination of genetic and bioinformatics approaches to reduce the number of candidate suppressor mutations identified. First, the suppressor is backcrossed to the un-mutagenized parental strain to reduce the number of unlinked induced mutations present. Second, we sequence DNA from pools of confirmed homozygous suppressors plants (*sup/sup* genotypes), non-suppressed siblings from the same family (+/+ genotypes) and the original low-fertility background. Data from the *sup/sup* sample will contain only the suppressing allele while the +/+ and background data will carry only the wild-type allele for the gene causing suppression. By identifying sequence variants that are unique to the *sup/sup* sample and filtering with additional criteria, we produce a very short list of candidate sequence variants to quickly and easily confirm the causative mutation.

2 Materials

2.1 Plant Material and Growth Supplies

1. ~2,000 EMS-treated seeds of the low-fertility Arabidopsis genotype for which suppressors are sought (*see* **Note 2**).
2. Flats.
3. Potting mix suitable for Arabidopsis.
4. 32-Cell flat inserts.
5. 96-Cell flat inserts.
6. Stakes and twist ties.
7. Coin envelopes for seed storage.
8. Growth chamber or room suitable for Arabidopsis (*see* **Note 3**).
9. Fine forceps (e.g., Dumont #5).

2.2 Sequencing DNA Preparation

1. DNeasy Plant Mini Kit (Qiagen) and required reagents.
2. 3 M sodium acetate, pH 5.2.
3. 100% ethanol.
4. TE buffer: 10 mM Tris pH 8.0, 1 mM EDTA.

2.3 Bioinformatics Resources

1. Access to a computer server or cluster capable of handling the significant volumes of data generated by Illumina sequencing.
2. Reference Genome: The *Arabidopsis thaliana* top level genome sequence can be downloaded from: ftp://ftp.ensemblgenomes.org/pub/plants/release-41/fasta/arabidopsis_thaliana/dna/
3. Software: In addition to Microsoft Excel, the steps in this protocol are performed using the open-source tools listed in Table 1.

Table 1
Tools for data analysis

Tool	URL
Bowtie2 [5]	http://bowtie-bio.sourceforge.net/bowtie2/index.shtml
FastQC	http://www.bioinformatics.babraham.ac.uk/projects/fastqc/
Freebayes	https://sourceforge.net/projects/acoros/files/acoros/stable/pool/main/f/freebayes/
R Studio	https://www.rstudio.com/products/rstudio/download/
Samtools	http://www.htslib.org/
SnpEff [6]	http://snpeff.sourceforge.net/index.html

3 Methods

3.1 Screening for Suppressors

1. Sow the mutagenized seeds on soil, distributing them evenly across nine flats containing 32-cell inserts for an average density of approximately seven seeds per cell.
2. As the plants begin to bolt, stake all plants growing in one cell together, securing them with a twist tie to limit pollen transfer between adjacent cells. Allow the M1 plants to self-fertilize and collect all seeds from one cell into a single seed envelope forming an M2 pool (*see* **Note 4**).
3. Sow seed from each pool into a 96-cell flat with about two seeds per cell. Following germination, thin and transplant seedlings as required so that each cell contains one plant. It is generally most convenient to sow M2 pools for screening over multiple rounds as space and manpower allow.
4. As the M2 plants begin setting seed, screen each individual for increased silique length and seed set compared to the background strain (*see* **Note 5**). Screening will likely require multiple rounds as the flowering time will vary. Discard non-suppressed plants. Stake each individual candidate suppressor. When dry, collect seed from all candidate suppressors individually, noting the source pool for each.
5. Sow ~50 seeds from each candidate suppressor (now the M3 generation) into 24 cells of a 96-cell flat. When flowering, phenotype for suppression as above and genotype suppressors for the background mutation(s) causing low fertility to confirm that they are not wild-type contaminants. For the suppressors that prove phenotypically stable and genetically clean, chose one individual per original pool to progress further (*see* **Note 6**).

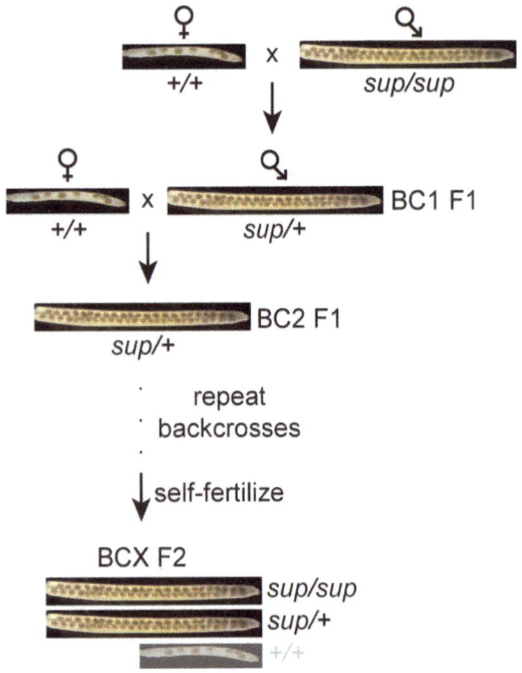

Fig. 2 Backcrossing the suppressor reduces the number of superfluous mutations. Mutagen treatment will produce not only the desired *sup* mutation, but other random mutations as well. Backcrossing the suppressor to the parental background will dilute these superfluous mutations. Fortunately, a suppressor with improved pollen fertility will have a suppressed phenotype in the F1 generation. Therefore, the next backcross can be done directly in this generation. After the chosen number of backcrosses have been completed (*see* **Note 8**), the F1s self-fertilize, forming the F2 generation. Most F2s are expected to be suppressed (either *sup/sup* or *sup/+*) since the non-suppressed (+) pollen will not transmit well

3.2 Generating the Sequencing Samples

1. Backcross the suppressor to the starting genetic background (Fig. 2). Using the suppressor as the male parent will increase cross seed yield and favor transmission of the *sup* allele (*see* **Note 7**).
2. Remove open flowers, small buds, and the inflorescence meristem from an inflorescence of the background genotype, leaving 2–4 of the most mature buds.
3. Using fine forceps, emasculate each bud, taking care not to damage the pistil.
4. Pollinate the stigma of each emasculated bud with pollen from a young, open flower of the suppressor, applying as much pollen as the stigma will accept.
5. Label the crossed inflorescence with the parent plant information and allow the seeds to mature over the next few weeks.

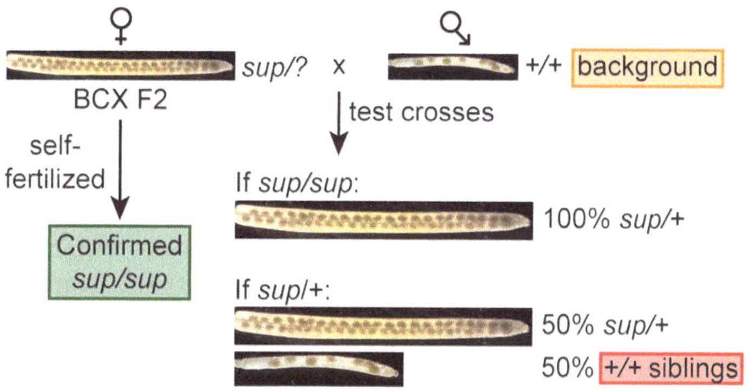

Fig. 3 Source of tissue samples for DNA sequencing. The backcross F2 generation will be composed primarily of suppressed individuals which are either *sup/sup* or *sup/+* genotypes. Determining the genotype of a given plant requires a test cross with the background genotype. Here, the suppressed parent should be used as the female parent to allow the non-suppressed (+) allele to transmit. The test crosses will either produce all suppressed progeny if the parent was *sup/sup*, or an equal mixture of suppressed and non-suppressed progeny if the parent was *sup/+*. The non-suppressed siblings produced by the *sup/+* crosses are the source of the non-suppressed sibling (*+/+*) tissue for sequencing (red box). The *sup/sup* tissue pool comes from the self-fertilized seeds of F2 plants that were confirmed to be homozygous based on the test cross results (green box). The final sequencing tissue sample comes from the un-mutagenized background genotype (orange box)

6. When the crossed siliques are nearly ripe, collect them and allow the seeds to finish drying on the bench.
7. Sow the backcross F1 seed. When the F1s are flowering, screen for suppression and backcross as above using a suppressed F1 as the male parent. Repeat the backcross step as many times as desired (*see* **Note 8**).
8. When the desired number of backcrosses have been carried out, allow the F1s to self-fertilize forming the F2 generation.
9. Sow the F2 seeds to produce a minimum of 32 F2 plants (*see* **Note 9**).
10. When flowering, identify suppressed F2s. The suppressed plants will be a mixture of *sup/sup* and *sup/+*genotypes.
11. Cross each suppressed F2 individual as a female to the background genotype (*see* **Note 10**). When ready, collect both the cross seeds and self-fertilized seeds from each F2 parent, carefully tracking the parent individual.
12. Sow the test cross seeds. When the plants are setting seed, score each individual for suppression. Categorize the parent F2 as either *sup/sup* (if all test cross plants are suppressed) or *sup/+* (1:1 suppressed to non-suppressed test cross plants, Fig. 3).

For statistical confidence, we require a minimum of seven scored test cross progeny for a given F2 parent (*see* **Note 11**).

13. The test cross progeny from *sup*/+ parents will include non-suppressed (+/+) sibling plants. Collect a total of 100 mg of young leaf tissue from these non-suppressed plants to serve as the non-suppressed sibling (+/+) DNA sample. Preventing suppressor contamination in this pool is critical, so avoid collecting tissue from any questionable individuals.
14. For each confirmed *sup/sup* F2 parent, sow self-fertilized seed (now the F3 generation). Following germination, genotype each F3 population using the background mutation genotyping assay and discard any F3 populations showing wild-type contamination.
15. Collect a total of 100 mg of pooled tissue from the confirmed *sup/sup* seedlings, using roughly equally amounts of tissue for each individual F3 population.
16. Sow seeds of the low-fertility genetic background and collect 100 mg of seedling tissue for the background DNA sample.
17. Extract DNA from the pooled +/+ tissue, the pooled *sup/sup* tissue and the background genotype tissue using the Qiagen DNeasy Plant Mini Kit following the manufacturer's instructions.
18. To concentrate the purified DNA, add 20 μl of 3M sodium acetate, pH 5.2 and 600 μl of ethanol to the 200 μl of eluted DNA. Mix by inversion and freeze at −20 °C overnight.
19. Spin DNA samples at 4 °C for 30 min at maximum speed in a standard microcentrifuge.
20. Gently decant the supernatant and allow the tube to dry. Resuspend the DNA pellets in 30 μl of TE.
21. Submit the purified DNA samples for whole-genome library preparation and sequencing (*see* **Note 12**).

3.3 SNP-Mapping Pipeline

The programs used in this pipeline are open-source and can be downloaded from the sources listed in Table 1. Alternatively, if you have access to a server and computer cluster with the software installed, then no downloads and installations are necessary, other than loading programs when needed. The command scripts are designed to be typed into a command-line interpreter program—otherwise known as a shell (e.g., "Terminal" for Mac and Linux, or a Unix shell emulator, like Cygwyn, for Windows).

The SNP mapping pipeline described here consists of several main steps: (1) mapping sequence reads to a reference genome, (2) calling and filtering variants to identify suppressor-specific mutations, and (3) filtering variants based on additional criteria including variant type, sequencing depth, and variant frequency.

It is important to note that most steps generate output files which will be used in subsequent steps. Prepare a work space by creating an empty folder that will serve as the working directory in which multiple subdirectories will be created to store files.

1. Create a new subdirectory called "tair10ref" to store the fasta file of the reference genome sequence. Download the most recent Arabidopsis genome assembly, "Arabidopsis_thaliana.TAIR10.dna.toplevel.fa.gz" and transfer it to the "tair10ref" folder.
2. Create a subdirectory named "raw_reads," download the raw sequencing reads, and store them in this folder. As an example, here are two "mates" of raw reads from the same sample/pool of DNA; "R1" and "R2" in the file name denotes each mate for a single sample. The ".gz" indicates that the file was compressed.
   - 112369_ATCACG_S1_L004_R1_001.fastq.gz
   - 112369_ATCACG_S1_L004_R2_001.fastq.gz
3. The first action to perform on sequence read files is a quality check using FastQC. Run the following command to perform quality analysis on your raw reads.

```
fastqc Raw_reads/112369_ATCACG_S1_L004_R1_001.fastq.gz
```

   This will generate an output filed named "112369_ATCACG_S1_L004_R1_001_fastqc.zip". Download and open this file to unpack it and open the .html file. This will bring up the FastQC report page. Analyze each section listed in the summary (*see* **Note 13**).

4. Indices for the reference genome must be generated before performing alignment. To do this, load Bowtie2 [5] and run the following command from your working directory.

```
bowtie2-build
tair10ref/Arabidopsis_thaliana.TAIR10.dna.toplevel.fa.gz
tair10ref/TAIR10
```

   This will generate multiple files with ".bt2" extensions. This command specifies that all the .bt2 files will have the base name "TAIR10," and they will be directed to the "tair10-ref" folder. The reference genome and the .bt2 index files must be present in the same subdirectory to perform alignment.

5. Create a new subdirectory called "aligned." Run the following command to align the sequence reads to the reference genome. This will create an output file named "11269.sam", which will be directed to the "aligned" folder. Be sure to include "-1" or "-2" before each file name to denote that these files contain

paired-read mates, otherwise the reads will be aligned without taking pairing into account.

```
bowtie2 -x tair10ref/TAIR10 -1 Raw_reads/112369_AT-
CACG_S1_L004_R1_001.fastq.gz, -2
Raw_reads/112369_ATCACG_S1_L004_R2_001.fastq.gz -S aligned/
11269.sam
```

6. Convert SAM files to BAM files. SAM and BAM files contain the same information, but SAM files are a human-readable text format and BAM files are a machine-readable binary version. The BAM format is used for downstream analysis. Create a subdirectory called "bams." Run the following command to convert the SAM file to a BAM file named "11269.bam" and direct the output file to the "bams" folder.

```
samtools view aligned/112369.sam > bams/112369.bam
```

7. Create a subdirectory named "sorted_bams." Run the following command to sort a BAM file and create an output file named "112369sorted.bam" directed to the "sorted_bams" folder.

```
samtools sort bams/112369.bam -o sorted_bams/112369sorted.bam
```

8. Create a new subdirectory named "bams_rmdup" Run the following command to remove PCR duplicates from a sorted .bam file. This command generates an output file named "112369rmdup.bam," which will be directed to the "bams_rmdup" folder.

```
samtools rmdup sorted_bams/112369sorted.bam bams_rmdup/
112369rmdup.bam
```

9. Variant-calling tools (including Freebayes) require an index for a .bam file. Run the following command to create an index for a .bam file. Although it is not written in the command, the output file will be directed to the subdirectory containing the .bam file to be indexed, and the output file will have the same name plus ".bai" at the end (e.g. the above command will generate a new file called 112369rmdup.bam.bai that will appear in the "bams_rmdup" folder). Leave the files together in the same subdirectory.

```
samtools index bams_rmdup/112369rmdup.bam
```

3.4 Identify Sequence Variants and Filter for Candidate Causative Mutations

1. Create a new subdirectory called "VCFs" to store output files. To call variants using Freebayes, run the following command.

```
freebayes-f  tair10ref/Arabidopsis_thaliana.
TAIR10.dna.toplevel.fa bams_rmdup/112369rmdup
.bam > VCFs/112369.vcf
```

The .vcf files can be opened with Microsoft Excel and will contain a large header at the top with the variants listed below. Details of the variants are specified in different columns, of which, the following are the most important (*see* **Note 14**):

- the chromosome that a variant was found in is listed in the "CHR" column.
- the position of the variant on the chromosome are listed in the "POS" column.
- the nucleotide(s) present in the reference genome in the "REF" column.
- the variant listed in the "ALT" column.
- other information, including the number of times the reference and variant sequences were counted in the sample, read coverage, and sequencing quality is listed in the "INFO" column. The header contains key information to interpret data listed in the INFO column.

2. Annotate the suppressor variants (*see* **Note 15**). Knowing the possible consequences for a given variant on gene function will be useful data for later analysis. Here we use SnpEff to predict the effect of the sequence variants [6]. Though several filtering steps remain, it is convenient to annotate the variant effect now, since SnpEff takes a .vcf file as input and the file format will be altered in later steps. This step is only required for the *sup/sup* data and can be skipped for the +/+ and background data. Using the below command, SnpEff will automatically install the required Arabidopsis genome database, annotate the .vcf file with a new data column and save the results as a new file "112368ann."

```
java -Xmx4g -jar snpEff.jarArabidopsis_thaliana 112369.vcf
>112369ann.vcf
```

3. For downstream analysis the information in the INFO column must be split into individual columns. Open the annotated .vcf file in Microsoft Excel and preform the following steps:
   (a) Delete the header.
   (b) Highlight the INFO column.
   (c) In the "Data" tab, click "Text to Columns."

(d) Select "Delimited" and set delimiters as "Semicolon."

(e) Save as a .csv file.

4. Remove common variants. When all files have been processed, there will be two pairs of .csv files for each sequenced suppressor (the *sup/sup* and the +/+ pools) along with the .csv file for the genetic background. The causative mutation will be present in the *sup/sup* sample but absent from the background and the +/+ samples. In this step, we remove variants from the *sup/sup* sample that also occur in the +/+ sample and/or the background sample (*see* **Note 16**). The following steps are performed with R Studio.

   (a) Move the .csv files to the working directory.

   (b) Install the "DBI" and "dplyr" packages if not already installed.

   (c) Import the .csv files for comparison using the "Import Dataset" function with the "no header" option selected.

   The following commands takes the annotated suppressor variant list (data frame "112369ann") and removes any lines which share the same chromosome number (column V1), position (column V2), and alternate sequence (column V5) as any variant in the +/+ pool (data frame "112370"). We further subtract the non-mutagenized background variants ("112371") from the intermediate result ("112369ns") to produce the final list of unique *sup/sup* variants ("u112369"). The resulting list is saved as a new .csv file. The variant number is typically reduced by ~40% following this step.

```
library(dplyr)
112369ns <- anti_join('112369ann','112370', by= c("V1", "V2",
"V5"))
u112369 <- anti_join('112369ns','112371', by= c("V1", "V2",
"V5"))
write.csv(u112369, file = "u112369.csv")
```

5. Determine variant frequency in the unique variant list:

   (a) Open the unique *sup/sup* variant .csv file in Excel.

   (b) Find and highlight the column with information that says "RO=n," where n is the number of times a read included the reference nucleotide and use the find and replace tool to delete the "RO=" from each cell.

   (c) Repeat these steps for the alternate nucleotide observation column with the "AO=n" information to remove the "AO=."

   (d) Insert a new column and label it "AO freq." Divide the values in the "AO" column by the sum of the values in the

"RO" and "AO" column for each row. These values are the frequencies with which a variant was found in that sample.

6. Using the Filter tool, you can now filter the file to show only the variants meeting the filtering criteria:
   (a) To filter by allele frequency, highlight the "AO freq" column, select Filter, and filter for values greater than or equal to your cutoff value. We generally use 0.8.
   (b) To filter by sequencing depth, filter the "AO" column. We generally require a minimum of five reads.
   (c) There is also a column containing the type of change, "type=n," where n refers to a type of variant, such as a single nucleotide polymorphism (SNP), insertion, etc. Since EMS typically induces transition mutations, filter for "type=SNP."

   Following these filtering steps, the total number of candidate variants will be significantly reduced, generally to <2% of the starting unique variant number. Most of these variants will occur in one chromosomal region which includes the causal mutation and those linked to it. The remaining few dozen candidates can be further filtered based on their annotated effect. Missense or non-sense mutations occurring in pollen-expressed genes near the center of the cluster are the most likely causative mutations.

4 Notes

1. We do not screen for enhancers (mutants with reduced pollen fitness and lower seed set) since they would be under a strong male-transmission disadvantage, making their recovery difficult. Though sterile mutants are a frequently recovered class in any mutagenesis screen, such sterility is likely caused by an unrelated sporophytic mutation.
2. The induction of Arabidopsis mutants by seed treatment with ethyl methanesulfonate (EMS) has been described several times previously [7–9]. We generally treat seeds of the starting genotype with 0.2% EMS for 16 h. It is advisable to check for segregation of albino seedlings in the M1 generation as an indicator of mutagenesis success.
3. Fertility is influenced by many factors, both genetic and environmental. Therefore, growth conditions should be as consistent and as favorable as possible. We have found that high air movement is a major contributing factor to contamination by wild-type pollen. Pollen contamination can be limited by using

lower airflow growth areas or by excluding flowering wild-type plants from the growth area.

4. The M1 plants may be screened directly for increased seed set. However, the M1s will be chimeras with sectors carrying different mutations so only some of their siliques may be carrying the suppressive mutation, which can complicate screening. If the M1s are screened, it is still advisable to screen the M2s of any pools that did not produce any suppressor candidates.
5. Older plants are generally easier to screen since they will have more siliques available for comparison. The choice of which siliques to consider is also important. The first few flowers produced by a plant should be avoided for phenotyping purposes as these often have low fertility, even in wild-type plants. We also avoid scoring siliques produced late in life or from tertiary branches as these often have low fertility.
6. Multiple unique suppressor mutations may exist in a single M2 pool. Unfortunately, due to the haploid nature of gametophytic mutants, complementation tests are not possible and allelic mutants can only be recognized after their cloning. Therefore, we progress a single candidate suppressor per pool as a reasonable compromise between total number of suppressors isolated and the highest likelihood of recovering independent mutations, and therefore novel alleles and genes.
7. We assume here that suppression is due to an improvement in pollen fertility. This has always been the case in our experience, but it is possible that female suppression pathways may exist. To determine whether suppression is due to improved fertility of the pollen itself or through compensation by the female tissue, perform reciprocal test crosses with the suppressor plants and the background genotype. If the suppressor acts by improving pollen fertility, the crossed siliques will have high seed set when the suppressor is used as the male parent (similar to the suppressor self-fertilized siliques), but low seed set when it is used as the female parent (similar to the self-fertilized background genotype).
8. We generally backcross for four generations, but this number may be reduced (or increased) as desired. If time is a constraint, reducing the number of backcrosses is preferable to eliminating the test cross step. Much of the power to filter candidate mutations comes from filtering based on the allelic frequency in the *sup/sup* data. If the suppressors sequenced were not confirmed homozygotes, the stringency of this filtering must be significantly reduced to account for heterozygous suppressors.
9. A *sup/sup* F2 plant is homozygous for the causative mutation, but may be homozygous for unrelated mutations as well. By

pooling F2 individuals we reduce the allele frequency of non-causative mutations. We expect about half of the suppressed individuals used for test crosses will prove homozygous. We target a minimum of ten homozygous F2s to include in the sequencing DNA pool, but have used as few as five with good success. We start with an F2 population of 32 plants, allowing for some non-suppressed F2s, some test crosses which fail to produce enough progeny to screen and other potential problems.

10. Due to the male-transmission advantage expected for a suppressor, the test crosses *must* be done using the suppressors as the female parent. However, the low male fertility of the background genotype will reduce the total seed yield from crosses when it is used as the male parent. Therefore, more flowers may need to be pollinated than is typical to ensure sufficient cross seed yield.
11. From a *sup/+* female parent, each individual test cross progeny has a 50% chance of inheriting the non-suppressive (+) allele and appearing non-suppressed. The probability of all progeny of sample size N inheriting the + allele by chance is calculated as 0.5^N. For seven individuals, the overall likelihood is $0.5^7 = 0.0078$ or 0.78%. In practice, we sow all cross seed and discard any crosses for which fewer than seven individuals can be scored. We also generally discard test crosses if the segregation ratio is not clearly one of the expected classes (100% suppressed or 50% suppressed).
12. We have used both in-house DNA sequencing core facilities and commercial sequencing operations to both prepare the libraries and carryout the sequencing with good success. We typically sequence the libraries as paired-end 150 bp reads using an Illumina HiSeq 4000 sequencer, pooling 16 libraries per lane with a target sequencing depth of 20- to 30-fold per sample [10]. The specific type of sequencing run is not critical and single-end sequencing and/or different read lengths can be used with only minor adjustments to data analysis. If a different read type, length or sequencer will be used, Illumina offers an online sequence coverage calculator to determine how many libraries may be pooled per lane (https://support.illumina.com/downloads/sequencing_coverage_calculator.html).
13. A tutorial to guide beginners through interpreting FastQC reports is available at the FastQC link in Table 1.
14. The total list of sequence variants will be long (in our experience ~40,000 per sequenced line) and includes many classes of variants (e.g., induced mutations, sequencing errors, polymorphisms between the sequenced line vs. the genome release,

etc.). Therefore, the variant number should not be taken as an indication of mutation rate.

15. The read mapping and variant calling steps are computationally too intensive for a standard personal computer and will require use of a server or cluster. However, all of the remaining steps after variant calling can be done on a standard laptop or desktop computer.

16. In this step, any variant that is found in the non-suppressed sibling (+/+) sample or the background genotype sample will be removed from the suppressor variant list. Even a small amount of suppressor contamination in these samples will eliminate the causative mutation since no account is made for the frequency or the quality of the variant. If the analysis does not yield promising candidate mutations, this is the most likely point of failure. For a less stringent approach this step can be skipped leaving a longer list of candidate variants.

References

1. Christensen CA, Subramanian S, Drews GN (1998) Identification of gametophytic mutations affecting female gametophyte development in Arabidopsis. Dev Biol 202:136–151
2. Lalanne E, Michaelidis C, Moore JM, Gagliano W, Johnson A, Patel R, Howden R, Vielle-Calzada JP, Grossniklaus U, Twell D (2004) Analysis of transposon insertion mutants highlights the diversity of mechanisms underlying male progamic development in Arabidopsis. Genetics 167:1975–1986
3. Johnson MA, von Besser K, Zhou Q, Smith E, Aux G, Patton D, Levin JZ, Preuss D (2004) Arabidopsis hapless mutations define essential gametophytic functions. Genetics 168:971–982
4. Pagnussat GC, Yu HJ, Ngo QA, Rajani S, Mayalagu S, Johnson CS, Capron A, Xie LF, Ye D, Sundaresan V (2005) Genetic and molecular identification of genes required for female gametophyte development and function in Arabidopsis. Development 132:603–614
5. Langmead B, Salzberg S (2012) Fast gapped-read alignment with Bowtie 2. Nat Methods 9:357–359
6. Cingolani P, Platts A, Wangle L, Coon M, Nguyen T, Wang L, Land SJ, Lu X, Ruden DM (2012) A program for annotating and predicting the effects of single nucleotide polymorphisms, SnpEff: SNPs in the genome of Drosophila melanogaster strain w1118; iso-2; iso-3. Fly 6:80–92
7. Weigel D, Glazebrook J (2006) EMS mutagenesis of *Arabidopsis* seed. Cold Spring Harb. Protoc 2006:pdb.prot4621
8. Dinh TT, Luscher E, Li S, Liu X, Won SY, Chen X (2014) Genetic screens for floral mutants in Arabidopsis thaliana: enhancers and suppressors. In: Riechmann J, Wellmer F (eds) Flower development. Methods in molecular biology (methods and protocols), vol 1110. Humana, New York, NY
9. Qu LJ, Qin G (2014) Generation and Identification of Arabidopsis EMS mutants. In: Sanchez-Serrano J, Salinas J (eds) Arabidopsis protocols. Methods in molecular biology (methods and protocols), vol 1062. Humana, Totowa, NJ
10. James GV, Patel V, Nordström KJV, Klasen JR, Salomé PA, Weigel D, Schneeberger K (2013) User guide for mapping-by-sequencing in Arabidopsis. Genome Biol 14:R61

Chapter 8

Workflow to Characterize Mutants with Reproductive Defects

Jennifer A. Noble and Ravishankar Palanivelu

Abstract

Reverse genetics approaches for characterizing phenotypes of mutants in a gene of interest (GOI) require thorough genotyping and phenotypic analysis. However, special challenges are encountered when a GOI is expressed in reproductive tissues: a variety of assays are required to characterize the phenotype and a mutant may show sporophytic and/or gametophytic defects in male and/or female reproductive tissues, which are structurally and functionally intertwined. Here, we present a streamlined workflow to characterize mutants with reproductive defects, primarily using Arabidopsis as a model, which can also be adapted to characterize mutants in other flowering plants. Procedures described here can be used to distinguish different kinds of reproductive defects and pinpoint the defective reproductive step(s) in a mutant. Although our procedures emphasize the characterization of mutants with male reproductive defects, they can nevertheless be used to identify female reproductive defects, as those defects could manifest alongside, and sometimes require, male reproductive tissues.

Key words Seed set, Male gametophyte and sporophyte, Transmission efficiency, Pollen, Pollen tube, Pistil, Ovules, Pollen tube–pistil interactions, Aniline blue staining, Transmission efficiency, Fertility

1 Introduction

The rise of comparative genomics and next-generation sequencing technologies have increased the need for reverse genetics approaches to characterize the function of a gene of interest (GOI). In angiosperms, reproductive mutants can be challenging to characterize, as a flower is comprised of four different types of tissues: female sporophyte, female gametophyte, male sporophyte, and male gametophyte [1–3]. Resolving the function of a GOI in reproduction requires multiple analyses, including expression and mutant analysis. Cell-specific expression datasets are beginning to be beneficial to confirm expression of a GOI [4–8]; however, mutant analysis is still required to determine the role of the gene in those cells during sexual reproduction of angiosperms.

Anja Geitmann (ed.), *Pollen and Pollen Tube Biology: Methods and Protocols*, Methods in Molecular Biology, vol. 2160, https://doi.org/10.1007/978-1-0716-0672-8_8, © Springer Science+Business Media, LLC, part of Springer Nature 2020

Here, we created a pipeline of experiments that can be used to begin analyzing mutants with reproductive defects. We start with identification of GOIs functioning in reproduction and obtaining mutants in Subheading 3.1. In Subheading 3.2, we describe how to score seed set in *Arabidopsis thaliana* siliques and explain in Subheading 3.3, how that assay can be used to examine a series of crosses to determine if the mutation is primarily affecting the male and/or female sporophyte and male and/or female gametophyte. In the final two sections, we describe assays that could be used to follow-up on the results from Subheading 3.3. In Subheading 3.4, we describe using transmission efficiency, where the transmission of the transgene is analyzed in the progeny, to confirm the gametophytic defects caused by the mutation. In the final Subheading 3.5, we describe using aniline blue staining to identify which post-pollination step is defective in the mutant of a GOI. In summary, the methods described here can be used to characterize reproductive mutants and gain insights into the reproductive function of a GOI.

2 Materials

2.1 Seed Set Analysis

Materials and Equipment

1. Microscope slides (75 × 25 × 1 mm).
2. Forceps.
3. Scissors.
4. Stereoscope with a bright light source.
5. Double-sided tape (¾").
6. Syringe needles (27 G).
7. 1 mL syringe.

2.2 Transmission efficiency

Materials and equipment

1. Microscope slides (75 × 25 × 1 mm).
2. Forceps.
3. Scissors.
4. Stereoscope with a bright light source.
5. Double-sided tape (¾″).
6. 1.5 mL microcentrifuge tubes and microcentrifuge tube rack.
7. Gas sterilization chamber.
8. 100 mL graduated cylinder.
9. 250 mL beaker.
10. Petri dishes (100 × 15 mm).

11. Growth chamber for growing Arabidopsis seedlings on plates.

Solutions

1. Gas sterilization solution (to be prepared right before sterilization and can be reused for 2 weeks):
   (a) To gas sterilize seeds, place open tubes of seeds labeled with pencil (not with a permanent marker) in a glass sterilization chamber.
   (b) Place a beaker with 95 mL of bleach in the chamber.
   (c) Carefully add 5 mL of 75% hydrochloric acid to the beaker, then immediately close the lid of the sterilization chamber.
   (d) Remove seeds from sterilization chamber after 3–5 h. Seeds are now ready to be plated. Left over seeds can be stored in a refrigerator for future use.
2. Murashige and Skoog (MS) plates with appropriate antibiotic:
   (a) In a flask, at least twice the volume of the media, add the following: 1/2× strength MS Salt Bases, 2% sucrose, 0.9% Bacto Agar, and 0.05% MES hydrate.
   (b) Dissolve in dH_2O, then adjust pH to 5.8.
   (c) Autoclave solution.
   (d) Before pouring solution into petri dish plates, add appropriate amount of antibiotic to the MS agar and swirl to mix.
   (e) Plates could be used immediately or stored up to 3–6 months at 4 °C.

2.3 Aniline Blue Staining

Materials and Equipment

1. Forceps.
2. Stereoscope with a bright light source.
3. Scissors.
4. 96-well plates, flat bottomed, ≥200 μL.
5. Plastic wrap or parafilm.
6. Resealable plastic bag (quart-sized).
7. Multi-channel pipette or single channel pipette.
8. 200 μL pipette tips.
9. Microscope slides (75 × 25 × 1 mm).
10. Coverslips (18 × 18 mm, #1.5).
11. Syringe needles (27 G).
12. 1 mL syringe.

13. Clear nail polish; we recommend "Hard As Nails" by "Sally Hansen," which does not react with mounting solution, form precipitates, or interfere with observation or staining.

Solutions

1. Fixative solution: glacial acetic acid and 100% ethanol (1:3 volume ratio).
2. Ethanol series: separate bottles of 70%, 50%, and 30% ethanol in dH_2O.
3. dH_2O.
4. Alkaline treatment solution (ATS): 8 M NaOH.
5. Decolorized aniline blue solution (DABS):
   (a) Prepare the 0.1% (w/v) aniline blue in 108 mM K_3PO_4 (pH 11) and store at 4 °C overnight.
   (b) Filter the solution using decolorizing active carbon powder and filter paper. Collect the filtrate in a dark container to avoid exposure to light.
   (c) Repeat step "b" with clean filter paper and decolorizing active carbon powder until the filtrate color is clear/faint light blue.
   (d) Once filtered, add glycerol to a final concentration of 2% glycerol.
   (e) Store the solution in the dark at 4 °C.
   (f) DABS solution can last at least 3 years, but its color may turn yellow over time. Color change does not reflect poor staining ability.
6. Mounting solution: DABS with a total concentration of 15% glycerol.

3 Methods

3.1 Obtain or Generate Mutants of a Gene of Interest with Decreased Expression in Reproductive Tissues

1. Confirm expression of GOI in flowers (i.e., anthers, pollen, pistil, or ovules) (*see* **Note 1**).
2. Obtain mutants from stock center or generate mutants using antisense or CRISPR technology in the GOI; genotype mutation in GOI using PCR or sequencing (*see* **Note 2**).
3. Perform RT-qPCR to confirm that there is decreased or no expression of the GOI in the mutant reproductive organ(s), compared to that in wild type, in which the GOI is typically expressed.

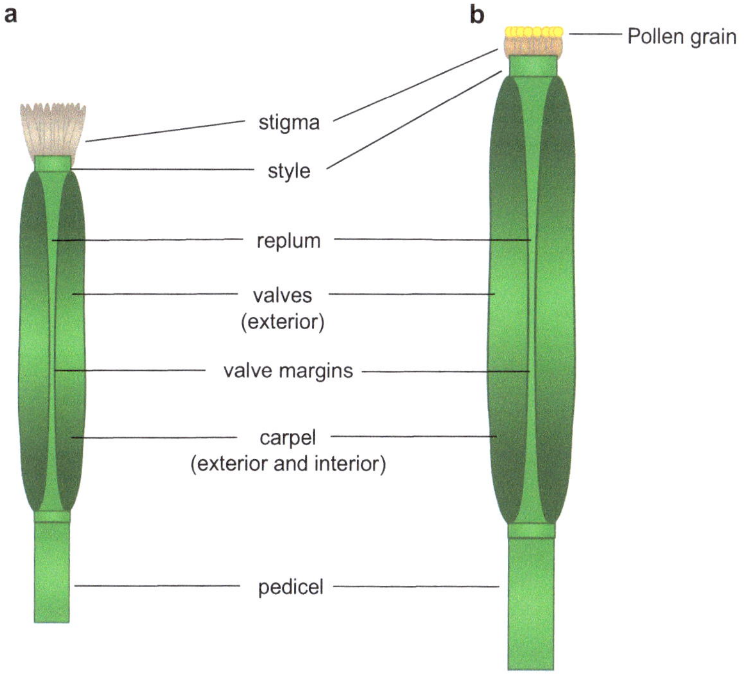

Fig. 1 Diagrams of *Arabidopsis thaliana* pistils. (**a**) An emasculated pistil, which is ready to receive pollen in a cross. (**b**) A pollinated pistil is relatively larger and has elongated to accommodate the developing seeds

3.2 Seed Set Analysis in a GOI Mutant Silique

1. Place a piece of double-sided tape on a clean microscope slide.
2. Excise a silique from the main stem of GOI mutants using scissors and/or forceps (*see* **Note 3**). Make sure to excise the silique with the pedicel (*see* Fig. 1), so that it can be used to handle and/or hold the sample during microsurgery in subsequent steps.
3. Place the silique on the double-sided tape on the microscope slide, ensuring that both carpel valves are visible with the replum facing upwards (*see* Fig. 1) and the pedicel is not on the tape. Press the silique down on the tape with forceps, once the silique is in the correct position.
4. Place the prepared sample under the stereoscope. The silique should be positioned horizontally with the majority of the silique within the field of view and in focus.
5. Using a needle attached to a 1 mL syringe, make four incisions (two at the ends of each carpel valve) perpendicular to the replum (*see* Fig. 1) [9].
6. Keeping the smooth slide of the syringe needle down (the suction hole in the needle facing you), use the syringe needle to cut open the carpel that is closest toward you, by making a horizontal incision just below the replum and continue to make

incision by moving from one end of the carpel valve to the other end, taking care to avoid the septum (*see* Fig. 1, *see* **Note 4**). Gently use the syringe needle to peel the carpel wall back and stick the removed carpel wall to the double-sided tape, so that you can return to the opened pistil rapidly, without letting it [9].

7. Rotate the slide 180°, so that now the unopened carpel is close to your dissecting hand.
8. Score ovules in the exposed carpel as unfertilized, viable, and aborted (*see* Fig. 2, *see* **Note 5**). Make sure to look for any unfertilized ovules, as they are smaller and may be obscured or hidden behind larger fertilized ovules. To avoid risking exposed ovules from drying, it is important to score the ovules in the opened carpel before opening the second carpel.
9. Repeat **step 6** for the unopened carpel. Rotate slide 180° and repeat **step 8**.
10. Repeat **steps 2–9** for a total of three siliques per plant. A seed set noticeably different than wild type might indicate one or more fertility defects (*see* Fig. 2, *see* **Note 6**).

3.3 Pinpointing Source of Reproductive Defects Using Crosses

1. To determine if fertility defects are caused by sporophytic or gametophytic defects, perform each of the following crosses at least three times and score seed set (Subheading 3.2) in crossed pistils (10–14 days after pollination) (*see* **Note 7**). Although determining the source of fertility defects is the first goal of these crosses, they can be used to perform additional assays, such as aniline blue staining described in Subheading 3.5, to further characterize the reproductive phenotype.
2. Choosing the right bud for emasculation and flower to supply pollen in a cross are essential to maintain consistency and to compare seed set from different crosses (*see* Fig. 3). Emasculate a "stage -1" bud, wait for 24 h, and pollinate it with pollen preferably from a flower at "stage 2" and/or "stage 3" (*see* Fig. 3). Perform crosses using buds in inflorescences in the primary and/or first two secondary branches. Avoid performing crosses on young (just bolted) or older (with too many branches and about to cease flowering in a few days) plants. Label each cross by listing the female and male parents on a piece of labeling tape that is placed immediately below the crossed pistil, after removing older flowers and siliques to make room for the tape label. When multiple genotypes are involved in a crossing experiment, errors can be avoided by designating a distinct colored tape for each genotype.
3. A day after performing the crosses, inspect each cross, and remove older flowers above and younger buds below the crossed pistil, to prevent them from being mistaken for a

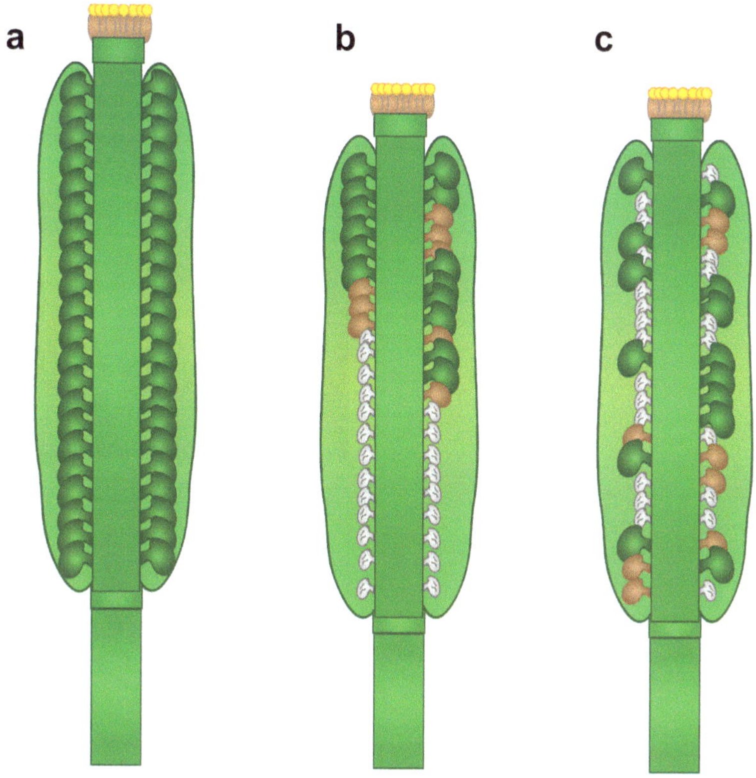

Fig. 2 Scoring seed set in *Arabidopsis thaliana* siliques. (**a**) Wild-type siliques typically show full seed set and each developing seed is relatively big, healthy, plump, and green. (**b**) Mutant siliques exhibiting male reproductive defects tend to show siliques with unfertilized ovules (shown as shriveled and gray) toward the bottom of the silique, as mutant pollen tubes are unable to grow the full length of the pistil and fail to fertilize the ovules in the bottom of a pistil. However, the seed set pattern alone may not be sufficient to distinguish the male gametophytic defect from the male sporophytic defect. (**c**) Mutant siliques exhibiting female gametophytic defects tend to have unfertilized ovules randomly in a silique, as ovules containing a wild-type or mutant female gametophyte are randomly distributed and fail to be fertilized in a heterozygous mutant pistil or randomly fail to be fertilized in a homozygous mutant pistil. (**b**, **c**) Post-fertilization defects can cause aborted seeds (shown as brown and flat) and point to recessive sporophytic defect(s) in embryo and/or endosperm development when homozygous seeds are produced by the fertilization of mutant ovule by a mutant pollen tube. Siliques with decreased seed set caused by the presence of unfertilized ovules and/or aborted seeds may be smaller than wild-type siliques, which is reflected in the diagram

Fig. 3 Selecting buds and flowers to serve as female and male parents in a cross, respectively. (**a**) Bird's eye view of an *Arabidopsis thaliana* inflorescence (Landsberg ecotype). The youngest bud with white petals protruding beyond the green calyx, is designated as "stage 0" and the oldest flower with white petals protruding beyond the yellow calyx is designated as "stage 5." The oldest bud that is entirely green is designated as "stage -1" (blue) should be chosen for emasculation. Twenty-four hours after emasculation, this emasculated pistil should be pollinated with pollen from a donor flower that is at "stage 2" (pink) and/or at "stage 3," at the time of pollination. As the flower matures, there is a change in the color of the calyx from green to yellow and an increase in the size of the flower. (**b**) A diagram showing *Arabidopsis thaliana* flowers and buds that tend to be in the corners of an imaginary triangle. The buds to be chosen for emasculation (stage -1, blue) and the flower that can serve as a male donor (stage 2, pink) are indicated

crossed pistil at the time of harvest. About 24–36 h after pollination, a successfully crossed pistil with fertilized ovules tends to accumulate anthocyanin pigments in the ovary wall.

4. The suggested crosses below will help characterize the fertility defect(s), if the mutant allele is recessive. If not, *see* **Note 8**. In the crosses below, the mutant allele is designated as "*m*," and the wild-type allele is indicated as "+." The pistil parent is listed first and denoted by the symbol (♀), while the pollen donor is listed second and denoted by the symbol (♂):
   (a) *m*/*m* ♀ × +/+ ♂.
   (b) +/+ ♀ × *m*/*m* ♂.
   (c) *m*/+ ♀ × +/+ ♂.
   (d) +/+ ♀ × *m*/+ ♂.
   (e) *m*/*m* ♀ × *m*/+ ♂.
   (f) *m*/+ ♀ × *m*/*m* ♂.
   (g) *m*/+ ♀ × *m*/+ ♂.
5. If a, c, e, f, and g show the phenotype, it indicates that the female reproductive organ is defective.
6. If b, d, e, f, and g show the phenotype, it indicates that the male reproductive organ is defective.
7. If only "a" shows the phenotype, it indicates that the female sporophyte is defective.
8. If only "b" shows the phenotype, it indicates that the male sporophyte is defective.
9. If "c" shows the phenotype, then it indicates that the female gametophyte is defective (*see* **Note 9**).
10. If "d" shows the phenotype, then it indicates that the male gametophyte is defective (*see* **Note 10**). In the event "d" does not show a phenotype, it should not be immediately assumed that there is no male gametophytic defect. Since such crosses are performed with unlimited pollen, the number of wild-type pollen from a heterozygous anther is still in vast excess of wild-type ovules in the pistil and they can fertilize all of the ovules and thus mask the defective mutant pollen tube growth. To rule out this possibility, limited pollination experiments need to be performed by depositing 20–30 pollen grains from a heterozygous mutant anther. If the seed set phenotype manifests in limited pollination experiments as well, then the phenotype is not attributable to mutant pollen being outcompeted by wild-type pollen; instead, these results point to inherent defects in mutant pollen function.

11. If selfed mutant pistils, but none of the reciprocal crosses (a–d), show the phenotype, then the mutant can be designated "self-sterile," which can be characterized by performing crosses "e" and "f".
12. Of all the listed crosses, if "e" alone shows the phenotype, it indicates that the interactions between male gametophyte and female sporophyte are defective [10].
13. Of all the listed crosses, if "f" alone shows the phenotype, it indicates that interactions between male sporophyte and female gametophyte are defective (*see* **Note 11**).
14. If e, f, and g show the phenotype along with selfed crosses but not in a, b, c, and d, it indicates that interactions between the male gametophyte and the female gametophyte are defective.
15. If no phenotype is observed in any of these crosses or selfed mutant pistils, it indicates that the single mutation in the GOI is not sufficient to cause the phenotype (*see* **Note 12**).

3.4 Transmission Efficiency to Examine Gametophyte Function

Measuring transmission efficiency (TE) of a mutation in the progeny, after the mutation is given an opportunity to be transmitted to the seed through one or both gametophytes, can help confirm gametophytic defects revealed by crosses listed in Subheading 3.3. If one or both gametophytes is defective in function due to the mutation, then the gametophyte will be defective in fertilization and will not produce a seed or generate seeds at a reduced rate. Consequently, frequency of mutation in the progeny is significantly decreased compared to the expected frequency.

1. A rapid way to measure TE is to use the resistance marker that is only associated with the mutation in GOI, this method is listed below. If no resistance marker is available *see* **Note 13**.
2. Gametophyte function is being tested in this assay. Hence, it is imperative to perform the following set of reciprocal crosses from the list of crosses mentioned in Subheading 3.3 (Fig. 4):
   - *m*/+ ♀ × +/+ ♂,
   - +/+ ♀ × *m*/+ ♂.
3. Perform four replicates of each crosses or enough replicates to score at least 200 seedlings from each type of cross.
4. Harvest the seeds from these crosses individually (typically 21–31 days after crossing, when siliques are brown and senesced). Harvesting individual siliques will come handy in tracking down errors in crossing, such as accidental crosses with a different parent or auto-selfing by the plant.
5. Desiccate siliques for 7 days after harvesting.
6. Surface sterilize seeds, then plate on 1/2× strength MS media plates with the appropriate antibiotic.

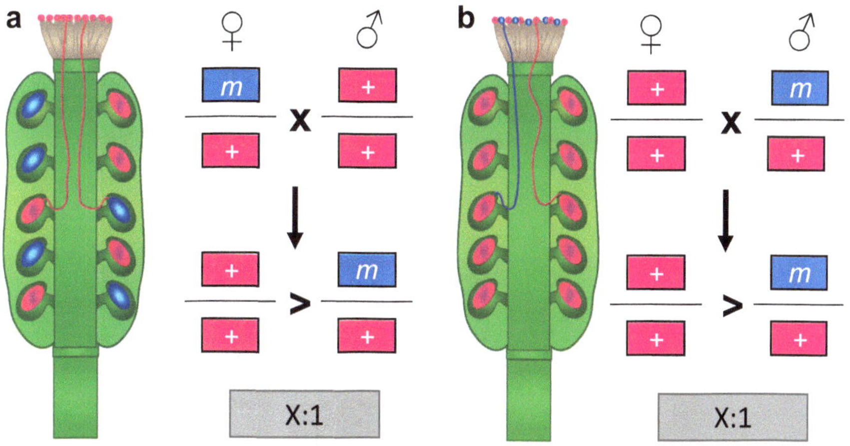

Fig. 4 Reciprocal crosses between wild type and heterozygous mutant to identify male or female gametophytic defects. (**a**) A heterozygous pistil pollinated with wild-type pollen (magenta) results in fertilization of either the wild-type (magenta) or the mutant (blue) female gametophyte. If the mutant female gametophyte is defective, then it is less likely to be fertilized by the wild-type pollen tube and results in significantly reduced frequency of heterozygous individuals in the progeny compared to wild-type siblings (*X*:1, where X is significantly greater than 1). (**b**) A wild-type pistil pollinated with heterozygous pollen (magenta) results in growth of either wild-type (magenta) or mutant (blue) pollen tubes. If the mutant pollen tube is defective, then most wild-type ovules are fertilized by wild-type pollen tubes. This will result in a significantly reduced frequency of heterozygous individuals in the progeny compared to wild-type siblings (*X*:1, where *X* is significantly greater than 1)

7. Stratify the seeds on plates in dark at 4 °C for 3–4 days.
8. Place in a plate growth chamber (example, Percival growth chamber) with the desired light condition (*see* **Note 14**).
9. 10–12 days later, score seedlings for antibiotic resistance (*see* **Note 15**).
10. To calculate transmission efficiency, divide the number of resistant seedlings by the number of sensitive seedlings. Significant reduction from the expected TE of 1 point to defective gametophyte function (Fig. 4).
11. In case of male gametophytic defects (Fig. 4b), a second round of TE experiments involving limited pollination crosses between the same parents should be performed. If the transmission of the mutation through the male gametophyte could not be restored even after limited pollination of mutant pollen on wild-type pistils, then it points to inherent defects in mutant pollen function.

3.5 Aniline Blue Staining of Pistils to Observe PT–Pistil Interactions

Aniline blue staining of selfed and manually pollinated pistils can pinpoint the reproductive step(s) that is (are) defective in the mutant (*see* **Note 16**). A key advantage of this protocol is that it does not rely on transgenic marker expression (such as YFP or GUS) in pollen tubes to visualize their growth in pistils. Thus, it can also be used to study pollen tube growth in any species or

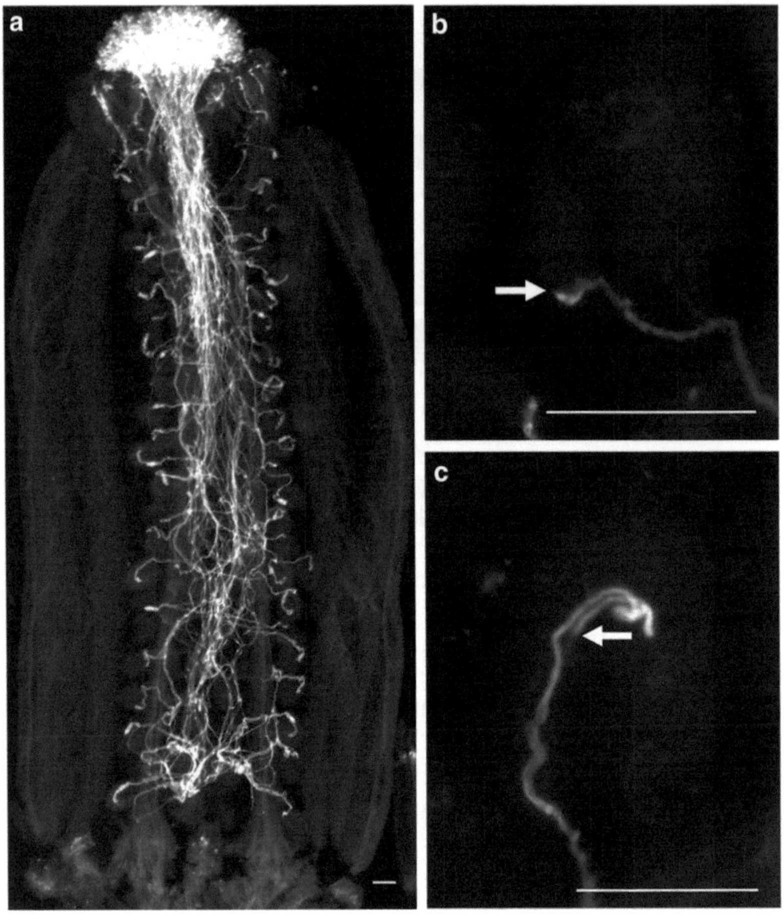

Fig. 5 Aniline blue staining assay to observe pollen tube–pistil interactions, including interspecific crosses. (**a**) Pollen tube growth in an *Arabidopsis thaliana* pistil pollinated with *Sisymbrium irio* pollen. Fluorescent staining shows that pollen tubes have germinated on the stigma, travelled down the transmitting tract, and entered the ovary chamber. Once there, some pollen tubes have entered the ovules and underwent normal pollen tube reception, while a majority either failed to enter the ovule or showed pollen tube reception defects inside the ovule. (**b**) A close-up view of an *A. thaliana* ovule interacting with an *A. thaliana* pollen tube showing successful pollen tube reception (burst of the tube), which is manifested as pollen tube growth arrest in ovules in the micropylar end of the ovule, where the synergid cells are located (white arrow). Pollen tube burst is not visible in such instances because aniline blue stains the callose in pollen tube cell walls and not the cytoplasm. After pollen tube burst, the wall disintegrates and only the cytoplasm is released into the synergid cell, which is not stained by aniline blue. (**c**) A close-up view of an ovule showing pollen tube reception defect, which is visualized as pollen tube overgrowth in the micropylar end of the ovule. White arrow marks the tip of the pollen tube showing pollen tube reception defect. Scale bar (**a**–**c**) = 10 mm

interspecific crosses (*see* Fig. 5a). This protocol is a modified version of the standard *A. thaliana* aniline blue protocol [11]. Modifications made include instructions for high-throughput processing of pistils, mounting modifications to ensure samples are positioned to score PT–ovule interaction defects, and incorporation of crosses for our recommended workflow.

1. Based on the results from Subheading 3.2 and/or Subheading 3.4, perform corresponding crosses from those listed in Subheading 3.3. Summarized below are the crosses from Subheading 3.3 that need to be performed for identifying the pollen tube growth defects underlying the corresponding reproductive defects. To quantify the pollen tube behaviors in pistils, perform indicated crosses in at least three pistils per plant that is used as a female parent in the cross:
   (a) If the mutation is male sporophytic, perform cross "b."
   (b) If the mutation is male gametophytic, perform crosses "b" and "d."
   (c) If the mutation is female sporophytic, perform cross "a."
   (d) If the mutation is female gametophytic, perform crosses "a" and "c."
   (e) In self-sterile mutants, perform crosses of selfed mutant pistils.
   (f) In cases where your mutant is defective in the interactions between male gametophyte and female sporophyte, perform crosses of selfed mutant pistils and "e."
   (g) In the rare case of a mutant with failed interactions between the male sporophyte and female gametophyte, perform crosses of selfed mutant pistils and "f."
   (h) If the mutant shows a failed interaction between the male gametophyte and the female gametophyte then perform crosses of selfed mutant pistils and e, f, and g.
2. Collect crosses 24 h after pollination in a 96-well plate for high-throughput processing. (If you have a small number of pistils (<12) to process, *see* **Note 17**).
3. Add 200 μL of fixative solution to each well, seal the plate lid with plastic wrap or parafilm and place in a resealable plastic bag to prevent evaporation. Keep pistils in solution at room temperature until pistils turn colorless (minimum 2 h minimum to a maximum of 48 h). Fixed pistils can be stored at 4 °C at this step while awaiting further processing.
4. Discard fixative solution via multi-channel or single channel pipette, taking care to avoid damaging or picking up pistils accidentally that get stuck to pipette tips. Having no more than three pistils per Eppendorf tube or two pistils per well in a 96-well plate also will help in not damaging the pistils during processing.
5. Pipette 200 μL of 70% ethanol to wash pistils, wait 10 min, and discard the solution.
6. Add 200 μL of 50% ethanol to wash pistils, wait 10 min, and discard the solution.

7. Add 200 μL of 30% ethanol to wash pistils, wait 10 min, and discard the solution.
8. Add 200 μL of dH_2O to wash pistils, wait 10 min and discard the solution.
9. Add 200 μL of ATS (NaOH) solution, ensure pistils are submerged in solution, cover with plastic wrap to avoid evaporation, and incubate at room temperature overnight (*see***Note 18**).
10. Discard NaOH solution. Add 200 μL of dH_2O to pistils. Care must be taken in handling of pistils, as by this step they have become softened from NaOH treatment. Wait 10 min, then discard the solution.
11. Add 200 μL of DABS to each well. Ensure pistils are submerged. Cover plate with plastic wrap to avoid evaporation and then place at room temperature in the dark for 24 h.
12. After 24 h, samples can be mounted on slides using DABS with 15% glycerol or can be stored in a parafilm sealed plate in the dark at 4 °C.
13. To mount pistils on slides, clean slides and coverslips with 75% ethanol. Using forceps, pick the pistil by its pedicel from the 96-well plate. Remove excess DABS by gently sliding pistil up the side of the 96-well plate during removal.
14. Quickly place the pistil on a clean microscope slide under a stereoscope, so that both ovary valves are visible, with the replum in the middle.
15. Rapidly add 10 μL of mounting solution around the pistil, avoiding air bubbles. Position the pistil with a needle attached to a 1 mL syringe and pop air bubbles, if necessary.
16. Using forceps, quickly place a coverslip on top of the pistil and gently press. Applying slight pressure will separate the valves away from the replum, allowing ovules to be exposed even as most of them remain attached to the pistil and pollen tube growth in the pistil to the ovule is preserved. If too much pressure is applied, most of the tissues will be crushed. If enough pressure is not applied, the ovules are not exposed and continue to remain behind the ovary wall and the pollen tube growth and interactions with ovules are obscured. Therefore, care and practice should be taken to master this step, as it is what will allow consistent scoring of pollen tube–ovule interactions in the pistil.
17. Seal the slides with a clear nail polish in the following manner:
    (a) Place a dot of nail polish in each of the four corners of the coverslip and let it dry.
    (b) Seal edges with nail polish and let it dry.

(c) Repeat sealing the edges with another coat of nail polish and let it dry.

18. Keep samples in a dark microscope slide box until ready to view. If mounted correctly and maintained properly, samples can last at least 2 years.

19. To visualize aniline blue staining, observe in an epi-fluorescence microscope after filtering the UV light through recommended filter sets (UV-1A, UV-2A, and UV-2B), which have an excitation and emission spectrum, of 370 and 509 nm, respectively (*see* Fig. 5, *see* **Note 19**):

(a) Selfed crosses of wild type typically show normal pollen grain germination and pollen tube growth through the stigma, style, and transmitting tract [12]. Upon entering the ovary chamber, the pollen tube locates an ovule, enters through the micropylar end, arrests growth, and bursts to release the two sperm cells to complete double fertilization (Fig. 5b) [12]. All of these steps in pollen tube navigation could be observed in aniline blue-stained pistils, except for sperm release.

(b) If pollen grains do not germinate on the stigma, the mutant is defective in pollen tube germination in crosses between those parents.

(c) If pollen tubes fail to grow on the stigma or through the transmitting tract, the mutant shows failed interactions between pollen tubes and sporophytic pistil tissues.

(d) If pollen tubes do not exit the transmitting tract, then the mutant shows failed interactions in long-distance pollen tube attraction and/or failed interactions between pollen tubes and sporophytic pistil tissues.

(e) If the pollen tube enters the ovary chamber but does not grow toward an ovule, there may be failed interactions between female sporophytic and/or gametophytic tissues.

(f) If the pollen tube grows along the funiculus of an ovule but does not enter the micropyle end of the ovule, then pollen tube attraction is defective.

(g) If the pollen tube reception is defective, pollen tube will enter an ovule but fail to arrest growth and continue to coil within the female gametophyte (Fig. 5c).

(h) If multiple pollen tubes enter a single ovule, then the mutant shows a polytubey phenotype [13, 14].

(i) Additional experiments, involving marking the sperm with fluorescent markers, are required to determine if this phenotype is due to failure to release sperm cells [15] (*see* **Note 20**).

4 Notes

1. An excellent resource to check gene expression in *A. thaliana* and a variety of other plant species is available through The Bio-Analytic Resource (BAR) for Plant Biology https://www.bar.utoronto.ca/.
2. The Arabidopsis Information Resource (TAIR, https://www.arabidopsis.org/) and the Arabidopsis Biological Resource Center (ABRC, https://abrc.osu.edu) contain seed stocks of T-DNA insertion lines in *A. thaliana*. For studying male reproductive defects, Blue SAIL lines, if available in the GOI, are preferable, as these T-DNA insertion lines also carry the pollen and pollen tube expressed *pLAT52::GUS* transgene in the *quartet (qrt)* mutant background, which allows all meiotic products to remain as a single tetrad [16].
3. For seed set analysis of selfed siliques, we prefer to select siliques at positions between 6 and 10 on the main stem of plants with at least 20 siliques [17].
4. Take caution when cutting below the replum. Too deep of an incision will damage the ovules or dislodge the funiculus from the septum. Too shallow of an incision will not open the carpel.
5. Unfertilized ovules will remain small, shriveled, and remain light in color (*see* Fig. 2). Viable ovules will be green with mature embryos. Aborted ovules will be flat or misshaped, larger than unfertilized ovules and be brownish or white in color. If all ovules look white, select older siliques or wait until the plant is older before selecting more siliques to score, it is most likely due to selecting a pistil too young to score. Suitable siliques for scoring should have viable ovules at mid- to late-bent cotyledon stage [18, 19].
6. It is imperative to include wild-type plants that were grown besides the mutant plants as a control in all seed set scoring experiments.
7. Emasculate buds at 12c stage [20], wait 24 h for the pistil to mature, and then pollinate them manually. If the mutant is associated with an antibiotic resistance marker, determining whether fertility defects are sporophytic or gametophytic can be confirmed by analyzing transmission efficiency of the mutation (Subheading 3.4).
8. If the mutant allele is dominant negative and shows sporophytic defects, phenotyping in heterozygous individuals and recapitulating the phenotype by introducing the transgene into wild-type plants may be required.

9. This is the key cross to make this conclusion. Such a mutation will also cause a phenotype in crosses a, e, f, g listed in Subheading 3.3.
10. This is the key cross to make this conclusion. Such a mutation will also cause a phenotype in b, e, f, g listed in Subheading 3.3.
11. We want to note this type of self-sterile mutant is highly unlikely because by the time pollen tube (the male gametophyte) extends and interacts with the female gametophyte, there is no physical or other known influence of the male sporophyte on the pollen tube. The last derivative of the male sporophyte on the male gametophyte is in the pollen coat, which too gets left behind on top of the stigma as part of the "foot" formed when the pollen grain makes initial contact with the stigma. If seed phenotypes instead of merely seed set are being scored, then "e" and "f" crosses in conjunction with "c" and "d" crosses can be useful in testing the possibility that seed defect is dependent on the genotype of the zygote in the seed. For example, if neither "c" or "f" cross show a seed phenotype but the seed phenotype is only seen in "a" cross, then it indicates that the seed defect is caused by the mutant maternal sporophyte [21].
12. In such instances, perform other cytological or molecular analysis to uncover subtle defects [22] or examine if GOI is a member of a gene family to rule out redundancy (either for shared function or compensatory function) as the reason for lack of a phenotype [23, 24].
13. TE analysis should be performed only after confirming that the T-DNA insertion (and by extension, the resistance marker in the T-DNA) is only in the GOI and that the T-DNA insertion in the GOI is tightly linked with the reproductive phenotype. The former can be achieved by back crossing mutant to wild type and the latter can be established by performing co-segregation analysis, in which the T-DNA insertion in the GOI confirmed by genotyping is correlated with the reproductive phenotype. Any reproductive phenotype or genotype can be used to measure TE; however, it is important to ensure that what is being used to assess TE is tightly linked with the mutation in the GOI.
14. Constant light will encourage faster growth and early flowering; however, the plants may be stressed under these conditions. If stress is observed in wild-type plants grown on regular MS plates, check the amount of light that plants are exposed to using a light meter. The following conditions are sufficient to raise healthy plants: plates are incubated in 75–100 μmol m^{-2} s^{-1} of light in 24 h light at 21 °C until 10–14 days old, when they are scored; then resistant plants are

immediately transplanted into soil and grown at 16 h light (100–120 μmol m^{-2} s^{-1}) at 21 °C and 8 h dark at 18 °C as described [25].

15. Resistant seedlings will typically have true leaves, longer roots, and remain green. Sensitive seedlings will turn white, have little to no root growth, and not develop true leaves. We often see that SALK lines may have resistant plants that can be extremely sensitive to Kanamycin; in such instances, delay scoring resistance until 15–20 days after plating. In cases where antibiotic resistance is silenced (often seen with Kanamycin resistance in SALK lines from ABRC), genotyping will be required (*see* **Note 13**).
16. Aniline blue stains callose-rich tissues—in pollinated pistils, this includes pollen grains, pollen tubes, pistil vascular tissue, funiculi from ovules, and damaged tissues [26].
17. Processing steps can be adjusted for small sample sizes; instead of placing into a 96 well plate, add each pistil to a 1.0 mL microcentrifuge tube. Reduce volumes of all solutions in **steps 3–11** from 200 to 150 μL in Subheading 3.5.
18. When adding ATS, due to difference in viscosity of solutions, there is an increased chance of introducing air bubbles. It is important to remove any air bubble in the tube/well, so that pistils do not dry after accidentally getting trapped in the air bubble.
19. Slides should be scored within 24 h after mounting. Occasionally, slides that have been overexposed to UV light tend to become cloudy and obfuscate scoring. Longer DABS staining time (at least 24 h) may help increase the durability of a sample.
20. Crosses to a dual pollen tube and sperm cell marker such as *pLAT52::GFP, HTR10::RFP* [15] can be used to visualize defects in sperm cell release.

Acknowledgements

J.A. Noble was supported by the following: IGERT Comparative Genomics Program at the University of Arizona (Award ID: 0654435); NSF Graduate Research Fellowship: Grant DGE-1143953; the Boynton Graduate Fellowship from the School of Plant Sciences, University of Arizona; and the University of Arizona Graduate College Office of Diversity and Inclusion. Additional support for this work was provided by an NSF grant to R. Palanivelu (IOS-1146090).

References

1. Honys D, Reňák D, Twell D (2006) Male gametophyte development and function. In: Teixeira da Silva JA (ed) Floriculture, ornamental and plant biotechnology: advances and topical issues, vol 1. Global Science Books, London, pp 76–87
2. Sundaresan V, Alandete-Saez M (2010) Pattern formation in miniature: the female gametophyte of flowering plants. Development 137:179–189
3. Palanivelu R, Tsukamoto T (2012) Pathfinding in angiosperm reproduction: pollen tube guidance by pistils ensures successful double fertilization. Wiley Interdiscip Rev Dev Biol 1:96–113
4. Rutley N, Twell D (2015) A decade of pollen transcriptomics. Plant Reprod 28:73–89
5. Qin Y, Leydon AR, Manziello A, Pandey R, Mount D, Denic S, Vasic B, Johnson MA, Palanivelu R (2009) Penetration of the stigma and style elicits a novel transcriptome in pollen tubes, pointing to genes critical for growth in a pistil. PLoS Genet 5:e1000621
6. Leydon AR, Weinreb C, Venable E, Reinders A, Ward JM, Johnson MA (2017) The molecular dialog between flowering plant reproductive partners defined by SNP-informed RNA-sequencing. Plant Cell 29:984–1006
7. Swanson R, Clark T, Preuss D (2005) Expression profiling of Arabidopsis stigma tissue identifies stigma-specific genes. Sex Plant Reprod 18:163–171
8. Wuest SE, Vijverberg K, Schmidt A, Weiss M, Gheyselinck J, Lohr M, Wellmer F, Rahnenführer J, von Mering C, Grossniklaus U (2010) Arabidopsis female gametophyte gene expression map reveals similarities between plant and animal gametes. Curr Biol 20:506–512
9. Johnson MA, Kost B (2010) Pollen tube development. In: Hennig L., K-hler C. (eds) Plant Developmental Biology, Methods in Molecular Biology (Methods and Protocols), vol 655. Humana Press, Totowa, NJ, pp 155–176
10. Palanivelu R, Brass L, Edlund AF, Preuss D (2003) Pollen tube growth and guidance is regulated by POP2, an Arabidopsis gene that controls GABA levels. Cell 114:47–59
11. Mori T, Kuroiwa H, Higashiyama T, Kuroiwa T (2006) GENERATIVE CELL SPECIFIC 1 is essential for angiosperm fertilization. Nat Cell Biol 8:64–71
12. Johnson MA, Harper JF, Palanivelu R (2019) A fruitful journey: pollen tube navigation from germination to fertilization. Annu Rev Plant Biol 70:809–837
13. Kasahara RD, Maruyama D, Hamamura Y, Sakakibara T, Twell D, Higashiyama T (2012) Fertilization recovery after defective sperm cell release in Arabidopsis. Curr Biol 22:1084–1089
14. Beale KM, Leydon AR, Johnson MA (2012) Gamete fusion is required to block multiple pollen tubes from entering an Arabidopsis ovule. Curr Biol 22:1090–1094
15. Hamamura Y, Saito C, Awai C, Kurihara D, Miyawaki A, Nakagawa T, Kanaoka MM, Sasaki N, Nakano A, Berger F et al (2011) Live-cell imaging reveals the dynamics of two sperm cells during double fertilization in *Arabidopsis thaliana*. Curr Biol 21:497–502
16. Palanivelu R, Johnson MA (2010) Functional genomics of pollen tube–pistil interactions in Arabidopsis. Biochem Soc Trans 38:593–597
17. Yuan J, Kessler SA (2019) A genome-wide association study reveals a novel regulator of ovule number and fertility in *Arabidopsis thaliana*. PLoS Genet 15:e1007934
18. Le BH, Cheng C, Bui AQ, Wagmaister JA, Henry KF, Pelletier J, Kwong L, Belmonte M, Kirkbride R, Horvath S et al (2010) Global analysis of gene activity during Arabidopsis seed development and identification of seed-specific transcription factors. Proc Natl Acad Sci 107:8063–8070
19. Goldberg RB, de Paiva G, Yadegari R (1994) Plant embryogenesis: zygote to seed. Science 266:605–614
20. Smyth DR, Bowman JL, Meyerowitz EM (1990) Early flower development in Arabidopsis. Plant Cell 2:755–767
21. Grover JW, Kendall T, Baten A, Burgess D, Freeling M, King GJ, Mosher RA (2018) Maternal components of RNA-directed DNA methylation are required for seed development in *Brassica rapa*. Plant J 94:575–582
22. Pereira AM, Nobre MS, Pinto SC, Lopes AL, Costa ML, Masiero S, Coimbra S (2016) "Love is strong, and you're so sweet"ω: JAGGER is essential for persistent synergid degeneration and polytubey block in *Arabidopsis thaliana*. Mol Plant. 9:601-614
23. Leydon AR, Beale KM, Woroniecka K, Castner E, Chen J, Horgan C, Palanivelu R, Johnson MA (2013) Three MYB transcription factors control pollen tube differentiation required for sperm release. Curr Biol 23:1209–1214

24. Liang Y, Tan Z-M, Zhu L, Niu Q-K, Zhou J-J, Li M, Chen L-Q, Zhang X-Q, Ye D (2013) MYB97, MYB101 and MYB120 function as male factors that control pollen tube-synergid interaction in *Arabidopsis thaliana* fertilization. PLoS Genet 9:e1003933

25. Kessler SA, Shimosato-Asano H, Keinath NF, Wuest SE, Ingram G, Panstruga R, Grossniklaus U (2010) Conserved molecular components for pollen tube reception and fungal invasion. Science 330:968–971

26. Śnieżko R (2000) Fluorescence microscopy of aniline blue stained pistils. In: Dashek WV (ed) Methods in plant electron microscopy and cytochemistry. Totowa, NJ, Humana, pp 81–86

Chapter 9

Internally Controlled Methods to Quantify Pollen Tube Growth and Penetration Defects in *Arabidopsis thaliana*

Devin K. Smith and Ian S. Wallace

Abstract

Double-fertilization in angiosperms requires precise communication between the male gametophyte (pollen), the female tissues, and the associated female gametophyte (embryo sac) to facilitate efficient fertilization. Numerous small molecules, proteins, and peptides have been shown to impact double-fertilization through the disruption of pollen germination, pollen tube growth, pollen tube guidance, or pollen tube penetration of the female tissues. The genetic basis of signaling events that lead to successful double-fertilization has been greatly facilitated by studies in the model organism *Arabidopsis thaliana*, which possesses a relatively simple reproductive physiology and a widely available T-DNA mutant seed collection. In this chapter, we detail methods for determining the effects of single gene loss-of-function mutations on pollen behavior through the creation of an internally controlled fluorescent hemizygous complement line. By transforming a single copy of the disrupted gene back into the homozygous mutant background, a precise endogenous control is generated due to the fact that pollen containing equal ratios of mutant and complemented pollen can be collected from a single flower. Using this experimental design, we describe multiple assays that can be performed in series to assess mutant pollen defects in germination, pollen tube elongation rate, and pistil penetration, which can be easily quantified alongside a "near-wild-type" complemented counterpart.

Key words Pollen germination, Double-fertilization, *Arabidopsis thaliana*, Pollen tube, Male gamete defects, In vitro fertilization, Fluorescence semi-in vivo (*SIV*) fertilization, Fluorescence microscopy, Pollen tube penetration

1 Introduction

Double-fertilization is a highly coordinated process involving the male gametophyte (pollen), the female pistil tissues, and the female gametophyte (embryo sac), which communicate through defined cell-specific mechanisms that ensure successful fertilization events [1–3]. Although various proteins, peptides, and small signaling molecules have been mechanistically implicated in specific and critical roles during double-fertilization [4–8], many molecular factors that facilitate this cellular communication pathway remain to be elucidated.

Anja Geitmann (ed.), *Pollen and Pollen Tube Biology: Methods and Protocols*, Methods in Molecular Biology, vol. 2160, https://doi.org/10.1007/978-1-0716-0672-8_9,

Double-fertilization begins when a pollen grain lands on the apical stigmatic surface of the pistil, hydrates, and produces a sperm cell-trafficking structure called a pollen tube [9]. The pollen tube is the fastest growing cell in all plants and functions to deliver two sperm cells to a distant ovule buried deep within the pistil [10]. As the pollen tube rapidly penetrates through the pistil tissues, the female tissues provide the growing pollen tube with nutrients and chemical guidance cues that ensure its successful arrival at an unfertilized ovule, where the pollen tube ruptures to release its two sperm cells into the female gametophyte [4–8, 11]. *Arabidopsis thaliana* provides an ideal system to investigate the genetic and cellular bases of double-fertilization, due to the wide availability of genetic tools, self and cross compatibility, and T-DNA insertional mutants that allow investigators to genetically dissect this process [12, 13]. Phenotypic analyses of these T-DNA mutant lines have revealed a plethora of genes involved in the regulation of pollen tube growth, pollen tube chemical guidance, and sperm cell fusion [14, 15]. Additionally, Arabidopsis is an ideal model organism for reproductive investigations, due to its generation time and pollen-pistil self-compatibility, which have facilitated the study of pollen tube behavior resulting from genetic loss-of-function in addition to testing different forms of stimuli, including chemical, electrical, and gravitational [10, 16–20].

While Arabidopsis represents a strong model system for investigating pollen–pistil interactions, certain technical challenges are presented. Pollen viability can be strongly influenced by the developmental stage of harvested floral organs or growth conditions of the parent plant [21]. Additionally, in vitro growth conditions of pollen or associated pistil organs can substantially influence critical growth parameters, such as pollen tube germination or growth rate [21, 22]. Finally, the investigation of pollen trajectories through the pistil as well as associated fertilization events can be difficult to quantify using traditional methods, such as aniline blue staining, and often these experiments require limited pollination conditions, which can be difficult to robustly reproduce between laboratories. Therefore, it would be ideal to internally control various pollen growth or pollen–pistil interaction assays to circumvent some of these technical challenges.

In this chapter, we detail simple techniques for investigating and quantifying Arabidopsis T-DNA mutant pollen defects in germination, pollen tube elongation, and pollen tube penetration related to single gene disruption. Precise analysis of these pollen phenotypes is facilitated by using a hemizygous complement construct that provides a reliable endogenous experimental control. First, the hemizygous complement line is created through Agrobacterium-mediated transformation of a complementation construct encoding the gene of interest, an herbicide marker, and a fluorescent tag into the homozygous mutant T-DNA background. T_1 transgenic lines are identified by herbicide selection, and after reaching sexual maturity, these lines are screened for single transgene insertion into the mutant genome by in vitro pollen germination. Germination of transgenic pollen under an epifluorescence microscope facilitates the visualization of complemented pollen tubes, which are fluorescent, and original mutant pollen that are nonfluorescent. Simple quantification of fluorescent vs. nonfluorescent pollen tubes is used to identify T_1 lines that are hemizygous. Using these genetic resources, pollen germination frequency and pollen tube elongation rate of the mutant and complemented line can be determined simultaneously by in vitro germination over a defined time course. Additionally, we describe a method to assess pollen penetration defects using hemizygous pollen in a semi-in vivo (*SIV*) fertilization assay. Emasculated homozygous *male sterility 1* (*ms1*$^{-/-}$) mutant pistils are pollinated with hemizygous pollen and incubated for 1 h before the pistil is dissected and transferred to an agarose growth medium pad on a glass microscope slide. Four hours after pollination, pollen tubes exiting the bottom of the style can be visualized using an epifluorescence microscope. By determining the number of pollen tubes per genotype exiting the bottom of each style, one can quantitatively evaluate pollen tube penetration defects if the pollen tube population ratio of 50:50% fluorescent, complemented versus nonfluorescent mutant pollen tubes is altered.

2 Materials

Deionized water (sensitivity of 18 MΩ cm) is recommended for preparation of all solutions. A set of p10, p20, p200, and p1000 pipettes and their respective sterile, disposable tips will be needed. In addition, sterile consumables that are RNase, DNase, and DNA free including 1.5 mL microcentrifuge tubes, 15 mL conicals, and 50 mL conicals will be required.

2.1 Selection of Hemizygous Complement Lines

1. Floral dipped *Arabidopsis thaliana* seed [23].
2. Seed cleaning solution: 3% (v/v) sodium hypochlorite, 0.1% (w/v) sodium dodecyl sulfate.
3. Biosafety cabinet.
4. Arabidopsis Growth Medium with appropriate herbicide/selection marker: 1/2× Murashige and Skoog salts, 10 mM MES-KOH pH 5.7, 1% (w/v) sucrose, 1% (w/v) phytoagar + appropriate concentration of herbicide.
5. Seed Germination Growth chamber (e.g., Percival, CU41L4).
6. Potting soil.
7. Germination flat containing insert with holes.

8. 4″ square pots with drainage holes.
9. Humidity dome (3.8″ tall).
10. Growth chamber (e.g., Percival, AR-66L).

2.2 Validation of Hemizygosity

1. Fine forceps.
2. Liquid Pollen Germination Medium (PGM): 2.5 mM $CaCl_2$, 0.005% (w/v) boric acid, 2.5 mM KCl, 0.5 mM $MgSO_4$, 10% (w/v) sucrose, pH 7.5 (*see* **Note 1**).
3. PGM-Agarose pad: 1 mM $CaCl_2$, 0.002% (w/v) boric acid, 1 mM KCl, 0.2 mM $MgSO_4$, 10% (w/v) sucrose, pH 7.5, 1.5% (w/v) low melting point agarose.
4. Glass slides (25 × 75 × 1 mm).
5. Kimwipes® (11 × 21 cm).
6. Square petri dishes (100 × 100 × 15 mm).
7. Wooden barbeque skewers/plastic drinking straws and lab tape (optional, *see* **Note 2**).
8. Rectangular glass coverslips (24 × 50 mm, 0.13–0.17 mm thickness).
9. Epifluorescence microscope with GFP filter (or fluorescence filter compatible with expression vector tag of choice), 20× objective, and software package to process z-stacked images as well as stitch images together.
   - A Keyence BZ-X700 epifluorescence microscope equipped with Keyence GFP Filter Cube (Ex 470/40, Em 525/20, Dm 495), Nikon CFI Plan Apochromat Lambda 20× objective (NA 0.75), and Keyence BZ-X Analyzer software that was used to generate images associated with this chapter.

2.3 In Vitro Germination Assay

1. Confirmed hemizygous complement plant lines at flowering stage.
2. Materials, **items 1–9**, from Subheading 2.2.
3. ImageJ software.

2.4 Semi-In Vivo (SIV) Penetration Assay

1. Confirmed hemizygous complement plant lines at flowering stage.
2. Equal-age homozygous recessive *male sterility 1 (ms1*$^{-/-}$*) Arabidopsis thaliana* plant lines or *Arabidopsis thaliana* Col-0 wild-type plant lines as hemizygous complement plant line (*see* **Note 3**).
3. Materials **items 1–9**, from Subheading 2.2.
4. Luer-Lok stainless steel needle (19 G).
5. 10 mL Luer-Lok tip syringe.

3 Methods

The following instructions assume that a single male-specific gene in *Arabidopsis thaliana* is responsible for the reproductive defects observed in the homozygous T-DNA mutant line, which should be confirmed through reciprocal crosses. Additionally, the protocols assume that the homozygous T-DNA mutant has been successfully transformed with a plant expression vector possessing an herbicide resistance gene as well as the causative gene expressed in pollen translationally fused to a GFP tag, and that the unconfirmed transgenic seed has been collected (*see* **Note 4**).

3.1 Selection and Growth of Transgenic Lines

1. In four to five 1.5-mL sterile microcentrifuge tubes, add no more than 0.1 mL of unconfirmed floral dipped seed (*see* **Note 5**).
2. Add 500 μL of Seed Sterilization Solution to each tube and briefly vortex to completely resuspend seed in solution.
3. Incubate tubes at 25 °C for 20 min.
4. During incubation, ready biosafety cabinet (or sterile hood) by turning on blower and spraying surface down with 75% (v/v) ethanol.
5. Following incubation, move working environment into the biosafety cabinet and aspirate Seed Sterilization Solution from each tube.
6. Add 1 mL of sterile diH_2O to each tube and wait for the seed to resettle to the bottom of the tube.
7. Aspirate the sterile diH_2O.
8. Repeat **steps 6** and 7 four more times for a total of five diH_2O washes.
9. Following the fifth wash diH_2O aspiration, finish the sterilization process by adding 1 mL diH_2O to each tube.
10. Incubate seed at 4 °C in the dark for 48 h.
11. Transfer stratified seed and selection growth media plates to prepped biosafety hood (*see* **Note 6**).
12. Seed can be easily plated on selection media using a p1000 pipette and sterile tips to gently draw up seed and aspirate onto the media (*see* **Note 7**).
13. Selection plates with seed should be grown horizontally in a seed germination growth chamber set to 24 °C long-day light conditions for 7–10 days to select for T_1 transformants (*see* **Note 8**).

14. Following 7–10 days, transplant 12 herbicide-resistant T_1 seedlings into 4″ square pots filled with loose, moist potting soil (maximum four seedlings per pot).
15. Place transplanted seedling pots into closed bottom flat equipped with drainage insert.
16. Fit humidity dome over flat and place flat in Arabidopsis growth chamber set to 24 °C long-day light conditions.
17. Allow seedlings to grow for two days with humidity dome and then remove.
18. Water seedlings with tap water by filling bottom base of flat with water until it reaches approximately 1 in. up from the base of the pots every third day until seedlings begin to bolt and then as needed following. Allow soil to soak up water for 10–30 min and then drain water from flat to prevent fungal and insect contamination.

3.2 Determining Hemizygosity in T_1 Transgenic Lines

The following confirmation method assumes that the transplanted T_1 lines have reached flowering stage. The following protocol will be used to determine which lines have integrated a single copy of the complement construct.

1. Prepare for making Pollen Germination Media (PGM)-Agarose microscope slides by laying out 12 microscope slides on a clean working surface. Additionally, gather a p1000 pipette and tips (*see* **Note 9**).
2. In a 50 mL falcon tube, prepare components of PGM-Agarose to a final volume of 20 mL (*see* **Note 10**).
3. Place the uncapped falcon tube in a 250 mL beaker (to hold upright) and microwave the mixture in short 10 s burst until the agarose has dissolved completely.
4. Pipette approximately 750 μL of hot PGM-Agarose onto each slide and allow medium to solidify at room temperature.
5. To avoid drying out the media pad, prepare a humidity chamber using a square petri dish with a dampened Kimwipe®, and place the slides face up in the petri dish immediately after the media has completely solidified (~5 min) (*see* **Note 11** and Fig. 1).
6. Using a single media slide per T_1 line, identify a flower that has been open for about 1 day and detach it from the remainder of the plant at the base of the floral pedicel.
7. Release the flower into the palm of your hand and using forceps, pick up the flower just above the sepal while applying slight pressure forcing the flower to fan out allowing for easier access to the stamens (*see* Fig. 2).

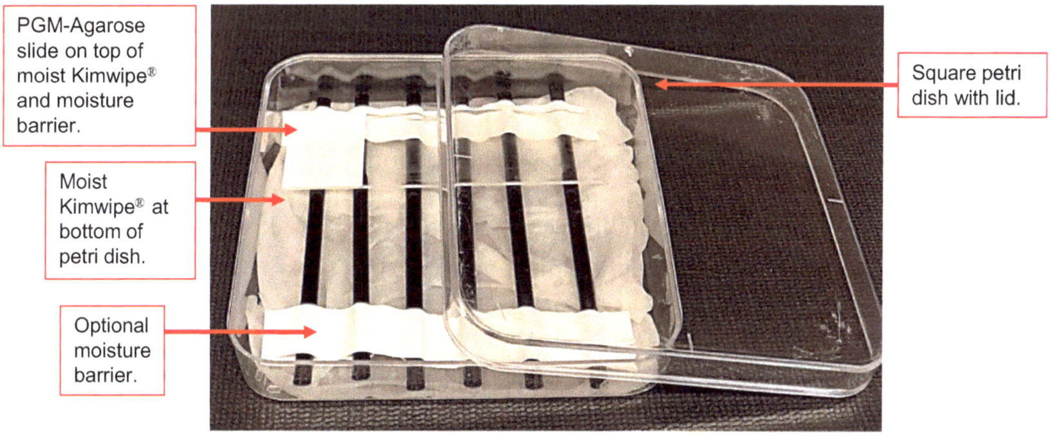

Fig. 1 Humidity chamber setup. A humidity chamber is used throughout the assays in this chapter to prevent slides with PGM-Agarose from drying out. The chamber is composed of two basic materials: (1) square petri dish and (2) moistened Kimwipe®. An optional, but highly recommended barrier between the moist Kimwipe® and media slide was created using coffee straws and lab tape

Fig. 2 Preparing Arabidopsis flowers for in vitro pollen germination. (**a**) *Arabidopsis thaliana* flower at peak sexual maturity. (**b**) Using fine forceps, apply slight pressure just above the sepal to allow for the floral structures to fan out facilitating greater access to the male reproductive structures (stamens). Maintain this pressure while simultaneously *gently* brushing/blotting the flower to release the pollen onto a PGM-Agarose media slide for in vitro germination assays

8. *Gently* brush or blot the flower across one-third of the media pad to transfer pollen grains (*see* **Note 12**).
9. Repeat **steps 6–8** for two more flowers each using another third of the media pad, for a total of three individual pollen samples from the same T_1 line per slide. Additionally, germinate a control slide with nonfluorescent line pollen, such as wildtype Col-0 or the original homozygous T-DNA mutant.

10. *Immediately* following complete pollination for a single line, return the slide to the humidified chamber, set a timer for 4 h, and allow the pollen on the slide to germinate at 25 °C in the dark for 4 h (*see* **Note 13**).
11. Following the 4-h incubation, prepare each slide for imaging by gently pipetting ~20 μL of liquid, room temperature PGM in small droplets across the agar surface of the slide.
12. Allow liquid medium to slightly soak in to PGM-Agarose pad for approximately 2 min before *very gently* applying coverslip to slide (*see* **Note 14**).
13. Individually image each slide at 20× magnification under brightfield and GFP excitation with an epifluorescence microscope using the z-stack setting to capture the entire depth of the field of view. A minimum of three images per line should be acquired for analysis (*see* **Note 15**).
14. After all images are acquired, use desired image processing software to create a z-stack maximum projection for each image and then merge the individual brightfield and GFP maximum projection images to allow for both fluorescent and nonfluorescent pollen tubes to be visualized.
15. To determine if the T_1 line has integrated a single copy of the complement vector into its genome, count the number of fluorescent and nonfluorescent pollen tubes observed for each line. Truly hemizygous lines that will be used for further analysis should possess a ratio of approximately 50:50% ± 2% fluorescent, complemented pollen tubes to uncomplemented, nonfluorescent mutant pollen tubes (*see* **Note 16**).

3.3 In Vitro Pollen Germination

The following protocol assumes a hemizygous complement line has been generated. The methods described in this section will allow for the observation of morphological abnormalities as well as images that can be used in Subheading 3.4 to calculate pollen germination frequency and pollen tube elongation rate between complemented and uncomplemented, mutant pollen. To generate these data, a time course will be implemented, and images will be acquired at 1, 2, 4, and 6 h after pollination.

1. Prepare seven PGM-Agarose slides as previously described in Subheading 3.2 **steps 1–5**.
2. For each time point, germinate at least one PGM-Agarose slide with hemizygous pollen from three flowers as previously described in Subheading 3.2 **steps 6–8** for confirmation of T_1 lines. Additionally, germinate a single slide with wildtype Col-0 pollen for each time point (*see* **Note 17**).
3. *Immediately* following complete pollination of a slide, transfer it to a humidified chamber to prevent media dry out. Place the

humidified chamber containing the pollinated slide in a dark environment at 25 °C and start timer (*see* **Note 18**).

4. After the necessary amount of time has elapsed for each time point, prepare the slides for imaging, perform microscopy, and process images as previously described in Subheading 3.2 **steps 11–14**. Make sure each final image is exported with a representative scale bar.

3.4 Analysis of In Vitro Germination

The following methods will describe how to determine pollen germination frequency and pollen tube elongation rate between complemented and uncomplemented, mutant pollen. ImageJ software [24] will be used for pollen tube length measurements and is recommended for calculating pollen germination frequency. Determining differences in overall pollen tube morphology is subjective unless overtly obvious and tools for quantifying such abnormalities will not be discussed in this chapter. Nonetheless, morphological information should not be disregarded as it may shed light on the potential cause of an underlying pollen tube defect and could be important for suggesting further experiments to elucidate the abnormality.

3.4.1 Determining In Vitro Pollen Germination Frequency

1. To determine the in vitro germination frequency of complemented versus uncomplemented, mutant pollen, open a single image in ImageJ (*see* **Note 19**).
2. For each image captured per time point, count the number of fluorescent, complemented pollen tubes and nonfluorescent, uncomplemented tubes (*see* **Note 20**).
3. For each time point, sum the number of all complemented pollen tubes and uncomplemented pollen tubes counted. Next, calculate percent germination of each pollen genotype using the following equation for a specific time point (*see* **Note 21**):

$$X\ \text{pollen genotype}\%\text{germination} = \frac{\text{number of } X\ \text{pollen tube genotype}}{\text{total number of all pollen tubes}} \times 100.$$

4. Repeat **steps 2–3** for each image captured per time point to determine the percent gemination of complemented and uncomplemented, mutant pollen. A germination frequency of approximately 50% per pollen genotype would be expected for mutants with unaffected germination. Deviations in these percentages, specifically a higher percentage observed for complemented pollen or a temporal delay in germination observed for uncomplemented, mutant pollen would suggest abnormalities with pollen grain hydration or problems with pollen tube production (*see* **Note 22**).

3.4.2 Determining In Vitro Pollen Tube Elongation Rate

1. To determine the in vitro elongation rate of complemented versus uncomplemented pollen tubes as well as wildtype Col-0 control, open a single image in ImageJ.
2. Set the pixel size within the image to a known length by selecting the straight-line icon (fifth icon in), and as precisely as possible, tracing over the scale bar within the photo by clicking and dragging the cursor.
3. In the ImageJ tool bar, select the *Analyze* dropdown menu and click *Set Scale.*
4. Change two of the image scale parameters, (a) *Known Distance* and (b) *Unit of Length*, to the value of the scale bar and make sure *Global* is unchecked. Select *OK* to exit menu and finish setting the scale for the image.
5. Right click on the straight-line icon and select *Segmented line* (*see* **Note 23**).
6. Start by individually measuring the lengths of all pollen tubes of either fluorescent, complemented pollen or nonfluorescent, uncomplemented, mutant pollen by double clicking on one end of the desired pollen tube to be measured and continue clicking to trace along the body of the tube. Double click at the opposite end of the tube to end the line tracing (*see* **Note 24**). The line drawn should run the entire length of the pollen tube from tip to emergence at pollen grain.
7. In the ImageJ tool bar, select the *Analyze* dropdown menu and click *Measure.* Alternatively, pressing the "M" key performs the same function. A second screen in table format should appear containing a value for the length of the line traced over the pollen tube in the units of image scale bar.
8. Continue measuring the pollen tube lengths of the genotype chosen until all have been analyzed in the image by repeating **steps 6–7**.
9. Record all lengths for the specific pollen genotype in the image before moving on to measure the second pollen tube genotype.
10. Repeat **steps 1–9** for each image, including Col-0 control images. *It is critically important to reset the scale parameters for each new image opened in ImageJ before continuing to measure pollen tube length.*
11. After all pollen tubes per genotype per time point have been measured, an XY graphical representation of pollen tube length over time should be constructed to display the growth rate of complemented, uncomplemented, and wildtype pollen tubes over the 6-h time course. Differences in elongation rate over time should be determined through statistical analysis, such as an unpaired t-test.

3.5 Fluorescence Semi-In Vivo (SIV) Fertilization Assay

The following protocol is used to investigate a potential mutant pollen penetration defect using a partially native semi-in vivo (*SIV*) setup. Performing hemizygous pollination of an emasculated pistil and then dissecting the pollinated stigma-style from the remainder of the pistil body facilitates the observation of pollen tubes emerging from the cut pistil bottom. Mutant pollen defective at penetrating effectively through the initial pistil tissues are expected to be observed at lower frequency than the complemented pollen [25].

1. Emasculate at least 10 *ms1*$^{-/-}$ pistils from flowers that have been open for approximately 1 day (*see* **Note 25**).
2. Additionally, prepare 3 PGM-Agarose slides as previously described in Subheading 3.2 **steps 1–5**.
3. Identify a hemizygous complement line flower that has been open for approximately 1 day and *gently* detach it from the remainder of the plant at the base of the floral pedicel using fine forceps.
4. Reposition the flower into nondominant hand and grip flower at base of receptacle.
5. With nondominant hand, hold flower between the thumb and index finger and with dominant hand, use fine forceps to fold back petals to expose the pistil and stamens (*see* Fig. 3a, b).
6. Remove pistil using fine forceps to expose remaining stamens.
7. Gather hemizygous stamens with fine forceps at base of the filaments and dissect the stamens by *gently* separating them in a

Fig. 3 Stamen dissection for fluorescence *SIV* pistil pollination. Demonstration of hemizygous flower stamen dissection to fertilize emasculated pistil for fluorescence *SIV* assay. (**a**) Hemizygous flower grasped between thumb and index fingers of nondominant hand. Using fine forceps in dominant hand, petals are folded back and wedged between nondominant fingers. (**b**) Exposed pistil and stamens with petals folded back. Image shown with flower held by forceps for demonstration purposes, while in actual practice, flower should remain grasped in fingers as in panel (**a**). (**c**) Using fine forceps, remove pistil from remaining flower. Finally, gather stamens with forceps and detach from flower in a single motion. Transfer hemizygous pollen atop anther to emasculated acceptor pistil

single motion from the remaining flower (*see* **Note 26** and Fig. 3c).

8. Transfer the hemizygous pollen atop the anthers of the grasped filaments onto the stigma of an emasculated $ms1^{-/-}$ pistil by gently brushing across the stigmatic papillae (*see* Fig. 3c).
9. Repeat **steps 3–8** until the $ms1^{-/-}$ pistil has been completely pollinated by hemizygous pollen (*see* **Note 27**).
10. Allow pollinated pistils to remain attached to the remainder of the plant at room temperature for 1 h following pollination. Place in area where the plant will not be disturbed and has low airflow.
11. Just before the end of incubation, position a strip of double-sided tape on a glass microscope slide (*see* Fig. 4a).
12. Following incubation, remove pollinated pistils from the plant by dissecting with fine forceps at the base of the pedicel.
13. Using a stereoscope, transfer dissected pistils to slide prepared with tape by positioning the pistils such that only the pedicel and lower half of pistil are adhered to tape. Make sure the pistil is fixed facing up such that both sides of the ovary and pistil septum are flat against the slide (*see* Fig. 4b, c).
14. With a Luer-Lok stainless steel needle (fitted with syringe for gripping), cut pistil as evenly as possible just below the style, and *immediately* transfer pollinated stigma-style to PGM-Agarose slide (*see* **Note 28** and Fig. 4d).

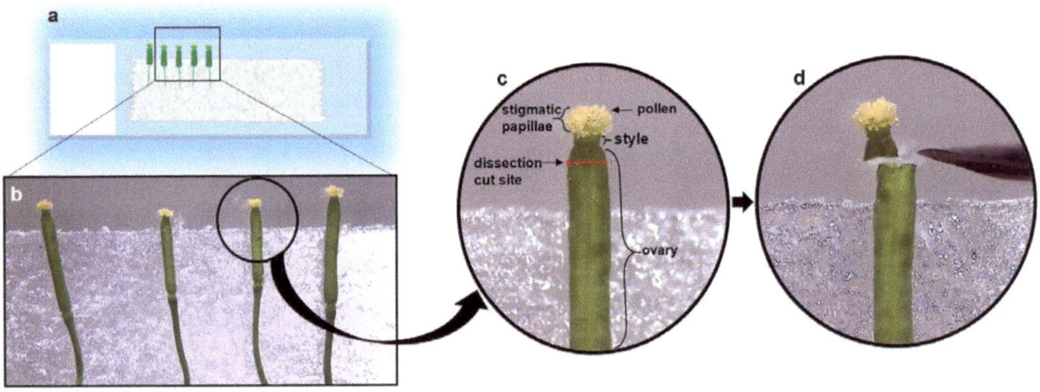

Fig. 4 Fluorescence *SIV* pistil dissection diagram. (**a**) Illustration of slide setup for pistil dissection. Double-sided tape is applied to glass slide. One hour after pollination (HAP), pistils are detached from the remaining plant at base of pedicel and transferred to adhesive. (**b**) Image of pistils transferred to glass slide 1 HAP. (**c**) Closeup of pistil on adhesive with important anatomical features denoted as well as dissection cut site (red dashed line). (**d**) Pistil cut with 19 G needle (rightmost object) just below style to sever the pollinated stigma-style from the remaining pistil. The stigma-style is then transferred to a PGM-Agarose slide and fertilization allowed to continue in a humidified chamber in darkness for 3 h before imaging

15. Repeat the dissection and transfer process to the PGM-Agarose slide for remaining pollinated pistils.
16. After all pistils have been transferred to the PGM-Agarose slide, *immediately* place slide in a humidified chamber and allow semi-in vivo fertilization to continue for 3 additional hours in the dark.
17. Four hours after pollination, prepare slides for imaging as previously described in Subheading 3.2 **steps 11–12**.
18. Individually image the bottom of each pistil to capture all emerging tubes at 20× magnification under brightfield and GFP excitation with an epifluorescence microscope using the z-stack and image stitching setting to capture the entire image depth and field of view.
19. After all images are acquired, use desired image processing software to stich images together as well as create a z-stack maximum projection for each image. Finally, merge the individual brightfield and GFP maximum projection stitched images to allow for both fluorescent and nonfluorescent tubes emerging from the bottom of the cut style to be visualized (*see* Fig. 5).

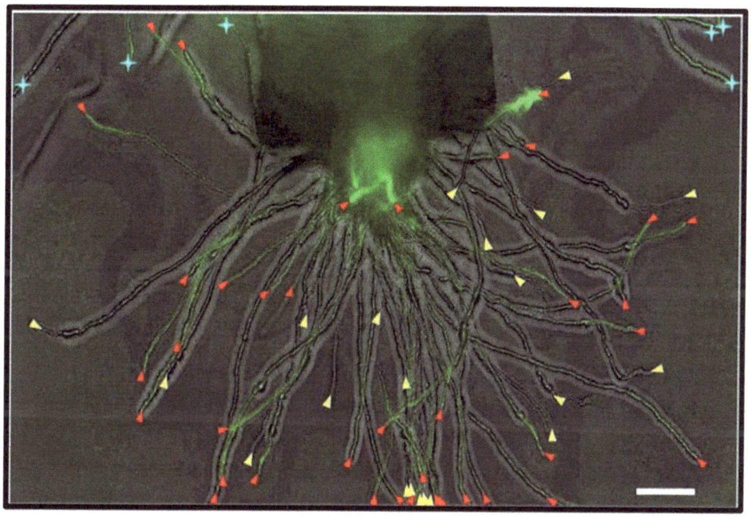

Fig. 5 Fluorescence *SIV* fertilization assay with hemizygous pollen. 20× GFP and brightfield z-stack maximum projection overlay image depicting pollen tubes (PTs) emerging from the bottom of a *SIV* fertilized *ms1*$^{-/-}$ style 4 h after pollination with hemizygous compliment pollen. Yellow carrots indicate uncomplemented, nonfluorescent PTs, while red carrots indicate complemented, fluorescent PTs. The blue stars indicate PTs that did not emerge from the bottom of the dissected stigma-style and should be excluded from further analysis. Bar = 100 μm

20. To determine if the mutant pollen of interest is defective at pistil penetration, count the number of fluorescent, complemented pollen tubes and nonfluorescent, uncomplemented, mutant pollen tubes emerging from the bottom of each pistil. An emerging pollen tube composition >50% of complemented pollen tubes would suggest the presence of a penetration defect for the mutant pollen (*see* **Note 29**).

4 Notes

1. A 2× Liquid Pollen Germination Medium stock solution without sucrose can be stored without contamination at 4 °C for a year (5 mM $CaCl_2$, 0.01% (w/v) H_3BO_3, 5 mM KCl, and $MgSO_4$, pH 7.5–7.8). Adjusting the pH of this solution is tricky and requires very slow addition of either 0.1 M NaOH or acetic acid. Before storing, pass solution through a 0.2 μm sterile disposable filtration unit to maintain sterility.
2. Storage and handling of slides is much easier if a barrier between the moist Kimwipe® and slide is made using wooden barbeque skewers or coffee stirrer straws. This barrier allows you to pick up a slide much easier and prevents the media pad from detaching/slipping off the slide following storage.
3. Homozygous recessive *male sterility 1 (ms1)* mutant lines (available from the Arabidopsis Biological Resource Center seed stock CS75; www.abrc.osu.edu) are used in each setup as the female pollen recipients because they do not produce viable pollen and can be emasculated at peak fertility the same day of assay setup [26]. Note that *ms1* mutants are maintained as heterozygotes, and homozygous mutants must be screened by inspecting siliques for elongation. Furthermore, the *ms1* mutant decreases the time and effort required for assay setup as emasculation of male-fertile flowers pre-pollen maturity requires more time for emasculation and additionally necessitates a wait period of a day or more for the pistil to further mature and become receptive to pollination.
4. The transgenic construct should have different herbicide resistance marker than the T-DNA line if the T-DNA insert contains a functional herbicide resistance gene. Typically, hygromycin-resistant binary plasmids are optimal because they do not interfere with the herbicide resistance markers of SALK, SAIL, or WiscDSLox T-DNA mutants. A suite of hygromycin selective Gateway-compatible plant expression vectors [27] can be purchased from Addgene (www.addgene.org). If the fluorescence signal associated with subcellular localization of the protein of interest is weak or complex, a LAT52-driven green fluorescent protein reporter can be integrated into

the complementation construct as a separate locus. While this modification requires additional plasmid engineering, it can provide a strong signal to distinguish complemented vs. mutant pollen tubes in the functional assays presented here.

5. Make sure you do not sterilize all of the unconfirmed, floral dipped seed and keep a reserve in case the T_1 lines do not generate a hemizygous line. Additionally, it is always recommended to have wildtype Col-0, *ms1*, and original, homozygous T-DNA mutant equal age plants as the transgenic line growing to be used as experimental controls and to provide plant material for crosses. Verified hemizygous lines can also be maintained by backcrossing to the original homozygous mutant. Thus, ~15–20 seed from each mentioned line should also be processed as described in Subheading 3.1.
6. For convenience, selection media can be made in large, round petri dishes (150 mm × 15 mm). Sow out 0.15 mL of seed per large selection plate.
7. Seed can be evenly dispersed on the media surface by using a p1000 pipette tip to gently spread seed out. Ideally, seed should be plated such that no other seed is touching.
8. True transgenic seedlings can be identified effectively after 7–10 days by checking for root penetration into the selection media.
9. Due to the low melting point agarose in the pollen germination media, you must work as fast as possible to pipette the media onto the slides before it solidifies or becomes too thick to pipette.
10. Five mL of PGM-Agarose makes about 3–4 media slides.
11. Slides can be stored at 4 °C for about 3 days in the humidified chamber as long as the edges of the square petri dish/humidity chamber are wrapped securely with Parafilm to lock in moisture and prevent media drying. If using slides that have been stored, make sure to allow them to come up to room temperature prior to germinating pollen or transferring *SIV*'s to the media pad.
12. Pollen germination is much more efficient when the pistil is dissected from the remainder of the flower following pollen transfer to the media slide and gently placed near the pollen grains.
13. After pollinating a single slide, it is wise to begin a timer. On average it takes roughly 2–5 min to prep a single slide, which will allow you the same amount of time to image each slide. Remember, imaging each of the 12 slides will take time and staggering the timepoints will allow for a more accurate 4-h image as well as time between images.

14. It is very important to be as gentle as possible when applying the coverslip to the PGM-agarose surface as pollen tubes will rupture if the coverslip is applied too abruptly or pressed down into the agarose.
15. Adjust the exposure by first optimizing the setting using a T_1 line and then checking to make sure the image is not over-exposed by observing the nonfluorescent control line. Reduce the exposure time if the control line appears fluorescent. Make sure to record fluorescence exposure time as well as any other important setting parameters.
16. If a hemizygous line is not obtained in the T_1 generation, two options are available to overcome this issue, both of which can be performed at the same time.
    (a) Sterilize more reserved, unconfirmed floral dipped seed and repeat Subheadings 3.1 and 3.2.
    (b) Perform back crosses using pollen from individual T_1 lines displaying the highest percentage of complemented, fluorescent pollen tubes with the original homozygous mutant T-DNA line. Due to the genetic complementation in the T_1 line, the uncomplemented, mutant pollen existing within the T_1 pollen mixture should be outcompeted to fertilize ovules by the complemented pollen population, resulting in mostly hemizygous seed generated from the back cross. Repeat Subheadings 3.1 and 3.2 to verify individual lines for hemizygosity.
17. The control slides with Col-0 pollen will be used to compare pollen tube elongation rates with the hemizygous pollen over the time course. Due to the negative effects of imaging and coverslip application forces on pollen tubes (e.g., bursting, growth arrest, exposure to light), germinating pollen for each timepoint on separate slides is recommended as it produces a more accurate sampling of pollen tube elongation rate over the time.
18. A small, light proof cardboard box with a removable lid, such as a microcentrifuge freezer box, is useful for storing germinating pollen slides and additionally eases transport of samples to the microscope.
19. Using the paintbrush or pencil icon in the ImageJ toolbar to indicate pollen tubes that have been counted greatly aids in this analysis.
20. Pollen grains that have not produced a pollen tube should be disregarded from this analysis as pollen grains alone may display auto-fluorescence on an epifluorescence microscope that is not related to containing the fluorescent complement construct.

21. The variable of "*X* pollen genotype" in the given equation relates to the pollen genotype being analyzed—either fluorescent, complemented pollen or nonfluorescent, mutant pollen.
22. Defects with pollen grain hydration can be further distinguished by observing pollen grain morphology, as the grain is distinctly different in size and shape when hydrated [28].
23. To verify that the scale has been set correctly, retrace over scale bar and select the "M" key on computer keyboard. If the scale parameters have been properly adjusted, the length measurement output should be the length of the image scale bar.
24. To zoom into a specific region of the image being analyzed, simply place cursor over the desired area and press the "+= " key. To un-zoom the image, press the "—- " key.
25. If Col-0 pistils are to be used as the pollen recipients instead of *ms1* pistils, select closed floral buds lacking petal emergence for emasculation. Prune away any open flowers and immature floral buds in the surrounding area of the flowers to be emasculated as well as siliques existing about 1 inch below on the same stem. Sterilize fine forceps with 70% (v/v) ethanol and let dry before removing all immature anthers and petals from the floral bud leaving only the naked pistil. Return Col-0 plants with emasculated pistils to growth chamber and position plant such that emasculated pistils do not come in contact with other plants or surfaces. Allow emasculated pistils to continue to develop for at least 24–48 h. Before hemizygous pollination, observe the pistil for browning and pistil elongation as these physiologies are indicative of poor pistil health and accidental pollination during the emasculation process, respectively, and should not be used for further experiments.
26. Detaching filaments from the flower must be performed in a single dissection. Grasp as many anthers as possible without accidentally including any other plant material.
27. Complete pollination can be checked by observing the pistil under a stereoscope or by using a magnifying head visor to observe the stigma displaying a yellow appearance.
28. The dissected stigma-style can be easily transferred to the PGM-Agarose slide using the 19 G needle and *gently* touching the cut surface (which is sticky). It is important to not further cut or damage the stigma-style during the transfer process with the needle. Position the bottom of the cut style to angle downward (~30° angle) toward the PGM-Agarose pad to allow for emerging pollen tubes to grow along surface of media. Position decapitated pistils in center of PGM-Agarose pad and avoid surrounding edges for easier imaging.

29. Again, using the paintbrush or pencil icon present in the ImageJ toolbar to label the end of emerging pollen tubes with respective colors for each genotype greatly aids in quantification (*see* Fig. 5). It is important to only quantify pollen tubes that are emerging from the bottom of the style and not pollen tubes that fell off of the top of the stigma during transfer to the PGM-Agarose slide and germinated solely on the growth medium. If a penetration defect is detected following quantification, aniline blue staining of pistils can be used to further investigate pollen behavior in vivo, and a standard protocol for performing this technique is outlined by Kuroiwa et al. [29].

Acknowledgements

This work was supported by startup funds from the University of Nevada, Reno Department of Biochemistry and Molecular Biology to I.S.W. and a National Science Foundation graduate research fellowship to D.K.S.

References

1. Berger F, Hamamura Y, Ingouff M, Higashiyama T (2008) Double-fertilization – caught in the act. Trends Plant Sci 13 (8):437–443
2. Dresselhaus T, Franklin-Tong N (2013) Male-female cosstalk during pollen germination, tube growth and guidance, and double-fertilization. Mol Plant 6(4):1018–1036
3. Leydon A, Weinreb C, Venable E, Reinders A et al (2017) The molecular dialog between flowering plant reproductive partners defined by SNP-informed RNA-sequencing. Plant Cell 29(5):984–1006
4. Chae K, Lord EM (2011) Pollen tube growth and guidance: roles of small, secreted proteins. Ann Bot 108(4):627–636
5. Takeuchi H, Higashiyama T (2016) Tip-localized receptors control pollen tube growth and LURE sensing in Arabidopsis. Nature 531(7593):245–245
6. Bosch M, Cheung AY, Hepler PK (2005) Pectin methylesterase, a regulator of pollen tube growth. Plant Phys 138(3):1334–1346
7. Guan Y, Guo J, Li H et al (2013) Signaling in pollen tube growth: crosstalk, feedback, and missing links. Mol Plant 6(4):1053–1064
8. Huang Q, Dresselhaus T, Gu H et al (2015) Active role of small peptides in Arabidopsis reproduction: expression evidence. J Integr Plant Biol 57(6):518–521
9. Li H-J, Meng JG, Yang WC (2018) Multilayered signaling pathways for pollen tube growth and guidance. Plant Reprod 31(1):31–41
10. Cameron C, Geitmann A (2018) Cell mechanics of pollen tube growth. Curr Opin Genet Dev 51:11–17
11. Palanivelu R, Tsukamoto T (2012) Pathfinding in angiosperm reproduction: pollen tube guidance by pistils ensures successful double-fertilization. Wiley Interdiscip Rev Dev Biol 1 (1):96–113
12. Azpiroz-Leehan R, Feldmann KA (1997) T-DNA insertion mutagenesis in Arabidopsis: going back and forth. Trends Genet 13 (4):152–156
13. Krysan PJ, Young JC, Sussman MR (1999) T-DNA as an insertional mutagen in Arabidopsis. Plant Cell 11(12):2283–2290
14. Johnson MA, von Besser K, Zhou Q et al (2004) Arabidopsis hapless mutations define essential gametophytic functions. Genetics 168(2):971–982
15. Beale KM, Johnson MA (2013) Speed dating, rejection, and finding the perfect mate: advice from flowering plants. Curr Opin Plant Biol 16 (5):590–597
16. Koornneef M, Meinke D (2010) The development of Arabidopsis as a model plant. Plant J 61(6):909–921

17. Agudelo C, Packirisamy M, Geitmann A (2016) Influence of electric fields and conductivity on pollen tube growth assessed via electrical lab-on-chip. Sci Rep 6(1):19812
18. Chebli Y, Geitmann A (2015) Live cell and immuno-labeling techniques to study gravitational effects on single plant cells. Methods Mol Biol 1309:209
19. Palanivelu R, Brass L, Edlund AF et al (2003) Pollen tube growth and guidance is regulated by POP2, an Arabidopsis gene that controls GABA levels. Cell 114(1):47–59
20. Higashiyama T, Yang W (2017) Gametophytic pollen tube guidance: attractant peptides, gametic controls, and receptors. Plant Physiol 173(1):112–121
21. Boavida L, McCormick S (2007) Temperature as a determinant factor for increased and reproducible in vitro pollen germination in *Arabidopsis thaliana*. Plant J 52(3):570–582
22. Rodriguez-Enriquez MJ, Mehdi S, Dickinson HG et al (2013) A novel method for efficient in vitro germination and tube growth of *Arabidopsis thaliana* pollen. New Phytol 197 (2):668–679
23. Clough S, Bent A (1998) Floral dip: a simplified method for Agrobacterium-mediated transformation of *Arabidopsis thaliana*. Plant J 16(6):735–743
24. ImageJ (1997) U. S. National Institutes of Health, Bethesda, MD. http://imagej.nih.gov/ij/. Accessed 23 Apr 2019
25. Smith DK, Jones DM, Lau JBR et al (2018) A putative protein O-fucosyltransferase facilitates pollen tube penetration through the stigma-style interface. Plant Physiol 17:2804–2818
26. van der Veen JH, Wirtz P (1968) EMS-induced genic male sterility in *Arabidopsis thaliana*: a model selection experiment. Euphytica 17(3):371–377
27. Nakagawa T, Suzuki T, Murata S et al (2007) Improved gateway binary vectors: high-performance vectors for creation of fusion constructs in transgenic analysis of plants. Biosci Biotechnol Biochem 71(8):2095–2100
28. Wang L, Clarke LA, Eason RJ et al (2017) PCP-B class pollen coat proteins are key regulators of the hydration checkpoint in *Arabidopsis thaliana* pollen–stigma interactions. New Phytol 213(2):764–777
29. Kuroiwa T, Higashiyama T, Kuroiwa H et al (2006) GENERATIVE CELL SPECIFIC 1 is essential for angiosperm fertilization. Nat Cell Biol 8:64–71

Chapter 10

Arabidopsis thaliana Pollen Tube Culture for Multi-omics Studies

Mario Costa, Jessy Silva, and Silvia Coimbra

Abstract

Pollen tubes have been key models to study plant cell wall elongation. *Arabidopsis thaliana,* although small, is a nice model, easy to grow and with a large set of studies to simplify result integration and interpretation. Pollen tubes may be used for gene expression essays, but also for biochemical characterization of the cell wall composition. However, pollen tube culture methods though seemingly straightforward have often a multitude of small technical details crucial for success, quickly deterring the more inexperienced and setting back experiments for months at the time. Here we propose a detailed method to set up easily a pollen tube culture routine in any lab, with a minimal set of equipment, to isolate and process pollen tubes for gene expression and/or cell wall biochemistry studies.

Key words Pollen tube, Easy method, Protein extraction, Cell wall, RNA extraction

1 Introduction

Reproduction in higher plants relies on the efficient delivery of the male gametes (sperm cells) to the receptive ovule by a specialized type of tip-polarized apical growth structure, the pollen tube [1].

This specialized growth type is present in nerve cells, root hairs, fungal hyphae, and pollen tubes and studies have shown that apical growth in all these models share common physiological characteristics [2, 3]. Animal neurons show growth rates of 7–13 nm/s [4], root hairs grow at rates of 10–40 nm/s [5, 6], and pollen tubes can achieve growth rates of 250 nm/s [7]. As the heterotrophic pollen tube creeps toward the receptive ovule guided by molecular cues to adjust the tip elongation direction [8, 9], an intense metabolism must fuel this growth, with a rapid turnover of cytoskeleton, proteins and an accelerated vesicle traffic to deliver membrane and cell wall material to the growing tip [10].

Due to this growth type, fast metabolism, and responsiveness to environmental cues, the pollen tube has been suitable for studies on polarization, signal transduction, cell-cell communication, ion

Anja Geitmann (ed.), *Pollen and Pollen Tube Biology: Methods and Protocols*, Methods in Molecular Biology, vol. 2160, https://doi.org/10.1007/978-1-0716-0672-8_10,

channel and ion flux activity, gene expression, cytoskeleton, cell wall structure and membrane dynamics, among others [11–14]. Knowing that about 35% of Arabidopsis cell wall synthesis genes are expressed in the pollen and pollen tube [15, 16], the pollen tube is a major model for the study of hormonal, enzymatic, or chemical effects on the plant cell walls [17]. In addition, an ever-increasing availability of cell wall mutant lines makes the pollen tube model furthermore appealing.

Arabidopsis thaliana, although being the leading model in plant science, is not the most frequently used species for studies of pollen germination and pollen tube growth [18]. Like other tricellular pollen species, Arabidopsis in vitro pollen germination has proven difficult [19]. Several more or less complex protocols for pollen tube culture media have been described [20–22] and even though some showed promising results with germination rates of 60–80%, they lacked reproducibility and consistency [23]. In addition, the lack of a standardized culture medium and growth method renders it difficult to isolate variables and compare the different studies. As shown by Costa in 2013 [24] the nutrient composition of the pollen germination medium and culture conditions influence pollen tube gene expression and metabolism to a significant extent. Therefore, the need for a standardized culture medium for the study of Arabidopsis pollen tube is evident. The optimization of composition, pH, and temperature of the culture medium together with the selection of the best flower stage for pollen harvest (best pollen density of the culture) by Boavida and McCormick [25], significantly improved the germination rate. Further improvement was introduced by an optimized pollen harvest method in liquid medium by Dardelle in 2010 [26].

The protocol proposed here describes a simple method to harvest Arabidopsis pollen and to culture pollen tubes in liquid medium. Additionally, several simple yet helpful details for characterizing the pollen tube growth are proposed. These detailed protocols can be combined to provide a more complete data set, or used individually. In vitro pollen tube germination and pollen tube elongation rates are considered the best indicator of pollen viability [27] and an important tool to detect physiological impairments that may interfere with pollen gamete delivery [28]. Gene expression studies by qPCR or RNAseq are excellent tools to help pinpoint the genetic origin of physiological disorders. Additionally, protein and cell wall composition analysis may provide the necessary complementary evidence to understand and describe the mechanics behind the observed physiological disorders of the pollen tube. The following protocol proposes a holistic in vitro study approach of the pollen tube, with optimized culture and harvest conditions for maximum purity and yields.

2 Materials

2.1 Equipment

1. 1.5 mL Eppendorf tubes.
2. 400 mm^2 flat bottom scintillation flasks with lid or parafilm.
3. Fine tip forceps.
4. Centrifuge.
5. 22 °C refrigerated incubation chamber.
6. Pestles.
7. 60 μm nylon mesh filter sheets.
8. 10 cm funnel.
9. Microscope.
10. Neubauer counting chamber.

2.2 Plant Material

Whether working with wild-type or mutant lines, plant material should be of the best quality possible and grown under stable conditions. For the purpose of pollen production, healthy plants with a good vegetative growth are strongly recommended. Plants grown under short day conditions for at least a few weeks are preferred. The plants should grow at 22 °C and relative humidity of 60%. Stress will affect pollen yields and introduce undesirable variables in the study. Sowing a maximum of five plants per 10 cm^2 pots. *Arabidopsis thaliana* grows best in well-drained soils with moderate watering; adding sand to the potting mixture will help to improve drainage. After bolting, most of the secondary inflorescences should be trimmed, and no more than four floral stems should be allowed to rise per plant. To maintain the strength of the floral meristem all tertiary branching and older flowers should be removed. This will keep the plant floral meristem active longer and reduces the wasteful allocation of nutrients to nonessential organs. An additional benefit is that it will make the flower harvest task swifter and less confusing. Also, the flower quality from secondary to tertiary branching drops exponentially. The main apical meristem and secondary meristems should provide each an average of four flowers (Fig. 1a); therefore, a strong healthy plant treated in this manner should provide about 50 flowers daily for 3 weeks. A single pot with five plants will thus generate up to 250 flowers. The flowers prepared by this method will look bigger than usual, produce bigger anthers, and thus produce more pollen.

2.3 Reagents

1. Pollen culture medium: 10% Saccharose, 5 mM KCl, 5 mM $CaCl_2$, 1 mM $MgSO_4$, 0.01% boric acid, pH 7.5 (Table 1).

   RNA extraction:
2. Purezol Bio-Rad.
3. Chloroform.

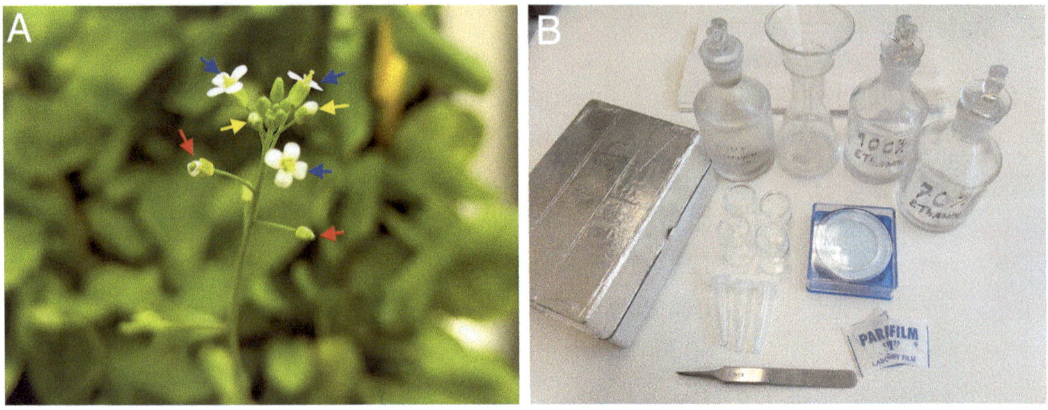

Fig. 1 Plants and materials. Good quality pollen is the key for success in this protocol (**a**). The flowers should be harvested freshly open (blue arrows). Pollen from older flowers (red arrows) and young closed flowers (yellow arrows) germinates poorly. The essential material for this protocol are a pair of fine tip tweezers, centrifuge tubes, and flat bottom glass flasks (**b**). Nylon filters and funnel are only required to filter out the pollen tubes for extraction of RNA, protein, or cell wall. A dark incubation box is useful to help maintain a stable temperature

Table 1
Pollen culture medium guide

Pollen tube culture medium composition					
			Direct weighing	Stock solutions	
	Concentration	Molar mass (g/mol)	For 200 mL of culture medium	Stock fold	For 50 mL stock
KCl	5 mM	74.55	0.076 g	100×	1.86 g
CaCl	5 mM	110.99	0.11 g	100×	2.78 g
$MgSO_4$	1 mM	120.4	0.024 g	100×	0.60 g
Boric acid	0.01 %(w/v)	n.a.	0.02 g	10×	0.05 g
Sucrose	10 %(w/v)	n.a.	20 g	n.a.	n.a.
Adjust to pH 7.5			Add ddH_2O to 200 mL		

Pollen tube culture medium should be prepared fresh. For preparing large volumes use direct weighing. To prepare small volume use stock solutions

4. Isopropanol.

   Cell wall extraction:

5. Absolute and 70% ethanol.
6. 1:1 Methanol-Chloroform.
7. Myo-Inositol.

8. 2:1 Methanol-HCl.
9. HMDS-TMCS.
10. Cyclohexane.

Light microscopy:

11. Methanol-free formaldehyde.
12. Glutaraldehyde.
13. Ethanol 10%, 30%, 50%, 70%, 90%, 96%, 100%.
14. LR-White medium grade epoxy resin.
15. Size 1 gelatin capsules.
16. *Protein extraction buffer*: 65 mM Tris–HCl pH 6.8, 1% SDS, 5% glycerol, 2.5% β-mercaptoethanol.

3 Methods

3.1 Pollen Tube Culture

This general protocol deals with the harvest of *Arabidopsis thaliana* pollen and initiation of the in vitro pollen tube culture in liquid medium (Table 1). Duration of culture and timing of pollen tube harvest method depend on the objective (Table 2).

In this method, pollen are harvested by dipping flowers in a liquid culture medium. After discarding undesired plant material, the pollen is pelleted by mild centrifugation and cultured at 22 °C in a 69 mm^2 flat bottom flask (*see* **Note 1**).

1. Collect 40 freshly opened flowers (Fig. 1a) per each 1.5 mL Eppendorf tubes (*see* **Note 2**).
2. Add 1 mL of pollen tube growth medium per tube (Fig. 2a).
3. Gently shake the tubes for 5 min. The solution in the tubes will turn cloudy and slightly yellow.
4. Remove the flowers with forceps and check the solution for contaminants, like petals, sepals, stamens, or other flower residues, under a stereoscope (*see* **Note 3**) (Fig. 2b, d).
5. Centrifuge the Eppendorf tubes at 3200 × *g* for 6 min and remove the supernatant with a pipette (Fig. 2c).
6. Add 200 μL of fresh culture medium to each tube and suspend the pollen pellets.
7. Group the pollen extracts of three tubes (120 flowers) in a 69 mm^2 flat bottom flask (Fig. 1e).
8. Gather any remaining pollen by rinsing the empty tubes with 150 μL of culture medium, and transferring to the culture flask (final culture volume of 750 μL) (*see* **Note 4**). Seal the tubes with parafilm to avoid evaporation or contamination during culture (Fig. 1f).

Table 2
Culture time for different objectives

Pollen tube culture time	
	Recommended
Pollen germination	5 h
Pollen tube elongation	5 h
mRNA extraction	6 h
Protein extraction	6 h
Cell wall extraction	Over 12 h

The time of culture will depend on the experiment objectives. For germination rate essays, the culture time should be between 4 and 6 h. Under 4 h some less responsive pollen grains may still germinate and over 6 h the percentage of pollen tubes bursting tends to increase. The same applies to the pollen tube cultures for gene expression studies, and the culture should be kept under 6 h to avoid any overcrowding, oxygen- or nutrient-related stress effects on gene expression. For cell wall harvest, the culture time can be extended beyond 12 h to maximize yields. If trying to combine and compare results from both essays, then stop the two cultures at the shortest culture time

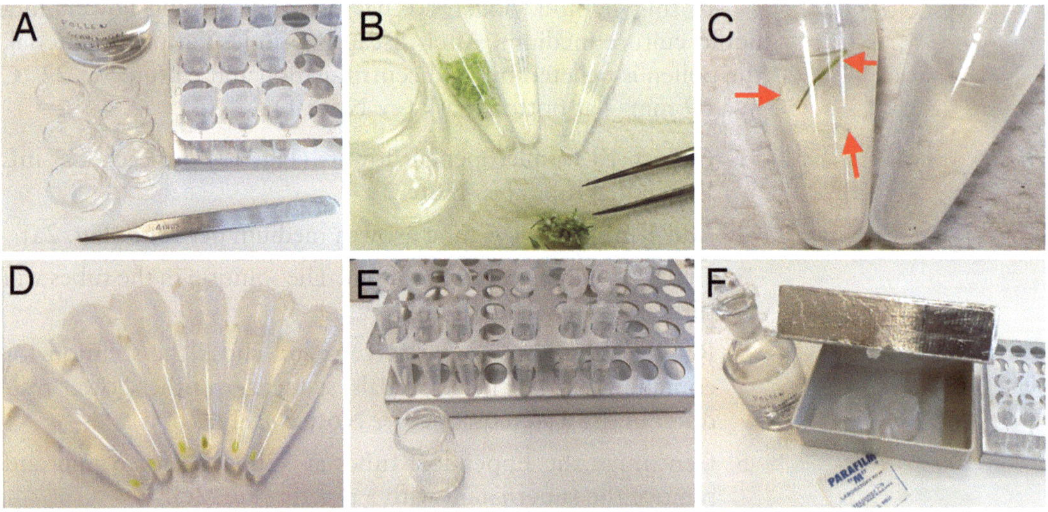

Fig. 2 Pollen harvest and pollen tube culture. This process will require fine tip tweezers, pollen tube germination media, centrifuge tubes, and flat bottom glass flasks (**a**). After transferring the harvested flowers to the centrifuge tubes and shaking, the solution will turn yellowish and the flowers are removed (**b**). Some flower residues (red arrows) may remain and will need to be removed using tweezers (**c**). Greenish yellow pollen pellets will form by the end of the mild centrifugation (**d**). These pellets are suspended in fresh culture media, and transferred to flat bottom glass flasks (**e**). The flasks are sealed and transferred to a dark culture chamber at 22 °C (**f**)

9. Incubate the culture in the dark at 22 °C, culture time will depend on the culture objective (Table 2).
10. Check the germination rate for each batch to ensure consistency (*see* **Note 5**).

3.2 Pollen Tube Culture Sub-protocols

The following inter-combinable protocols present simple ways to simultaneously analyze multiple important parameters for the study of pollen tube elongation.

3.2.1 Germination and Elongation Rate

Germination and elongation rates are the two simplest parameters to evaluate pollen fitness and culture conditions. In addition, they provide important support data for publication.

1. Stop cultures between 4 and 6 h.
2. Gently stir the culture flasks to homogenize the solution.
3. With a clipped 1000 μL pipette tips, transfer the solution to a 2-mL centrifuge tube.
4. Gently centrifuge at 3200 × *g* for 6 min, discard the supernatant, and suspend in 500 μL of fresh culture medium (*see* **Note 6**).
5. If a Neubauer counting chamber is available, make estimates for at least five samples.
6. If a Neubauer counting chamber is not available, transfer 10 μL drops to a microscope slide.
7. Count germinated pollen under a microscope at low magnification (*see* **Note 6**).
8. Pollen tube elongation essays will require an imaging system and measuring software (*see* **Note 7**).

3.2.2 Embedding for Immunohistochemistry

The study of the pollen tube cell wall by immunocytochemistry is an excellent complement to any biochemical essay (*see* Table 3 for useful antibodies). The following protocol describes a simple method for embedding of pollen tubes. Briefly, the pollen tube culture will be fixed in formaldehyde and pelleted by centrifugation. The resulting pellets will be trapped in an agarose matrix and then embedded in LR-white resin for light microscopy.

1. Stop the culture between 4 and 6 h by pipetting out the culture medium and replacing it with 1.5 mL of fixative solution (2% formaldehyde, 2.5% glutaraldehyde in pollen culture medium). Keep for 2 h at room temperature with mild agitation, followed by overnight at 4 °C.
2. Gently swirl the culture to suspend and transfer the pollen culture to a 2 mL centrifuge tube with a clipped 1000 μL pipette tip.
3. Centrifuge at 5000 × *g* for 5 min and discard supernatant.

Table 3
Useful cell wall-related *m*onoclonal *a*nti*b*odie*s* (MAbs)

Name	Affinity	Available at
LM2	Arabinogalactan protein	L
LM5	Pectins	L
LM6	Neoglycoprotein	L
LM7	Homogalacturonan	L
LM8	Xylogalacturonan	L
LM16	Processed Arabinan—RG-I	L
LM19	Homogalacturonan	L
LM20	Homogalacturonan	L/G
JIM5	Low methyl-esterified homogalacturonan	L/G
JIM7	Highly methyl-esterified homogalacturonan	L/G
JIM4	Arabinogalactan protein	G
JIM8	Arabinogalactan protein	L
JIM13	Arabinogalactan protein	L
JIM20	Extensin	G
MAC207	Arabinogalactan protein	G

All of these MAbs are obtained in rat and are unlabeled. For a comprehensive protocol for immunocytochemistry, *see* Coimbra et al. [29]
G CarboSource Georgia; *L* plant probes Leeds

4. Prepare a 2% low-melt agarose gel in $_{dd}H_2O$.
5. Pour the hot low-melt agarose gel into a small petri dish. After the temperature has cooled to 40 °C transfer the pollen tubes pellets to the agarose, apply a mild swirl and let set.
6. Once the agarose has set, with sharp scalpel cut blocks of gel containing pollen tube pellet of no more than 5 mm size, and transfer to glass vials.
7. Dehydrate the samples in an ethanol series (25%, 35%, 50%, 70%, 80%, 90%, and 100%), 15 min at each step, followed by two additional incubations in ethanol 100% for 10 min each.
8. Gradually replace ethanol by LR-white resin (3:1, 2:1, 1:1, 1:2, 1:3, ethanol:resin), incubating each step for 12 h.
9. Leave an extra 6 h in fresh LR-white resin and transfer the embedded blocks to size 1 gelatin capsules, fill with resin and hermetically close with the caps. Let polymerize at 55 °C for 48 h.
10. Make semi-thin sections with an ultramicrotome (*see* **Note 8**).

3.2.3 Preparation for RNA Extraction

Gene expression analysis is an important tool to understand cell mechanisms. Pollen tubes must be harvested early, as they tend to burst at later times (*see* Table 2). In this protocol, a variation of the phenol/chloroform method to isolate total RNA is used, after filtering out the pollen tubes from the culture by passing the culture through a nylon mesh filter.

1. Culture pollen tubes for 4–5 h.
2. Filter the cultures through a 60-μm mesh nylon filter (Fig. 3a), to discard ungerminated pollen (*see* **Note 9**). Rinse the filter with culture medium.
3. Pollen tubes will gather at middle of the nylon filter (Fig. 3b). With a pair of sharp forceps transfer them to an RNAse free tube (Fig. 3c).
4. Freeze de samples in liquid nitrogen and store until a significant quantity of pollen tubes is collected (*see* **Note 10**).
5. Grind the frozen pollen tubes to a fine powder in liquid nitrogen, add 1 mL of Purezol (Bio-Rad), homogenize, and let stand for 5 min at RT.
6. Add 200 μL of chloroform and mix vigorously for 15 s and let stand for 5 min at RT.
7. Centrifuge at 12,000 × *g* for 15 min at 4 °C.
8. Transfer the upper liquid phase to a new tube and mix in 500 μL of ice cold 2-propanol.
9. Let rest for 5 min on ice and centrifuge for 10 min at 12,000 × *g* at 4 °C.
10. Discard supernatant and wash the pellet with 1 mL of 75% cold ethanol by inverting the tube several times. Centrifuge for 5 min at 7500 × *g* at 4 °C.
11. Let the pellet air dry and dissolve in 50 μL of RNAse free water (*see* **Note 11**). Check for quality and DNA contaminants.

3.2.4 Preparation for Cell Wall Extraction

This protocol is designed to obtain a maximum yield of cell wall mass for GC-MS cell wall composition analysis. Pollen tubes will be cultured for over 16 h to maximize growth, thus increasing the yield of cell wall mass per batch.

1. Stop cultures by adding three volumes of absolute ethanol to each culture vial (Fig. 3d) and swirl slowly. Pollen tubes will tend to cluster forming a loose hank (Fig. 3e).
2. Filter the culture through a 60 μm mesh nylon filter.
3. Wash the filter by passing 70% ethanol through it to discard any remaining ungerminated pollen grains, salts, and sucrose (Fig. 3f).

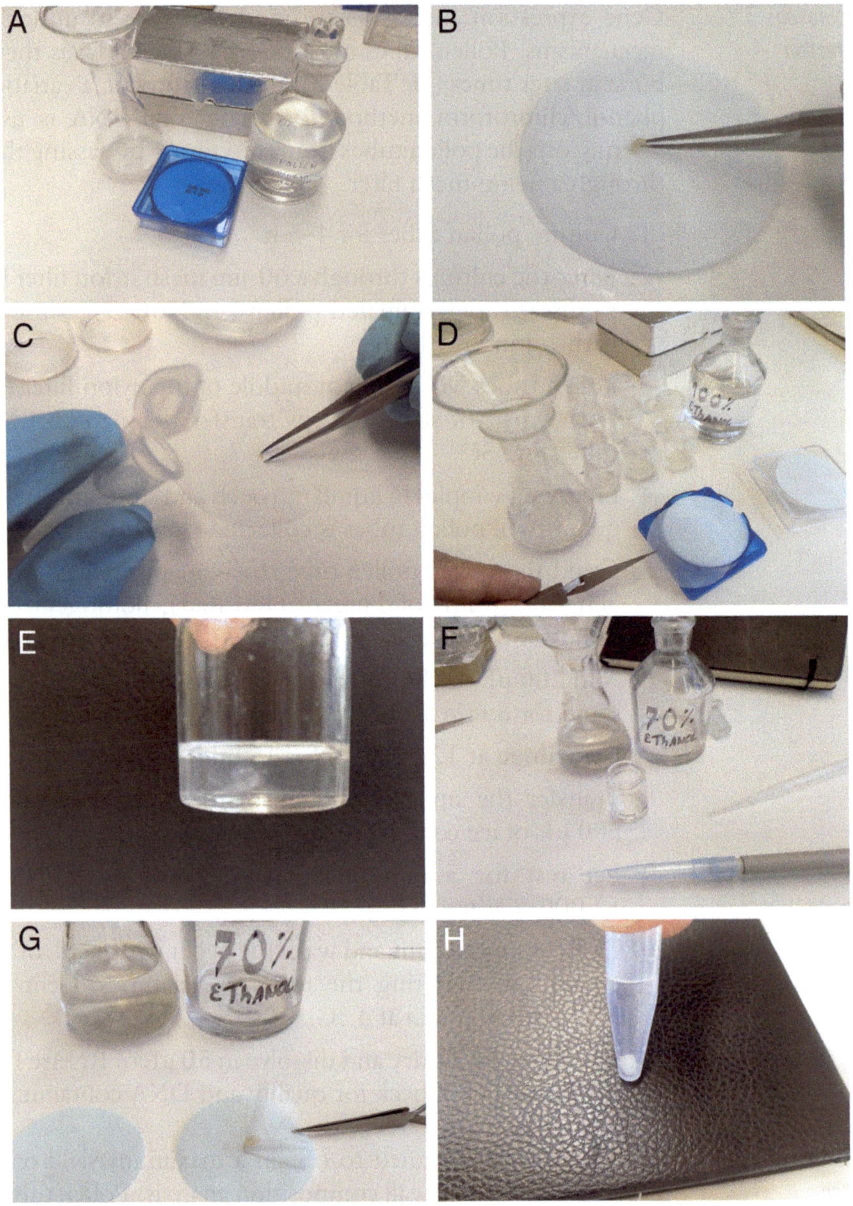

Fig. 3 Pollen tube isolation. The pollen tube culture may be used for various applications. Minor changes must be made to the protocol to maximize pollen tube recovery depending on the subsequent experiment. For RNA and protein extraction: the pollen tube cultures are filtered through a 60-μm nylon filter (**a**), and rinsed with culture media. The pollen tubes must be gently gathered with tweezers (**b**) and transferred to the respective extraction buffer (**c**). For cell wall extraction (**d**): 3 volumes of 100% ethanol is added directly to the culture. The pollen tubes will gather forming a loose hank (**e**). This solution is filtered through a 60-μm nylon filter (**f**), and the pollen tubes are rinsed with 70% ethanol. Pollen tubes will gather at the bottom of the filter (**g**). Pollen tubes can be stocked in 70% ethanol until the needed quantity is collected (**h**)

4. With a pair of forceps, slightly scrape the nylon filter to gather the pollen tubes (Fig. 3g).
5. Transfer the pollen tube bundles to a vial containing 70% ethanol and store at 4 °C until gathering sufficient material (12 mg per replicate) (Fig. 3h).
6. Grind samples in 500 μL of 70% ethanol. Incubate at 70 °C for 15 min. Centrifuge at 5000 × *g* for 10 min, remove supernatant, and mix in 1 mL of 70% ethanol. Repeat once.
7. Add 1 mL of methanol-chloroform (1:1 v/v), mix and centrifuge at 5000 × *g* for 10 min. Discard supernatant and let air dry. Add 2.5 μL (4.3^{-3} M Myo-Inositol) and freeze dry.
8. Add 250 μL of 2:1 methanol-HCl, transfer to a GC-vial and incubate for 16 h at 80 °C.
9. Mix in 200 μL of methanol and air dry. Repeat twice.
10. Add 200-μL HMDS-TMCS and incubate at 110 °C for 20 min.
11. Add 200-μL cyclohexane and let evaporate. Repeat twice.
12. Finally dissolve sample in cyclohexane 100 μL and analyze by GC-MS (*see* **Note 12**).

3.2.5 Preparation for Protein Extraction

For an integral study of pollen tube development, proteome analysis is an important tool. This protocol provides a simple method for a basic study of cytoplasmic proteins in a polyacrylamide gel.

1. Filter the pollen tube culture through a 60 μm mesh filter (Fig. 3a).
2. Wash out the pollen grains with culture medium.
3. Pollen tubes will collect at the bottom of the filter (Fig. 3b).
4. With a pair of sharp tweezers, transfer the samples to a sterile 1.5 mL centrifuge tube containing 250 μL protein extraction buffer (Fig. 3c).
5. Grind with a 1.5 mL centrifuge tube compatible pestle for 3 min, the solution should turn yellowish.
6. Centrifuge at 14,000 × *g* for 30 min.
7. Transfer the supernatant into a new centrifuge tube, add 250 μL of protein extraction buffer and repeat **step 6**.
8. The protein extract can be stored at 4 °C until use.
9. Load samples in a 30% acrylamide gel and stain with Coomassie blue (*see* **Note 13**).

4 Notes

1. If using 69 mm^2 flat bottom flasks (9.4 mm diameter), each batch of pollen tube culture will require 120 flowers. If using a different diameter culture vessel a ratio of 1.74 flower/mm^2 is desired.

   This proportion will ensure an average of 520 pollen grains/mm^2, which is close to the optimal pollen density for Arabidopsis pollen tube culture [16, 25, 26]. Flowers are collected in groups to keep the average concentration of pollen in culture stable.

   It is always a good idea to try working in batches. For example, we propose the use of 120 flowers batches distributed over three tubes with 40 flowers each. Always check if the number of flowers is sufficient before starting a new batch to avoid wasting time and flowers.

2. The collected flowers will tend to wither even in the centrifuge tubes. Withered flowers tend to shed less pollen. Therefore, proceed to the next step as quickly as possible. After addition of the culture medium, the pollen will start to hydrate almost immediately. Shake the tubes gently and continuously. The gentle movement will help improve the release of the pollen to the medium; by the end of the 5 min, the solution will look cloudy (Fig. 2b). Speed is important; the harvesting window should not extend over the 30-min mark in order to obtain a synchronous germination. It is also important to harvest a maximum of top quality pollen from freshly opened flowers (Fig. 1a). Young flowers in which anthesis has not occurred will not contribute any pollen. If the young anthers burst, the pollen will be immature and compete for resources with the mature pollen. The same applies to pollen collected from older flowers already pollinated, the little that remains usually germinates poorly affecting the overall pollen tube yields, in a similar manner as the immature pollen. Flowers that are freshly opened will both give maximum yields of pollen grains and the best pollen tubes.

3. The contamination of the culture with flower residues (Fig. 2c) alters the conditions of culture, not only through competition for oxygen and nutrients but also by introducing additional contaminants, thus greatly affecting the pollen tube culture yields. The foreign tissues may also affect the purity of the pollen tube RNA, protein and cell wall extracts.

4. Pollen tends to stick to the plastic wall of the tubes. The final wash by sequentially passing the 150 μL of culture medium through the three tubes will help to gather some of the remaining pollen. The pollen grains will hydrate and tend to sediment at the bottom of the culture flask. By gently swirling the culture flask, a cloud of pollen should be visible. If not, it may be

possible that the harvest has been ineffective. Mix the culture and check the concentration of pollen grains by pipetting 10 μL of the suspension to a Neubauer counting chamber. Alternatively, pipette 5 μL onto a slide and count all the pollen grains. The expected concentration for Colombia-0 wild type is of 50 ± 10 grains/μL. If below 50% of this value repeat the extraction process. After **step 6** the liquid in the centrifuge tube should be yellowish and cloudy. If this is not the case verify that a pellet is formed after centrifugation in **step 5**. If a pellet is not formed after **step 5** then, check **Note 2**.

5. To check germination rates gently swirl culture flasks to homogenize the culture. Pipette 10 μL samples of the culture to a slide and observe. Count the total number of pollen tubes and divide by the total number of objects (pollen tubes and ungerminated pollen grains) in five samples. The average germination rate of Colombia-0 wild-type pollen should normally be above 80%. Low germination rate, or pollen tube burst above 10% for cultures of up to 5 h, are indicative of a problem with the culture medium or culture parameter. Reduced germination and/or elongation rates are a trait of certain mutant lines. Checking that the germination rate is consistent between cultures and batches of the same lines may help to identify phenotypes. It is unusual but from time to time batches with significant divergent germination rates may occur. If they deviate significantly (over 20% variation) from the average germination rate for that specific line, discard it as it may generate artifacts in downstream applications.

   If the germination rate varies substantially in systematic manner, the culture media stock solutions should be checked for contaminants. Verify that the pH is perfectly adjusted to 7.5 with HCl. Make sure that the culture chamber temperature is stable at 22 °C.

   Glassware is a common source of contamination. Wash the glassware with a mild detergent, and rinse several times with deionized water, 100% ethanol and then deionized water again before letting it dry bottom side up. This should insure that no detergent or salts make it into the culture.

   Plastic ware, especially centrifuge tubes, is another common source of chemical contamination. Sadly, a simple wash won't solve the problem. If after checking culture parameter, medium and glassware, low Col-0 germination rates persist, consider using a different brand of centrifuge tubes.

   Yet another sign for problems are ruptured pollen grains. If appearing in great numbers accompanied by a low germination rate it may be a sign that centrifugation force was excessive. Observe a sample of pollen after centrifugation in **step 5** of Subheading 3.1. If damaged pollen appears, reduce the centrifugation speed. If pollen burst is only observed after some culture time has elapsed, then most probably it is a hydration issue, and the culture medium is hypotonic. Retry with a fresh batch of culture medium.

6. Pollen is considered to have germinated when the emerging pollen tube is longer than 1.5× length of the pollen grain. Pollen grain burst may occur and is very different from pollen tube emergence. Slides with reaction chambers can be used to make this task slightly easier. Calculate the germination rate dividing the germinated pollen average by the total number of pollen grains.
7. There are several free tools with the capacity of recording measurements; ImageJ is one of those versatile tools for microscopy (https://imagej.nih.gov/ij/download.html). Always use the same magnification and imaging resolution. Make measurements on unaltered images. Changing image resolution or sizing might cause the measuring tool to give inaccurate results.
8. LR-white resin is compatible with a wide range of water soluble stains. Also, it has low auto-fluorescence, making it perfect for fluorescence immunocytochemistry. A simple and efficient protocol for immunolocalization on LR-White sections of Arabidopsis tissues can be found in Coimbra et al. [29].
9. A minimum of four cultures should be pooled, as there are some losses during the filtering process. Some pollen tubes will stick to the filter mesh and will be unrecoverable, others will pass through the mesh.
10. Up to 10 ng can be purified from a single culture (120 flowers). For most microarray essays, replicates of 150 ng of RNA will be required (about 14 cultures, or 1680 Flowers per replicate). qPCR essays (*see* Table 4 for a list of tested qPCR primers) and RNASeq can be performed with as little as 1 ng. Consider these numbers before starting an experiment.
11. DEPC treatment is the most common solution to make RNAse free water. However, DEPC is toxic and some handling training is required. Other safer and easier solutions may be considered, as commercial RNAse free water may be purchased at low cost.
12. Culture medium carry over sucrose contamination is a frequent problem with this protocol, and will be manifested by a peak of glucose in chromatography. Make sure to rinse the pollen tubes thoroughly with 70% ethanol. Sugars from the pollen grain exine and intine may reduce the resolution of the test, especially if the pollen tubes are short. The pollen grain will (for the most part) remain attached to the pollen tube. It is therefore inevitable to see its influence in the final analysis. To create a

Table 4
qPCR primers

	AGI	Gene	Forward primer	Reverse primer
Reference genes	At4g27960	UBC9	GTTTCACCACCCTTTCTTC	AAATCCCACGATCAAATTCC
	At4g05320	UBQ10	GCTCCGACACCATTGACAAC	ACGCAGGACCAAGTGAAGAG
	At5g59370	ACT4	CACCAAACGAAGACGGAAGT	CATTGGGAGTGGAAATTGTTT
Cell wall-related genes	At1g71380	CEL3	TCTCCAAGTCATTGCTCTTCTTCC	CGTTGTCTCCTGCGTCATAGTAC
	At3g60570	EXPB5	GCAAACGGTGATGGGAACTTCG	GGACACGGCGGAGGTAAGC
	At1g71380	CEL3	TCTCCAAGTCATTGCTCTTCTTCC	CGTTGTCTCCTGCGTCATAGTAC
	At3g60570	EXPB5	GCAAACGGTGATGGGAACTTCG	GGACACGGCGGAGGTAAGC
	At5g42260	BGLU13	CCATTGGGACACACCGCAAAGC	ACACCTTCCTGGAGCCATTACACC
	At5g14380	AGP6	TTCTAAGTTAAGTCGTCCAC	GACATTTAGGTTTATATTTACTCC
	At3g01700	AGP11	CCACGTAATGTCAAGC	CAACAGGGGATGATGCTTTC
	At3g57690	AGP23	AATGGAGATGAAGAAGATTG	TGCAAGTAGTAGCTGAAG
	At5g40730	AGP24	CGGTTACAGATGGAGGAGGCACTC	AGGACACCACACGTAGCCACAATC
	At3g20865	AGP40	GCGGCTACAATGGAGTCTC	CAAGAAGGCGGAGAGTGATG
	AT4G35010	BGAL11	GGAATGGGAAAAGGGTTGAT	GCTCAGGCTTCACATTAGGC
	AT2G16730	BGAL13	AAACAAGGCGGTTTGAACAC	GGAAGTCCTCCATGAGTCCA
	AT3G01040	GAUT13	GAATTTTGCTGGGTTGTCGT	ACAACATCACCTCCCTACGC
	AT1G06780	GAUT6	CGGCTTAAGAGCATCACTCC	CCATCACTTTCAGCAGACGA
	At4g33240	PI4K	TCCTCCAGCCACACGCAGAATTG	GAAGCATAGCGGAAGCAAGCAACC
	At2g39890	PROT1	ATGGCGAGAGGCGGGTAC	GTGGTTGGCTAGAATGAATGTGAG
	AT1G69940	PPME1	ATGCCTACACCGAGATGACC	AGTTTACAAAGCGGGTGGTG
	AT2G47040	VGD1	ACTGGCCCAAGTCATTCAAC	CCACCGAGACCACAACTTTT

List of qPCR primers previously tested for useful cell-wall-related genes in pollen tube [24] base line a pollen cell wall extract can be used. Gather pollen by the method described in this chapter, using $_{dd}$water instead of culture media. At the end of **step 5** group all the pollen pellets and proceed as described in the cell wall extraction sub protocol starting at **step 6**.

13. A single culture of 120 flowers will yield an average of 80 μg of protein.

Acknowledgments

Our work was supported by EU project 690946 "SexSeed" (Sexual Plant Reproduction—Seed Formation) funded by H2020-MSCA-RISE-2015, and by the FCT project 02/SAICT/2017 POCI-01-0145-FEDER-027839. M.C. was supported by FCT Ph.D. grant SFRH/BD/111781/2015.

References

1. Lopes AL, Moreira D, Ferreira MJ, Pereira AM, Coimbra S (2019) Insights into secrets along the pollen tube pathway in need to be discovered. J Exp Bot 70:2979–2992
2. Malhó R, Serrazina S, Saavedra L, Dias FV, Ul-Rehman R (2015) Ion and lipid signaling in apical growth – a dynamic machinery responding to extracellular cues. Front Plant Sci 6:816
3. Cheung A, Wu HM (2008) Structural and signaling networks for the polar cell growth machinery in pollen tubes. Annu Rev Plant Biol 59:547–572
4. Gomes TM, Spitzer NC (1999) In vivo regulation of axon extension and pathfinding by growth-cone calcium transients. Nature 397:350–355
5. Galway ME, Heckman JW Jr, Schiefelbein JW (1997) Growth and ultrastructure of arabidopsis root hairs: the rhd3 mutation alters vacuole enlargement and tip growth. Planta 201:209–218
6. Wymer CL, Bibikova TN, Gilroy S (1997) Cytoplasmic free calcium distribution during development of root hairs of *Arabidopsis thaliana*. Plant J 12:427–439
7. Pierson ES, Miller DD, Callaham DA, van Aken J, Hackett G, Hepler PK (1996) Tip-located calcium entry fluctuates during pollen tube growth. Dev Biol 174:160–173
8. Johnson M, Harper J, Palanivelu R (2019) A fruitful journey: Pollen tube navigation from germination to fertilization. Annu Rev Plant Biol 70:809–837
9. Chebli Y, Kaneda M, Zerzour R, Geitmann A (2012) The cell wall of the arabidopsis pollen tube—spatial distribution, recycling, and network formation of polysaccharides. Plant Physiol 160:1940–1955
10. Campanoni P, Blatt MR (2007) Membrane trafficking and polar growth in root hairs and pollen tubes. J Exp Bot 58:65–74
11. Grebnev G, Ntefidou M, Kost B (2017) Secretion and endocytosis in pollen tubes: models of tip growth in the spot light. Front Plant Sci 8:154. https://doi.org/10.3389/s.2017.00154
12. Malhó R, Liu Q, Monteiro D, Rato C, Camacho L, Dinis A (2006) Signalling pathways in pollen germination and tube growth. Protoplasma 228:21–30
13. Onelli E, Moscatelli A (2013) Endocytic pathways and recycling in growing pollen tubes. Plan Theory 3:211–229
14. Sekereš J, Pleskot R, Pejchar P, Žárský V, Potocký M (2015) The song of lipids and proteins: dynamic lipid–protein interfaces in the regulation of plant cell polarity at different scales. J Exp Bot 66:1587–1598
15. Pina C, Pinto F, Feijó JA, Becker JD (2005) Gene family analysis of the Arabidopsis pollen transcriptome reveals biological implications for cell growth, division control, and gene

expression regulation. Plant Physiol 138:744–756

16. Costa M, Nobre M, Becker J, Masiero S, Amorim M, Pereira LG, Coimbra S (2013) Expression-based and co-localization detection of arabinogalactan protein 6 and arabinogalactan protein 11 interactors in Arabidopsis pollen and pollen tubes. BMC Plant Biol 13:7
17. Liepman AH, Wightman R, Geshi N, Turner SR, Scheller HV (2010) Arabidopsis—a powerful model system for plant cell wall research. Plant J 61:1107–1121
18. The Arabidopsis Genome initiative (2000) Analysis of the genome sequence of the flowering plant *Arabidopsis thaliana*. Nature 408:796–815
19. Taylor LP, Hepler PK (1997) Pollen germination and tube growth. Annu Rev Plant Physiol Plant Mol Biol 48:461–491
20. Fan LM, Wang YF, Wang H, Wu WH (2001) In vitro arabidopsis pollen germination and characterization of the inward potassium currents in arabidopsis pollen grain protoplasts. J Exp Bot 52:1603–1614
21. LiH LY, Heath RM, Zhu MX, Yang Z (1999) Control of pollen tube tip growth by a Rop GTPase-dependent pathway that leads to tip-localized calcium influx. Plant Cell 11:1731–1742
22. Boavida LC (2005). Genetic dissection of cell-to-cell interactions in sexual reproduction in *Arabidopsis thaliana*. s.l. : Ph.D. thesis, Departamento de Biologia Vegetal, Faculdade de Ciencias da Universidade de Lisboa, Lisbon
23. Johnson-Brousseau SA, McCormick S (2004) A compendium of methods useful for characterizing arabidopsis pollen mutants and gametophytically-expressed genes. Plant J 39:761–775
24. Costa M, Pereira LG, Coimbra S (2013) Growth media induces variation in cell wall associated gene expression in *Arabidopsis thaliana* pollen tube. Plan Theory 2:429–440
25. Boavida L, McCormick S (2007) Technical advance: temperature as a determinant factor for increased and reproducible in vitro pollen germination in *Arabidopsis thaliana*. Plant J 52:570–582
26. Dardelle F, Lehner A, Ramdani Y, Bardor M, Lerouge P, Driouich A, Mollet JC (2010) Biochemical and immunocytological characterizations of arabidopsis pollen tube cell wall. Plant Physiol 153:153–176
27. Shivanna KR, Linskens HF, Cresti M (1991) Pollen viability and pollen vigor. Theor Appl Genet 81:38–42
28. Hashida SN, Takahashi H, Kawai-Yamada M, Uchimiya H (2007) *Arabidopsis thaliana* nicotinate/nicotinamide mononucleotide adenyltransferase (AtNMNAT) is required for pollen tube growth. Plant J 49:694–703
29. Coimbra S, Almeida J, Junqueira V, Costa ML, Pereira LG (2007) Arabinogalactan proteins as molecular markers in *Arabidopsis thaliana* sexual reproduction. J Exp Bot 58:4027–4035

Chapter 11

Large-Scale Analysis of Pollen Viability and Oxidative Level Using H_2DCFDA-Staining Coupled with Flow Cytometry

Nicholas Rutley and Gad Miller

Abstract

Determining pollen viability and other physiological parameters is of critical importance for evaluating the reproductive capacity of plants, both for fundamental and applied sciences. Flow cytometry is a powerful high-performance high-throughput tool for analyzing large populations of cells that has been in restricted use in plant cell research and in pollen-related studies, it has been minimized mostly for determination of DNA content. Recently, we developed a flow cytometry-based approach for robust and rapid evaluation of pollen viability that utilizes the reactive oxygen species (ROS) fluorescent reporter dye H_2DCFDA (Luria et al., Plant J 98(5):942–952, 2019). This new approach revealed that pollen from *Arabidopsis thaliana* and *Solanum lycopersicum* naturally distribute into two subpopulations with different ROS levels. This method can be employed for a myriad of pollen-related studies, primarily in response to stimuli such as biotic or abiotic stress. In this chapter, we describe the protocol for H_2DCFDA staining coupled with flow cytometry analysis providing specific guidelines. These guidelines are broadly applicable to many other types of cellular reporters to further develop this novel approach in the field of pollen biology.

Key words Pollen, H_2DCFDA, Flow cytometry, Cell sorting, FACS, ROS, Viability, *Arabidopsis thaliana*, *Solanum lycopersicum*, Heat stress, Oxidative stress

1 Introduction

Successful fertilization is critically dependent on the fitness of the male gametophyte, which is determined by many developmental parameters such as the state of the mature pollen grains, dispersal, germination on the pistil, tube growth, and communication with the female gametophyte and the ambient conditions during these processes [1–3].

Pollen development begins in the anther locules when a diploid microspore mother cell undergoes meiosis to produce a tetrad of microspores. Each microspore divides mitotically once or twice (depending on the species), laying down a male germline encapsulated inside a larger vegetative cell (seen as the pollen grain). This is followed by a maturation stage when the pollen

Anja Geitmann (ed.), *Pollen and Pollen Tube Biology: Methods and Protocols*, Methods in Molecular Biology, vol. 2160,
https://doi.org/10.1007/978-1-0716-0672-8_11,

desiccates in preparation for dehiscing of the pollen sac and release of the mature pollen [3]. Upon landing on a receptive stigma, pollen hydrate and then germinate a tube and grow through the pistil toward the ovules in a complex and highly coordinated process [3]. Because these processes can be easily disrupted by environmental changes, pollen is extremely vulnerable to stresses such as high temperatures and low humidity, with even small increase in temperature causing pollen abortion or failure to germinate and grow a tube, resulting in reduced seed set and sometimes complete male sterility [4, 5].

A classical and extremely powerful way to evaluate pollen fitness is by out-crossing of pollen from a male-donor plant onto a female recipient, and to score allele segregation or seed set in the next generation [5]. However, this approach can be time consuming, labor intensive, and can be highly technical when for example genotyping the F1 requires PCR to determine allele segregation ratio. Estimating the ability of pollen to germinate tubes in vitro provides an accurate proxy for pollen fitness, however, this too is time consuming as it requires manual counting of hundreds of pollen by microscopy and can be challenging if a large number of plants are to be scored and if pollen density is different between individuals or genotypes (since pollen density itself has an impact of pollen germination percentage, in particular in Arabidopsis). As an alternative, a number of staining protocols have been commonly used to determine pollen viability, providing a quick and empirical assay to quantify pollen fitness. However, some of the most commonly used protocols such, Alexander's stain [6–8] and 2,3,5-triphenyl tetrazolium chloride [9] sometimes fail to distinguish between live and dead pollen in vitro [8, 10]. Another common tetrazolium salt, thiazolyl blue tetrazolium bromide (3-(4,5-dimethylthiazol-2-yl)-2,5-diphenyltetrazoliumbromide (MTT), which reports mitochondrial dehydrogenase activity by staining live pollen dark purple/black, is a highly reliable viability indicator [9, 10]. However, like the majority of histochemical staining procedures, MTT requires manual counting of pollen grains and the staining chemicals are cytotoxic which prevents any follow-up analyses on living cells.

In this chapter, we describe a simple-to-use recently developed approach for evaluating pollen viability on a population scale, which utilizes the general reactive oxygen species (ROS) probe dichlorodihydrofluorescein diacetate (H_2DCFDA) coupled with flow cytometry analysis [10].

H_2DCFDA is a cell-permeable molecule that becomes trapped in cells following hydrolysis by cytoplasmic esterases [11, 12]. Once the diacetate moiety is cleaved-off the non-fluorescent H_2DCF is free to react with ROS produced within the cells, oxidizing it to become green fluorescent dichlorofluorescein (DCF). Thus, only metabolically active cells are DCF-stained. In addition to being a reliable viability probe, H_2DCFDA is also a highly sensitive reporter of cellular ROS levels indicating the oxidative state of the cell. As such, two parameters can be measured when using H_2DCFDA: percentage of stained cells and median fluorescence intensity (MFI).

Large-scale analysis of single cells by flow cytometry is a high-throughput method providing robust and rapid quantification of optical and fluorescence parameters. Coupled with cell isolation using the fluorescence-activated cell sorter (FACS), it provides numerous applications for both basic and applied science. Accurate gating of cells can be informed by selective staining, cell type-specific reporters, and sorting.

Pollen grains make ideal cell-sized input for flow cytometry and FACS (diameters of mature ranging between 18 and 24 μm for Arabidopsis to 110 μm for lily [13], as flow cytometers have nozzles ranging from 70 to 125 μm in diameter.

Analyzing DCF-stained pollen of Arabidopsis (*Arabidopsis thaliana*), tomato (*Solanum lycopersicum*), and *Nicotiana benthamiana* in a flow cytometer revealed that pollen naturally distribute bimodally into a "low-ROS" subpopulation (having weak fluorescence bordering unstained pollen) and a "high-ROS" subpopulation [10]. The low-ROS and high-ROS subpopulations represent pollen with low and high metabolic activity, respectively. Purification of the two subpopulations by FACS demonstrated that under control conditions the ability of pollen to germinate a pollen tube is positively correlated with the level of ROS; in vitro germination of high-ROS Arabidopsis and tomato pollen was 35-fold higher than the low-ROS pollen [10].

This method is useful for measuring dynamic changes in the proportions of the unstained, low- and high-ROS subpopulations and in their MFI between control and treatments (e.g., exposure to stress conditions, hormones, antioxidants, etc.). ROS are generated as part of aerobic metabolism in cells but are also actively produced, and although harmful they serve as signaling molecules and play critical roles in development and stress response programs [14, 15].

Pollen, like any other cell type, can accumulate ROS within a range, beyond which it is not able to function properly [16]. Increasing ROS level of the high-ROS subpopulation as a consequence of oxidative stress increased pollen mortality [10]. On the other hand, quenching ROS level with increasing dosage of antioxidant can enhance pollen performance to a point; however, further increases will inhibit its normal activity, completely abolishing tube formation [10]. We strongly advise that quantification of DCF-stained pollen before and after treatment is assayed alongside tube formation capacity in vitro, at least for the first time an experiment is performed as experimental conditions and pollen type may have an independent effect on DCF-staining capacity.

Using the method detailed in this chapter requires access to a flow cytometry cell analyzer or FACS machine, which are nowadays found in core equipment units in many research institutes. Such units often rely on expert technical services provided by dedicated personnel with the relevant know-how of how to use the equipment properly. This is essential, at least initially, when applying DCF-stained pollen to flow cytometry analyses. Therefore, we do not provide instructions on operating flow cytometers or FACS machines nor on how to use flow cytometry analyses platforms such as FlowJo™ software. Rather, our goal in this book chapter is to provide basic guidelines and considerations for planning and conducting flow cytometry analysis and FACS of DCF-stained pollen, from the point of collecting pollen, sample preparation, flow cytometry analysis, and isolating subpopulations by FACS. These guidelines are also useful for optimizing flow cytometry analyses of pollen stained with any other fluorescent dye or reporter system.

2 Materials

1. Plant material: *Arabidopsis thaliana* plants at flowering stage with open flowers. For recommended numbers of flowers *see* Table 1.

Table 1
Pollen collection guide for flow cytometry and FACS

	Small-scale analysis		Bulk analysis and FACS	
Species	**Arabidopsis**	**Tomato**	**Arabidopsis**	**Tomato**
Sufficient for	One sample of 1 ml	One sample of 1 ml	2–3 samples of 3 ml	5 samples of 1.5 ml
Sampling	30 flowers	1 anther cone	40–50 inflorescences	6 anther cones
10% PGM volume[a]	1.5 ml	1.5 ml	10 ml	10 ml
Filtering	Modified column + collection tube, centrifuge for 5 min at 500 × *g*	Modified column + collection tube, centrifuge for 5 min at 500 × *g*	Doubled Miracloth over 50 ml tube	Doubled Miracloth over 50 ml tube

For small-scale flow cytometry analysis where only a small volume of filtered pollen suspension is required, pollen can be collected from a small number of Arabidopsis plants or an individual tomato plants. For FACS and for flow cytometry (bulk analysis), when larger volumes of pollen suspension are needed, pollen can be collected from a large number of Arabidopsis plants and one or more tomato plants. These two approaches use different volumes of 10% PGM and different filtering methods. *See* Fig. 2 for an illustration of the modified column and collection tube setup

[a]A small volume of 10% PGM is lost during sample handling

Tomato (*Solanum lycopersicum*) plants at flowering stage. For specific numbers of flowers *see* Table 1.

2. Arabidopsis pollen germination medium (10% Arabidopsis PGM): 10% w/v sucrose, 5 mM KCl, 5 mM $CaCl_2$, 1 mM $MgSO_4$, 0.01% w/v boric acid [17]. Adjust pH to 7.5 using 0.1 M NaOH after adding sucrose (*see* **Note 1**).

   Tomato pollen germination medium (10% tomato PGM): 10% w/v sucrose, 2 mM $Ca(NO_3)_2$, 2 mM $MgSO_4$, 1 mM KNO_3, 2 mM boric acid [18]. No adjusting of pH required (*see* **Note 1**).
3. 5 mM H_2DCFDA (Cayman Chemical, Ann Arbor, MI, USA) stock in ethanol, stored according to the manufacturer instructions.
4. Miracloth (Merck) or 40 μm nylon mesh.
5. FACS-compatible tubes (we use 5 ml round-bottom tubes or 15 ml centrifuge tubes).
6. 50 ml falcon tubes (must be unused to avoid particulate debris).
7. Modified plasmid miniprep spin column for sample filtration (*see* **Note 2**).
8. BD FACSAria™ III or any other FACS or optical flow cytometer analyzer that has FITC laser and filter set (488 nm laser line, BP filter 530/30).
9. FlowJo software or any other flow cytometry analysis platform.

3 Methods

Good experimental design is critical for successful flow cytometry and FACS experiments, especially since pollen yields from plants can be somewhat unpredictable and plants can only be sampled once or twice before flowering ends and senescence begins.

We use 10% PGM as the pollen suspension buffer in flow cytometry analyses as well as for isolating pollen by FACS, as this allows direct comparison between flow cytometry data and post-sort analysis of cells (e.g., correlating DCF-staining with in vitro pollen germination assays of sorted pollen subpopulations).

3.1 Experimental Design Considerations

The number of Arabidopsis flowers and tomato anther cones needed depends on the purpose of the experiment. In Table 1, we detail two approaches for collecting pollen for flow cytomtery and FACS; small scale for using a small number of plants and bulk scale typically needed for FACS to sort sufficient pollen. For a typical flow cytometry experiment using H_2DCFDA we aim to record fluorescence in 10,000 pollen grains and for sorting 10,000 pollen per sorting gate. Since a single tomato flower produces a large amount of pollen (~700,000 per flower; [19]), pollen from individual plants can easily be analyzed by flow cytometry and used for sorting experiments. Single plant analysis is more challenging for Arabidopsis since its flowers produce as little as 2000 pollen grains per flower [20]. For Arabidopsis flow cytometry ~30 open flowers are needed at a minimum to have sufficient pollen to work with (allowing for variability in pollen quality, loss of pollen during sampling, establishing gates, and technical issues at the flow cytometer), while large-scale sorting experiments need significantly more flowers. For sorting of Arabidopsis pollen we grow 48 plants to flowering stage, and divide this into four pools representing four biological replicates. Pollen from each replicate can be split into no treatment (i.e., control) and treatment groups if treatments are performed in vitro, such as heat stress or exposure to chemicals, such as antioxidants. For all flow cytometry and sorting experiments (in every experiment and for each genotype), an unstained sample is needed to serve as a negative control for distinguishing pollen autofluorescence from true DCF fluorescence.

As each sample can take several minutes to read in a flow cytometer, the order in which samples are read in the flow cytometer is an important consideration when using H_2DCFDA in order to keep variability to a minimum between replicates. This is because DCF-staining is not stable over a long period (tens of minutes). We recommend reading all samples that originate from the same bulk collection of pollen and performing normalization to the control sample. An example of read order:

Bulk 1: control temperature, in vitro heat stress.

Bulk 2: control temperature, in vitro heat stress.

Bulk 3: control temperature, in vitro heat stress.

The following method has been validated for BD LSRFortessa™ or BD FACSAria™ III machines but is applicable to almost any flow cytometer/FACS device.

We devised the protocol with the assumption that personnel experienced in flow cytometry and FACS will be supporting the user carrying out this method for the first time.

Flow cytometry and FACS require the user to establish gates that define the target population for analysis. Although significant post-run analyses can be performed (such as redrawing gates; since the machine records information on all events within the parameters) cells falling outside of the parameters will not be recorded.

As typical flow cytometers and FACS machines do not contain cameras to visualize the cells passing the detector, extra caution must be taken to make sure that the target cell types, and not other cells or tissue debris, are being analyzed. Therefore, when using flow cytometry for the first time the gating of pollen should be validated to make sure that events that fall within the gates are pollen. This can be done in one of three ways, a direct way using a pollen-specific reporter line such as pLAT52-H2B::GFP or pVCK-H2B::GFP [21] or indirectly using a dye that stains biological material (e.g., dichlorofluorescein) or sizing beads. A fourth way, if FACS is available, is to isolate pollen by cell sorting and validate the target populations post-sort by microscopy.

3.2 Sample Preparation (for Bulk Analysis and FACS)

1. Preparation of pollen suspension
   Arabidopsis:
   (a) Collect inflorescences containing open flowers and closed buds up to the ~20 ml mark in a 50 ml falcon tube.
   (b) Add 10 ml 10% Arabidopsis PGM and vortex for 2 min to release pollen into the medium.

   Tomato:
   (a) Separate six anther cones from newly open flowers. Cut each anther cone in half transverse and place into a 15 ml falcon tube.
   (b) Add 10 ml tomato 10% PGM at room temperature, and wash out pollen by gently squeezing the anthers against the side of the tube wall using a 5 ml pipette tip or a serological pipette, followed by vortexing for 2 min.
2. Filter the suspensions of Arabidopsis or tomato pollen by pouring it through folded Miracloth or a single layer of 40 μm nylon mesh into a clean 50 ml falcon tube.

3.3 Sample Preparation (for Small-Scale Analysis)

1. Preparation of pollen suspension
   Arabidopsis:
   (a) Pick ~30 open flowers from Arabidopsis plants and place inside a 2 ml centrifuge tube.
   (b) Add 1.5 ml of 10% Arabidopsis PGM and vortex for 2 min to release the pollen.

   Tomato:
   (a) Separate a single anther cone from a freshly opened flower. Cut the anther cone in half transverse and place into a 2 ml centrifuge tube.
   (b) Add 1.5 ml of 10% tomato PGM at room temperature and gently crush the anther against the side of the centrifuge tube using a 1 ml pipette tip, followed by vortexing for 2 min.
2. Transfer 800 μl of the Arabidopsis or tomato pollen suspension through a plasmid miniprep spin column modified for sample filtration (*see* **Note 2**; Fig. 1) and centrifuge for 5 min at 500 × *g*. Disassemble the cartridge and carefully pipette off

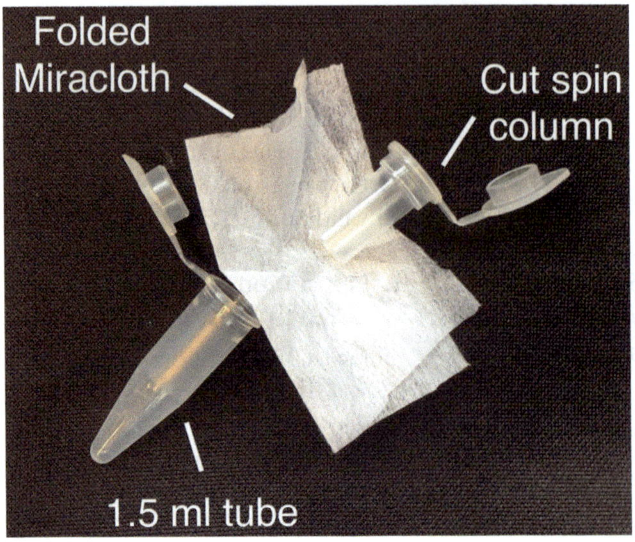

Fig. 1 Assembly of modified plasmid miniprep spin column for small-scale sample filtration

the liquid. Reassemble the column with a fresh piece of folded Miracloth and repeat with remaining volume of unfiltered pollen suspension.

3. Filter additional 10% PGM through folded Miracloth and use this filtered 10% PGM to resuspend the pollen pellet to a final volume of 1.5 ml of 10% PGM.

3.4 Pollen H_2DCFDA Staining

1. After cutting 2–3 mm off the end of a 1 ml pipette tip to widen the end, transfer 1.5 ml of pollen suspension to 2 ml microcentrifuge tubes (*see* **Note 3**). Leave pollen to hydrate in its PGM for 15 min at room temperature.
2. For H_2DCFDA staining, add 1.5 μl of 5 mM H_2DCFDA to 1.5 ml (final concentration of 5 μM) of pollen suspension, followed by inversion 5–7 times to mix. To a second pollen suspension add 1 μl of ethanol to serve as unstained control. Incubate at room temperature for 30 min.

3.5 Flow Cytometry Gating

1. To establish gating, first read the filtered unstained pollen sample in the BD FACSAria™ III using a forward scatter (FSC) neutral density filter of 2.0 and a 85 μm nozzle. Observe a FSC-area (FSC-A) vs. side scatter area (SSC-A) scatter plot (Fig. 2a). Filtering of sample preparation results in near-pure pollen preparations (*see* **Note 4**) however not all events that fall within an FSC-A vs. SSC-A scatter plot will be pollen; cells smaller than the filter pore size, broken or damaged cells and fibers carried over from sample processing might appear in the scatter plot. Gate the majority of pollen using a second density plot as a guide (Fig. 2b).

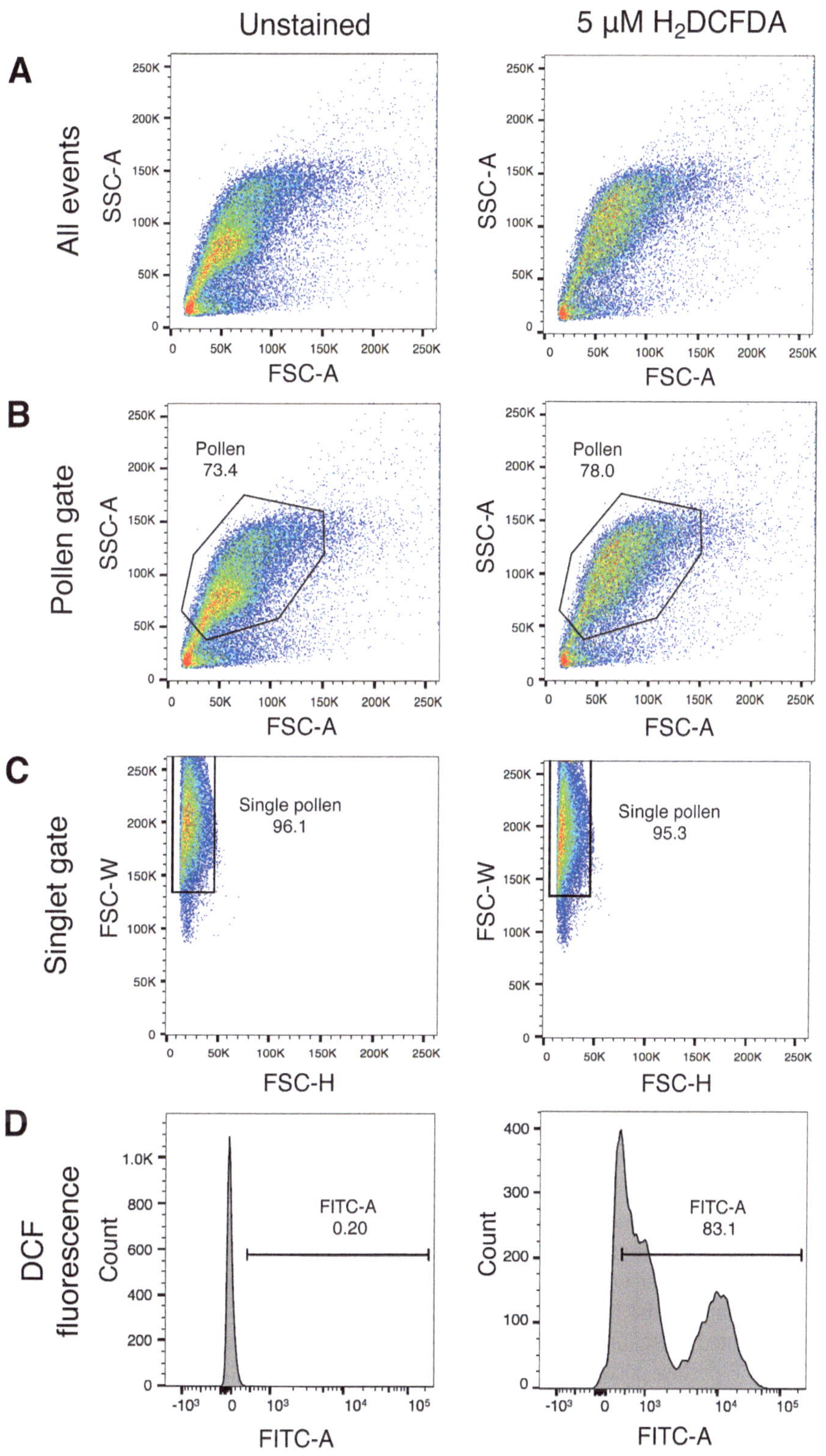

Fig. 2 Generalized workflow for gating DCF-positive Arabidopsis pollen for flow cytometry and fluorescence-activated cell sorting (FACS). Samples of unstained pollen (left) and pollen stained with 5 μM H_2DCFDA (right) in 10% pollen germination medium were read in a ACSAria™ III sorter. Gating to record DCF-positive pollen

2. Droplets containing single pollen must be defined from droplets containing more than one pollen grain. Since single pollen can be isolated based on signal duration (i.e., FSC-width), we draw a FSC-height (FSC-H) vs. FSC-width (FSC-W) scatter plot and draw a gate that encompasses the majority of the pollen and excluding outlier pollen along the FSC-W axis (Fig. 2c).
3. Using the FITC laser-filter set (488 nm laser line, BP filter 530/30), set a DCF fluorescence threshold gate using unstained pollen in order to discriminate pollen autofluorescence from true DCF-staining (Fig. 2d).
4. Once gating has been established, validate DCF fluorescence threshold gate by reading a H_2DCFDA-stained sample such as a positive control. Further gating of low-ROS and high-ROS subpopulations can be performed using this positive control sample.
5. Read all samples and record 10,000 DCF-positive pollen grains for the required gate. Depending on pollen density and flow rate, recording should be taken between 1 and 5 min. The recorded parameters enable scoring the number of DCF-negative (unstained) and DCF-positive (stained) pollen and additionally the two major FITC positive subpopulations (i.e., low-ROS and high-ROS), as well as evaluate the average DCF intensity of the entire stained pollen population and for each of the subpopulations.

3.6 FACS-Isolation of Pollen and Post-FACS In Vitro Germination

Larger amounts of pollen and large volumes of pollen suspension are required for sorting by FACS. Figure 3 outlines the major steps for FACS of DCF-stained pollen when sorting low- and high-subpopulations for in vitro germination. In FACS, pollen are sorted directly into the buffer needed for downstream processing such as TRIzol for RNA extraction or 10% PGM for post-sort in vitro germination on liquid medium.

Typically, we sort 10,000 pollen grains into 2 ml of TRIzol for RNA extraction, or 10,000 pollen grains into 2 ml of 10% PGM for post-FACS in vitro germination. Sorting 10,000 pollen using the

Fig. 2 (continued) was performed stepwise, first using the unstained sample: (**a**) all events were plotted in a scatter plot of forward scatter area (FSC-A; *x*-axis) and side scatter area (SSC-A; *y*-axis) which correspond to size and granularity. (**b**) Pollen were then gated using density as a guide. (**c**) Next singlet pollen were gated in a scatter plot of forward scatter height (FSC-H; *x*-axis) and forward scatter width (FSC-W; *y*-axis). (**d**) Singlet pollen were then gated to separate autofluorescence from true DCF-positive (DCF +ve) signal using the FITC-A laser-filter set. DCF-positive pollen were gated by drawing a gate to the right of the end of the histogram of the unstained sample. A positive control (such as a H_2DCFDA-stained sample of pollen from control conditions) was then run to validate gating. The number inset of each image refers to the percentage of events within the gate. Pseudocolor dot plots (**a–c**) and histograms of FITC-A fluorescence (**d**) were generated in FlowJo™ software for illustrative purposes

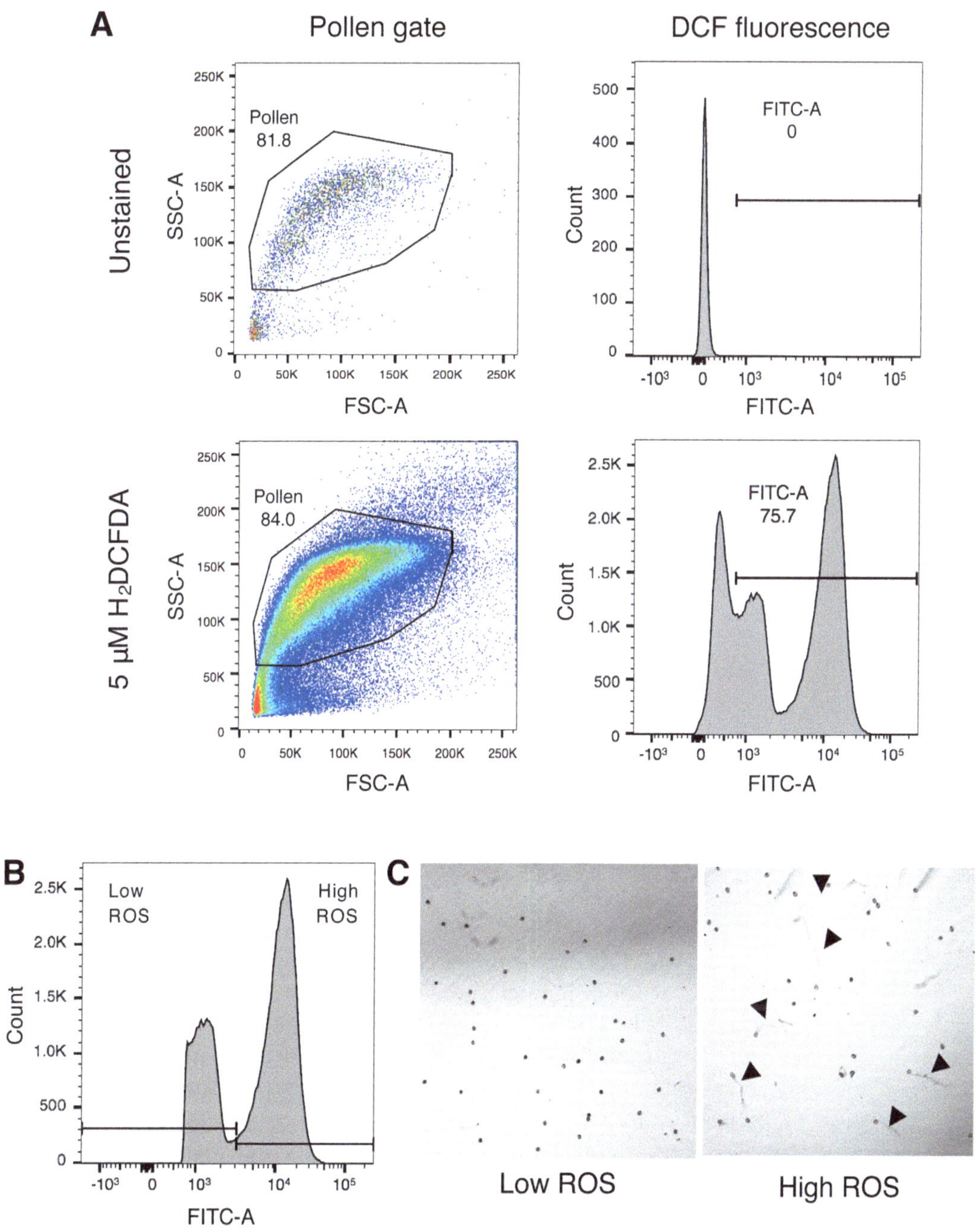

Fig. 3 Key features to look out for when performing FACS-isolation of H_2DCFDA-stained Arabidopsis pollen. (**a**) Run an unstained sample of pollen initially to establish pollen, singlet and true DCF fluorescence gates. It is important to also record all events when sorting in order to correlate any downstream findings with values from the FACS run. (**b**) An example of a gate setup to sort low-ROS and high-ROS subpopulations. (**c**) Representative images of low-ROS and high-ROS pollen sorted then germinated overnight in germination medium. Black arrowheads identify pollen tubes. Pseudocolor dot plots and histograms of FITC-A fluorescence in panel (**a**) were generated in FlowJo™ software for illustrative purposes

85 μm nozzle gives a sample volume of ~20 μl so by using 2 ml of buffer we keep the ratio of sample buffer to final buffer ratio very low.

1. Briefly vortex the sorted pollen sample then transfer 1 ml to a 1.5 ml microcentrifuge tube. Pellet the pollen by centrifugation at 500 × *g* for 5 min. Remove the supernatant and repeat with the remaining volume of sorted pollen suspension. The pellet of pollen might hardly be visible, but placing the tube inside the centrifuge at a standard orientation should facilitate pipetting without disrupting the pollen on the wall of the microcentrifuge tube.
2. Following this, the supernatant is removed, and the pollen is resuspended in 100 μl of fresh 10% PGM.
3. Tomato pollen can easily be germinated in wells of a 96-well plate covered with a lid. We germinate Arabidopsis pollen on 35 mm plates (*see* [10]). Briefly, the plates are first made adhesive by pipetting a droplet of 100 μl of polylysine onto the center of the plate and incubated for 10 min. This is then removed and the area where the drop was placed is washed three times in 10% PGM, each wash lasting 10 min. Pollen are then plated on the prepared area.
4. Pollen are incubated overnight at 22 °C and imaged the following day by microscopy.

4 Notes

1. For 10% Arabidopsis PGM, stock solutions of 100 mM KCl, 100 mM $CaCl_2$, 100 mM $MgSO_4$, 1% w/v boric acid, and ddH_2O can be prepared and autoclaved in advance and stored at room temperature. The medium itself must be made fresh daily. For 10% tomato PGM, the medium can be prepared in advance without sucrose, then autoclaved and stored at room temperature for 3–4 months. Add sucrose before use.
2. Take a used plasmid miniprep spin column and carefully cut off the end of the column containing the silica gel membrane. Clean the cut column in mild detergent and leave to dry, then assemble the filter by trapping a single piece of Miracloth between the cut column and a 1.5 or 2 ml centrifuge tube. *See* Fig. 2 as an example of how to assemble the filter.
3. Pollen rapidly sediments, so pollen should be resuspended by brief vortexing or pipetting up-and-down or inverting the microcentrifuge tube. Some liquid volume is lost during filtering leaving a smaller volume to work with from the starting volume of 10% PGM.
4. Filtering using Miracloth is a standard method to isolate pollen [22].

References

1. Carrizo Garcia C et al (2017) It is a matter of timing: asynchrony during pollen development and its consequences on pollen performance in angiosperms-a review. Protoplasma 254 (1):57–73
2. Firon N et al (2012) Water status and associated processes mark critical stages in pollen development and functioning. Ann Bot 109 (7):1201–1214
3. Suzuki G (2009) Recent progress in plant reproduction research: the story of the male gametophyte through to successful fertilization. Plant Cell Physiol 50(11):1857–1864
4. Rieu I et al (2017) Pollen development at high temperature: from acclimation to collapse. Plant Physiol 173(4):1967–1976
5. Zinn KE et al (2010) Temperature stress and plant sexual reproduction: uncovering the weakest links. J Exp Bot 61(7):1959–1968
6. Alexander MP (1969) Differential staining of aborted and nonaborted pollen. Stain Technol 44(3):117–122
7. Andreuzza S et al (2015) The chromatin protein DUET/MMD1 controls expression of the meiotic gene TDM1 during male meiosis in Arabidopsis. PLoS Genet 11(9):e1005396
8. Peterson R et al (2010) A simplified method for differential staining of aborted and non-aborted pollen grains. Int J Plant Biol 1: e13
9. Endo S et al (2013) A novel pollen-pistil interaction conferring high-temperature tolerance during reproduction via CLE45 signaling. Curr Biol 23(17):1670–1676
10. Luria G et al (2019) Direct analysis of pollen fitness by flow cytometry: implications for pollen response to stress. Plant J 98(5):942–952
11. Heslop-Harrison J et al (1984) The evaluation of pollen quality, and a further appraisal of the fluorochromatic (FCR) test procedure. Theor Appl Genet 67(4):367–375
12. Schuller-Levis GB, Sturman JA (1992) "Activation" of alveolar leukocytes isolated from cats fed taurine-free diets. Adv Exp Med Biol 315:83–90
13. De Storme N et al (2013) Volume-based pollen size analysis: an advanced method to assess somatic and gametophytic ploidy in flowering plants. Plant Reprod 26(2):65–81
14. Miller G et al (2010) Reactive oxygen species homeostasis and signalling during drought and salinity stresses. Plant Cell Environ 33 (4):453–467
15. Mittler R (2017) ROS Are Good. Trends Plant Sci 22(1):11–19
16. Halliwell B (2006) Reactive species and antioxidants. Redox biology is a fundamental theme of aerobic life. Plant Physiol 141(2):312–322
17. Boavida LC, McCormick S (2007) Temperature as a determinant factor for increased and reproducible in vitro pollen germination in *Arabidopsis thaliana*. Plant J 52(3):570–582
18. Firon N et al (2006) Pollen grains of heat tolerant tomato cultivars retain higher carbohydrate concentration under heat stress conditions. Sci Hortic 109(3):212–217
19. Pressman E et al (2002) The effect of heat stress on tomato pollen characteristics is associated with changes in carbohydrate concentration in the developing anthers. Ann Bot 90 (5):631–636
20. Yan J et al (2017) Arabidopsis pollen fertility requires the transcription factors CITF1 and SPL7 that regulate copper delivery to anthers and jasmonic acid synthesis. Plant Cell 29 (12):3012–3029
21. Grant-Downton R et al (2013) Artificial microRNAs reveal cell-specific differences in small RNA activity in pollen. Curr Biol 23 (14):R599–R601
22. Johnson-Brousseau SA, McCormick S (2004) A compendium of methods useful for characterizing Arabidopsis pollen mutants and gametophytically-expressed genes. Plant J 39 (5):761–775

Chapter 12

Obtaining Mutant Pollen for Phenotypic Analysis and Pollen Tube Dual Staining

Sheng Zhong, Zhijuan Wang, and Li-Jia Qu

Abstract

Mutant phenotype observation is the most useful and important method to study which biological process a gene-of-interest is involved in. In flowering plants, excessive pollen grains land and germinate on the stigma, then pollen tubes grow through the transmitting tract to reach the ovules, eventually enter the micropyle to complete double fertilization. *First*, for mutants whose homozygotes could not be obtained due to pollen tube defects, it is difficult to observe the defect phenotype since the pollen grains of different genotypes are mixed together. Here, we provide a detailed protocol to pick out mutant pollen grains from the heterozygous mutant plants in *Arabidopsis thaliana*. By using this method, we could obtain sufficient mutant pollen grains for phenotypic analysis. *Second*, it is difficult to compare the pollen/pollen tube behavior of two different genotypes/species in vivo in a same pistil. Here, we develop a new dual staining method which combines GUS staining with aniline blue staining. By using this method, we can analyze the competence of the two different pollen tubes in the same pistil.

Key words Pollen, Phenotypic observation, Eyelash pen, GUS staining, Aniline blue staining

1 Introduction

Double fertilization is the most essential link in flowering plants to connect haploid gametophyte generation and diploid sporophyte generation. Successful double fertilization depends on a series of tightly controlled processes, including development of both male and female gametophytes [1–4] and subsequent male–female interaction and fertilization [5]. Thus, studying these processes involved in plant sexual reproduction would eventually have a great impact and help for crop breeding.

In the past decade, a number of genes have been reported to participate in pollen/pollen tube–stigma interaction, pollen tube growth, and pollen tube guidance [6–11]. Loss of these genes results in lethality in recessive homozygous mutants [6]. Therefore, only heterozygous mutants could be obtained for these mutants. Because the heterozygous mutant plants produce pollen grains of

Anja Geitmann (ed.), *Pollen and Pollen Tube Biology: Methods and Protocols*, Methods in Molecular Biology, vol. 2160, https://doi.org/10.1007/978-1-0716-0672-8_12,

two different genotypes (i.e., wild type versus mutant) and these two different pollen grains are often mixed together, it is difficult to distinguish the homozygous pollen/pollen tubes from the wild type in those mutants with male defects in the processes after pollen–stigma interaction. For example, we found the male transmission efficiency was very low in *vps41* mutant, but we had not observed obvious defects of those pollen tubes in the pistil in vivo. Therefore, we need to pick out mutant pollen grains for phenotypic analysis [6].

In this chapter, we describe a method of how to obtain mutant pollen grains in those recessive homozygous male-generated lethal mutants. The first step is to make a complementation vector in which the gene-of-interest is fused with a fluorescence protein and is driven by a pollen/pollen tube-specific promoter [6, 12]. Then, screen a single insertional line in T1 generation of the transgenic plants. In the T2 generation, it is possible to obtain phenotype-complemented homozygous mutant with heterozygous allele for the fluorescent protein. In these plants, half of the pollen grains possess fluorescence signals (complemented, equivalent to wild-type pollen), while the other half of the pollen grains are not fluorescent (not complemented, mutant pollen). The two different pollen grains can be distinguished by their fluorescent signal under a fluorescence microscope and the mutant pollen grains can be collected to a glass slide. By using this method, we were able to obtain *vps41* mutant pollen grains, and characterized that while they could germinate normally, the pollen tubes exhibited a serious defect in penetrating the style [6]. In our lab, this method has also been successfully used to identify specific processes including pollen tube growth and guidance that are affected by our gene-of-interest. It is worth mentioning that we now can adopt CRISPR/Cas9 technology to generate homozygous mutant plants with fertility defects in the T1 generation [13–15], which makes phenotypic analysis of a sterile mutant much easier. The strategy, however, can only be used on those genes specifically involved in plant reproduction, because if the gene-of-interest is important in vegetative growth it would be difficult to obtain homozygous mutant plants by CRISPR/Cas9 technology. One additional note to add is that the eyelash pen described in the protocol can also be used to pick out pollen grains for other experiments, such as RNA sequencing [16].

We also provide a dual staining method that combines GUS staining [17] and aniline blue staining [18], both of which are among the most popular staining methods used in plants. Nowadays Promoter:GUS constructs are often used to transform wild-type plants to investigate the expression pattern of a gene-of-interest [14, 15, 19, 20]. For example, transferring LAT52pro:GUS into the plants would allow people to label pollen tubes in vivo with GUS staining, and to observe pollen tube growth and guidance in planta [21]. Aniline blue staining is another staining method to stain pollen tubes in the pistil, good for observing pollen tube behavior in vivo. However, neither of these two staining methods alone can distinguish behavior of two different pollen tubes in the same pistil.

Here we develop a new dual staining method, by combining GUS staining with aniline blue staining, so that LAT52:GUS pollen tubes can be distinguished from wild-type pollen tubes in the same pistil. To do this dual staining, the pistil is first pollinated with similar numbers of two different pollen grains. The pollinated pistil is first stained with GUS, to show LAT52pro:GUS pollen tubes. Then, the same pistil is stained with aniline blue, which would stain all the pollen tubes in the pistil. After dual staining, it is easy to distinguish the pollen tubes with GUS signal from those pollen tubes that are only stained by aniline blue. This dual staining method is specifically suitable to analyze competition between different pollen tubes in planta under a more natural condition. Notably, this dual staining method can be used to distinguish pollen tubes of two different genotypes as well as those of two different species [15].

2 Materials

2.1 Obtaining Mutant Pollen

2.1.1 Complementation Construct

1. Vectors: pDONRP4-P1R and pDONR221 (Invitrogen), pENTRY-TOPO-D (Invitrogen), pK7FWG0 (modified from pK7FWG2, from VIB-Ghent University, Belgium), pKm42GW,3 (VIB-Ghent University, Belgium).
2. Gateway kits: TOPO reaction (Invitrogen), BP reaction (Invitrogen), LR reaction (Invitrogen).
3. KOD-Plus (Takara).

2.1.2 Screen for Positive Transgenic Plants

1. LB medium: 1% Tryptone, 0.5% Yeast extract, 0.5% NaCl, 1.5% Agar.
2. Agrobacterium infection solution: 0.22% Murashige and Skoog (MS), 5% sucrose and 20–40 μL Silwet.

2.1.3 Tools

1. Eyelash pen (Fig. 1).
2. Straight tweezer.
3. Epifluorescence microscope (e.g., Olympus BX51) equipped with a UV filter set.
4. Stereomicroscope.

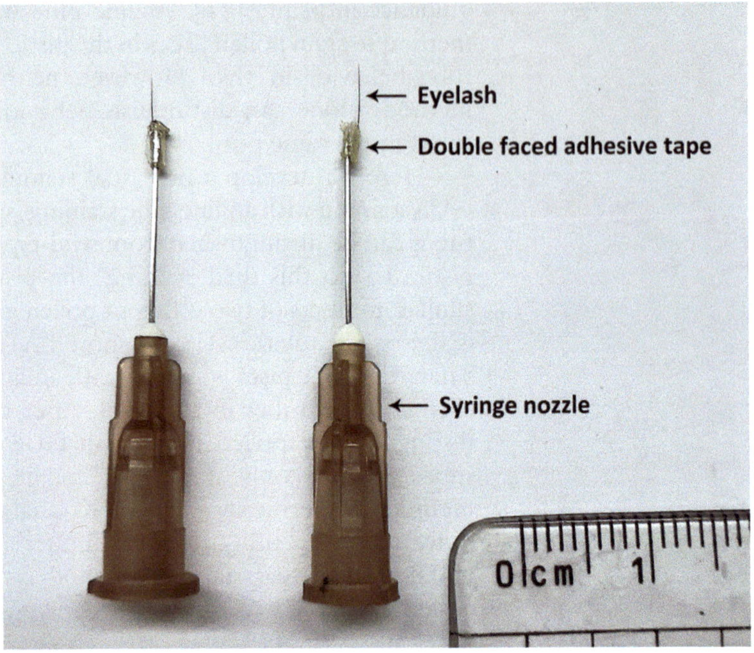

Fig. 1 The eyelash pen was handmade, using an eyelash and a syringe nozzle of 1 mL injection syringe

2.2 Pollen Tube Dual Staining

2.2.1 GUS Staining

1. Phosphate buffer: 69 mM $Na_2HPO_4 \cdot 2H_2O$, 15.7 mM $NaH_2PO_4 \cdot 2H_2O$, 0.5 mM $K_3Fe(CN)_6$, 0.5 mM $K_4Fe(CN)_6 \cdot 3H_2O$.
2. 2 mM GUS solution: 0.05217 g X-Gluc, 50 mL phosphate buffer.

2.2.2 Aniline Blue Staining

1. 8 M NaOH.
2. Aniline blue solution: 0.1% aniline blue, 108 mM K_3PO_4, adjust pH to 11 with H_3PO_4.

3 Methods

3.1 Obtaining Mutant Pollen

Two types of vectors can be generated to do the complementation experiments (Fig. 2a, b).

3.1.1 Generation of Complementation Vectors

1. To obtain the native promoter driving gene fused with GFP (Fig. 2a), a fragment containing the native promoter with the full-length genomic sequence (without stop codon) is amplified by PCR and cloned into pENTRY-TOPO-D or pDONR221 through TOPO reaction or BP reaction to make the intermediate vectors, respectively. The destination vector is then generated by LR reaction between the intermediate vector and pK7FWG0.

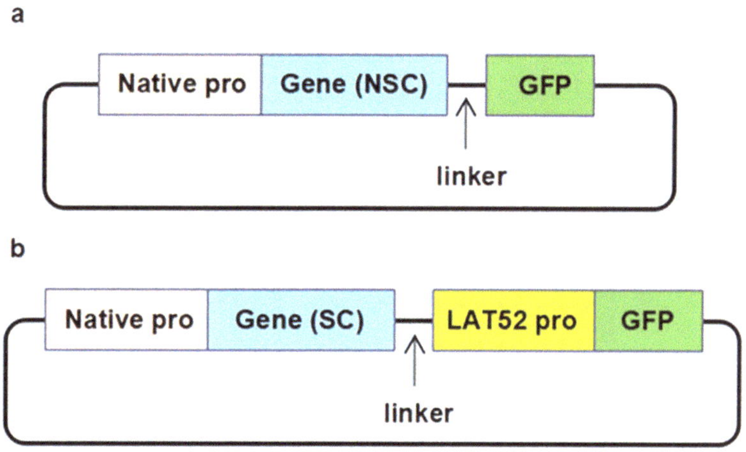

Fig. 2 Two types of complementation vectors. NSC, no stop codon; SC, stop codon; LAT52pro, a pollen-specific promoter; linker is produced by the LR reaction in gateway system

2. It is also recommended to make a second complementation vector if activity of the native promoter is not strong enough for pollen picking under the fluorescence microscope (Fig. 2b). First, a fragment containing the native promoter and gene-of-interest with stop codon is amplified by PCR and cloned into pDONRP4-P1R by BP reaction. Then the LAT52 promoter [22, 23] and the GFP gene are fused and cloned into pENTRY-TOPO-D through TOPO reaction. Finally, the destination vector is generated by LR reaction of the pDONR-Native promoter with gene-of-interest (with stop codon), pENTRY-LAT52 and pKm42GW3 by two-fragment LR reaction (*see* **Note 1**).

3.1.2 Obtaining Transgenic Plants

1. Transform the plants by Agrobacterium-mediated floral dip method [24].
2. Screen transgenic seeds on mediums with antibiotic resistance.

3.1.3 Picking Out Mutant Pollen Under the Fluorescence Microscope

1. Screen for single insertional transgenic line in the T1 generation (*see* **Note 2**).
2. Screen for rescued homozygotes with heterozygous GFP allele in the T2 generation (*see* **Note 3**).
3. Spread the mixed genotypic pollen grains on the glass slide (*see* **Note 4**) and take pictures with a fluorescence microscope using the GFP channel and GFP & Bright Field merged channels (Fig. 3a, b, *see* **Note 5**).
4. Use the eyelash pen to remove pollen grains with fluorescence (rescued pollen grains, equivalent to wild-type pollen grains) one by one under a stereoscope (*see* **Note 6**).

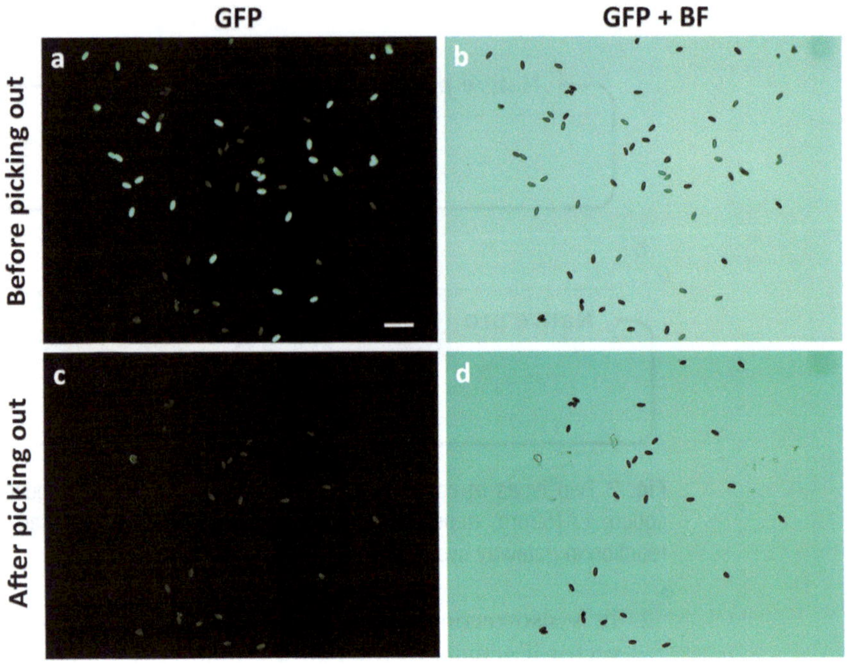

Fig. 3 Picking out mutant pollen grains under a fluorescence microscope. (**a**, **b**) Pollen grains from rescued homozygotes with heterozygous GFP allele in the single insertional transgenic plant. (**c**, **d**) The mutant pollen grains left on the glass slide. Scale bars = 100 μm

5. Using an epifluorescence microscope, confirm that the pollen grains left on the glass slide have only auto-fluorescence (these pollen grains are mutant pollen grains, Fig. 3c, d).

3.2 Pollen Tube Dual Staining

3.2.1 Homozygous LAT52pro:GUS Plant Confirmation

1. Pollinate wild type pistils with pollen grains from LAT52pro: GUS plants.
2. Screen plants with homozygous LAT52pro:GUS allele by GUS staining (*see* **Note 7**, GUS staining assay is described in **step 1–6** in Subheading 3.2.3).

3.2.2 Pollination with Two Different Pollen Grains

1. Pick up different pollen grains (Fig. 4a) by the eyelash pen.
2. Pollinate equivalent or proportionate pollen grains to the stigma, respectively (Fig. 4b–d, *see* **Notes 8** and **9**).

3.2.3 GUS Staining and Aniline Blue Staining

1. Put the pollinated pistils in a microcentrifuge tube.
2. Add 1 mL 90% acetone and put the tube on ice for at least 30 min (*see* **Note 10**).
3. Replace acetone with phosphate buffer twice, 15 min each (*see* **Note 11**).
4. Remove phosphate buffer, add 2 mM GUS solution to submerge the samples.

Fig. 4 Dual pollination by eyelash pen. (**a**) Eyelash pen with pollen grains collected. (**b–d**) Dual pollination step by step. (**b**) The stigma before pollination. (**c**) The stigma pollinated with one type of pollen grains. (**d**) The dual pollinated stigma. Scale bars = 100 μm

5. Vacuum the samples for 30 min, the samples will often start to sink to the bottom.
6. Incubate samples at 37 °C until the GUS signals appear.
7. Stop GUS staining with 70% ethanol.
8. Add 50%, 30% ethanol and ddH_2O in sequence, 15 min for each step.
9. Transfer samples into 8 M NaOH and incubate at room temperature overnight for tissue softening (*see* **Note 12**).
10. Remove 8 M NaOH thoroughly, add ddH_2O to wash the samples twice, 15 min each.
11. Remove ddH_2O, add aniline blue solution to stain the samples for more than 2 h in dark (*see* **Note 13**).

3.2.4 Sample Observation

Observe the samples under a fluorescence microscope (*see* **Note 14**). Take pictures using the BF channel, UV channel, BF and UV "merged channel".

4 Notes

1. To distinguish the complemented pollen (with GFP) from the non-rescued ones (without GFP), make sure that the GFP signal in the complemented plants is strong enough to be distinguished from the auto-fluorescence of dry pollen.

2. Single insertional lines can be judged by the fluorescence in T1 generation of the transgenic plants. There are two obvious characteristics of the single insertional line: (a) about half of the pollen grains have GFPsignals while the other half does not. (b) The intensity of the GFPsignals in those complemented pollen grains is approximately homogeneous.
3. Do not select the complemented plants with homozygous GFP alleles.
4. In order to pick out pollen grains more easily, spread the pollen grains on the glass slide as dispersedly as possible.
5. Choose appropriate exposure parameters to make sure that the auto-fluorescence of dry pollen can be observed.
6. It is alright if some mutant pollen grains are removed together with complemented pollen grains.
7. Homozygous LAT52pro:GUS plants can be judged by two characteristics: (a) each ovule is targeted by a GUS-stained pollen tube. (b) all the ovules have GUS signals at the micropyle.
8. If the eyelash cannot adhere the pollen grains, put the eyelash on a wet paper for a few seconds to eliminate electrostatics.
9. It is important to do the dual pollinations on each half of the stigma to guarantee an approximately same condition for the dual pollinations on the papilla cells.
10. To ensure good penetration of the fixing solution, pistils should be dissected at the ventral suture.
11. Keep the samples on ice during this step.
12. The samples must be incubated in NaOH for long enough, usually longer than 8 h.
13. After aniline blue staining, the samples can be stored at 4 °C for more than 1 week.
14. For the UV optical filter, an absorptive longpass filter should be used.

Acknowledgement

The research in the Qu laboratory is supported by grants from National Natural Science Foundation of China (31830004, 31620103903 and 31621001) and by the Peking-Tsinghua Joint Center for Life Sciences.

References

1. Borg M, Brownfield L, Twell D (2009) Male gametophyte development: a molecular perspective. J Exp Bot 60:1465–1478. https://doi.org/10.1093/jxb/ern355
2. Hafidh S, Fila J, Honys D (2016) Male gametophyte development and function in angiosperms: a general concept. Plant Reprod 29:31–51. https://doi.org/10.1007/s00497-015-0272-4
3. Drews GN, Yadegari R (2002) Development and function of the angiosperm female gametophyte. Annu Rev Genet 36:99–124. https://doi.org/10.1146/annurev.genet.36.040102.131941
4. Berger F, Twell D (2011) Germline specification and function in plants. Annu Rev Plant Biol 62:461–484. https://doi.org/10.1146/annurev-arplant-042110-103824
5. Zhong S, Qu L-J (2019) Peptide/receptor-like kinase-mediated signaling involved in male-female interactions. Curr Opin Plant Biol 51:7–14. https://doi.org/10.1016/j.pbi.2019.03.004
6. Hao L, Liu J, Zhong S, Gu H, Qu L-J (2016) AtVPS41-mediated endocytic pathway is essential for pollen tube-stigma interaction in Arabidopsis. Proc Natl Acad Sci U S A 113:6307–6312. https://doi.org/10.1073/pnas.1602757113
7. Muschietti JP, Wengier DL (2018) How many receptor-like kinases are required to operate a pollen tube. Curr Opin Plant Biol 41:73–82. https://doi.org/10.1016/j.pbi.2017.09.008
8. Liang X, Feng QN, Li S, Zhang Y (2018) Vacuolar trafficking in pollen tube growth and guidance. Plant Signal Behav 13:e1464854. https://doi.org/10.1080/15592324.2018.1464854
9. Li HJ, Meng JG, Yang WC (2018) Multilayered signaling pathways for pollen tube growth and guidance. Plant Reprod 31:31–41. https://doi.org/10.1007/s00497-018-0324-7
10. Mizuta Y, Higashiyama T (2018) Chemical signaling for pollen tube guidance at a glance. J Cell Sci 131. https://doi.org/10.1242/jcs.208447
11. Higashiyama T, Takeuchi H (2015) The mechanism and key molecules involved in pollen tube guidance. Annu Rev Plant Biol 66:393–413. https://doi.org/10.1146/annurev-arplant-043014-115635
12. Guan Y, Lu J, Xu J, McClure B, Zhang S (2014) Two mitogen-activated protein kinases, MPK3 and MPK6, are required for funicular guidance of pollen tubes in Arabidopsis. Plant Physiol 165:528–533. https://doi.org/10.1104/pp.113.231274
13. Ge Z, Zheng L, Zhao Y, Jiang J, Zhang EJ, Liu T, Gu H, Qu L-J (2019) Engineered xCas9 and SpCas9-NG variants broaden PAM recognition sites to generate mutations in Arabidopsis plants. Plant Biotechnol J. https://doi.org/10.1111/pbi.13148
14. Ge Z, Bergonci T, Zhao Y, Zou Y, Du S, Liu MC, Luo X, Ruan H, García-Valencia LE, Zhong S, Hou S, Huang Q, Lai L, Moura DS, Gu H, Dong J, Wu HM, Dresselhaus T, Xiao J, Cheung AY, Qu L-J (2017) Arabidopsis pollen tube integrity and sperm release are regulated by RALF-mediated signaling. Science 358:1596–1600. https://doi.org/10.1126/science.aao3642
15. Zhong S, Liu M, Wang Z, Huang Q, Hou S, Xu YC, Ge Z, Song Z, Huang J, Qiu X, Shi Y, Xiao J, Liu P, Guo YL, Dong J, Dresselhaus T, Gu H, Qu L-J (2019) Cysteine-rich peptides promote interspecific genetic isolation in Arabidopsis. Science 364:eaau9564. https://doi.org/10.1126/science.aau9564
16. Zhang J, Huang Q, Zhong S, Bleckmann A, Huang J, Guo X, Lin Q, Gu H, Dong J, Dresselhaus T, Qu L-J (2017) Sperm cells are passive cargo of the pollen tube in plant fertilization. Nat Plants 3:17079. https://doi.org/10.1038/nplants.2017.79
17. Jefferson RA, Kavanagh TA, Bevan MW (1987) GUS fusions: beta-glucuronidase as a sensitive and versatile gene fusion marker in higher plants. EMBO J 6:3901–3907
18. Martin FW (1959) Staining and observing pollen tubes in the style by means of fluorescence. Stain Technol 34. https://doi.org/10.3109/10520295909114663
19. Hou Y, Guo X, Cyprys P, Zhang Y, Bleckmann A, Cai L, Huang Q, Luo Y, Gu H, Dresselhaus T, Dong J, Qu L-J (2016) Maternal ENODLs are required for pollen tube reception in Arabidopsis. Curr Biol 26:2343–2350. https://doi.org/10.1016/j.cub.2016.06.053
20. Liu J, Zhong S, Guo X, Hao L, Wei X, Huang Q, Hou Y, Shi J, Wang C, Gu H, Qu L-J (2013) Membrane-bound RLCKs LIP1 and LIP2 are essential male factors controlling male-female attraction in Arabidopsis. Curr Biol 23:993–998. https://doi.org/10.1016/j.cub.2013.04.043
21. Lu Y, Chanroj S, Zulkifli L, Johnson MA, Uozumi N, Cheung A, Sze H (2011) Pollen

tubes lacking a pair of K+ transporters fail to target ovules in Arabidopsis. Plant Cell 23:81–93. https://doi.org/10.1105/tpc.110.080499

22. Twell DWR, Yamaguchi J, McCormick S (1989) Isolation and expression of an anther-specific gene from tomato. Mol Gen Genet 217:240–245. https://doi.org/10.1007/BF02464887

23. Muschietti JDL, Vancanneyt G, McCormick S (1994) LAT52 protein is essential for tomato pollen development. Plant J 6:321–338. https://doi.org/10.1046/j.1365-313X.1994.06030321.x

24. Clough SJ, Bent AF (1998) Floral dip: a simplified method for Agrobacterium-mediated transformation of *Arabidopsis thaliana*. Plant J 16:735–743. https://doi.org/10.1046/j,1365-313x.1998.00343.x

Chapter 13

Galvanotropic Chamber for Controlled Reorientation of Pollen Tube Growth and Simultaneous Confocal Imaging of Intracellular Dynamics

Firas Bou Daher and Anja Geitmann

Abstract

Successful fertilization and seed set require the pollen tube to grow through several tissues, to change its growth orientation by responding to directional cues, and to ultimately reach the embryo sac and deliver the paternal genetic material. The ability to respond to external directional cues is, therefore, a pivotal feature of pollen tube behavior. In order to study the regulatory mechanisms controlling and mediating pollen tube tropic growth, a robust and reproducible method for the induction of growth reorientation in vitro is required. Here we describe a galvanotropic chamber designed to expose growing pollen tubes to precisely calibrated directional cues triggering reorientation while simultaneously tracking subcellular processes using live cell imaging and confocal laser scanning microscopy.

Key words Directional cues, Galvanotropism, Galvanotropic chamber, Guided growth, Pollen tube, Tropic growth

1 Introduction

The pollen is a cell on a singular mission—to deliver paternal genetic material required for fertilization to the female gametophyte. To do so, the pollen forms a tube that has to find its way through different tissues comprising the stigma, the style, and the ovary. Once it reaches the ovary, the pollen tube has to identify an unfertilized ovule, pass through its micropyle to reach its final destination, the embryo sac. While traversing these tissues, the pollen tube responds to external signals and consequently changes its growth direction multiple times exhibiting tropic behavior [1, 2]. The guiding signals informing pollen tube growth direction are emitted by the sporophytic tissues of the pistil as well as the female gametophyte [3–18] and have to be sensed by the pollen tube to induce subcellular changes leading to growth reorientation [15, 19, 20].

Anja Geitmann (ed.), *Pollen and Pollen Tube Biology: Methods and Protocols*, Methods in Molecular Biology, vol. 2160, https://doi.org/10.1007/978-1-0716-0672-8_13, © Springer Science+Business Media, LLC, part of Springer Nature 2020

Although our understanding of the upstream players involved in the signaling processes regulating pollen tube growth has improved in recent years, our knowledge about the mechanisms involved in the execution of growth reorientation is poor. Being able to induce a controlled reorientation of the pollen tube in vitro while tracking subcellular changes in transgenic and/or mutant lines will help us shed more light on the cellular machinery underlying the tropic growth control and execution.

In vitro experiments for targeted redirection of pollen tube growth have been elegantly applied using as guidance cues proteins, ion gradients, electrical fields, and whole ovules [9, 12, 13, 15, 21, 22]. Many of these assays have been used to simply score in binary manner whether or not pollen tubes reorient upon exposure to the signal, but most assays are limited in terms of controlling exact timing and magnitude of the directional cue or do not offer the possibility to simultaneously monitor the cells at high optical resolution. For example, placing entire ovules near the path of in vitro growing pollen tubes does not inform on the nature or concentration of the guiding agents, or exposing the cells to diffusion-based chemical gradients does not specify the actual concentrations of the agent at cell level. This limitation has been addressed by a microfluidic device that is able to expose the growing tip of a pollen tube to two discrete concentrations and to tune the position of this extremely sharp chemical gradient to the spatial location of the tip [23, 24]. This device is complex to fabricate and use, however, and does not allow for any kind of high throughput testing required for efficient phenotyping.

As an alternative to a chemical turning signal, electrical fields have been used to trigger the directional cellular response, with the advantage of being able to easily control the signal in time and intensity. This has been done in macroscopic experimental set-ups [25–27] and based on MEMS technology [28]. However, these setups were either not amenable to simultaneous imaging or require advanced technology not available in most biology laboratories. Here we present a low-tech experimental device that can be used to investigate pollen tube tropic growth allowing exposure of multiple cells to an electrical field while performing high resolution confocal laser scanning imaging.

2 Material

2.1 Chamber Construction

1. Thin transparent plexiglass (e.g., from an old decontaminated electrophoresis chamber).
2. Electric drill (wear appropriate personal protective equipment including goggles or a face shield when using).
3. Superglue.

4. Clear silicon.
5. 22 mm square cover glass compatible with the microscope objectives used (typically #1.5).
6. 0.25 mm platinum wire.
7. Copper electrodes.
8. Direct current (DC) power supply (e.g., TEKPOWER HY1803D).

2.2 Pollen

Depending on the experiment, one of the following:

- *Camellia japonica* pollen: easily collected using a vibrating toothbrush.
- *Lilium longiflorum* pollen: scraped off the anthers.
- *Arabidopsis thaliana* pollen: most efficiently harvested using a vacuum cleaner [29].

All pollen is ideally used fresh, but can also be stored in gelatin capsules at −80 °C (Lilium) or −20 °C (other species) between harvest and use. Prior to freezing, pollen should be dried overnight in a desiccator over silica gel beads.

2.3 Growth Media

- *Camellia japonica*: 1.62 mM H_3BO_3, 2.54 mM $Ca(NO_3)_2$, 1 mM KNO_3, 0.81 mM $MgSO_4$, 8% sucrose (w/v).
- *Lilium longiflorum*: 1 mM KNO_3, 0.13 mM $Ca(NO_3)_2$, 1.62 mM H_3BO_3, 10% sucrose (w/v), and 5 mM MES (2-(N-morpholino)ethanesulfonic acid) buffer adjusted to pH 5.5.
- *Arabidopsis thaliana*: 1.62 mM boric acid, 1 mM $CaCl_2$, 1 mM $Ca(NO_3)_2$, 1 mM $MgSO_4$, 1 mM KCl and 18% sucrose (w/v) with a pH adjusted to 7 [29].

2.4 Imaging

1. Inverted microscope with appropriate objectives (e.g., 10×, 40×, 60×).
2. Immersion oil (if using oil immersion objective).

3 Methods

3.1 Galvanotropic Box

1. Use 0.5-mm-thick transparent sheet of plexiglass to build the chamber. Cut a rectangular piece to fit the microscope stage (here a 4.5 × 4 cm piece was used). This will be the base of the chamber (*see***Note 1**).
2. Make a rectangular cut in the center of the base with 1.5 cm sides to form a mini-chamber (Fig. 1a).

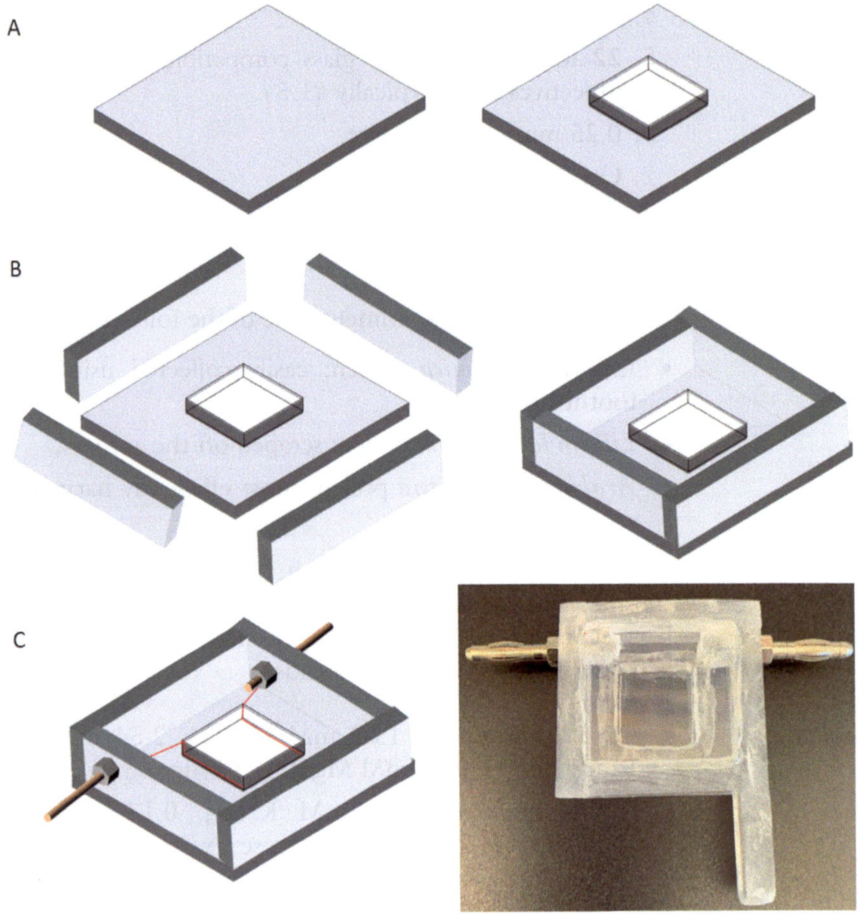

Fig. 1 Assembly of the galvanotropic chamber. (**a**) Preparation of the base of the galvanotropic chamber. (**b**) Addition of side walls. (**c**) Final chamber after attachment of wires and electrode ports

3. Attach four 1 cm high pieces of plexiglass to the base to form the big chamber (Fig. 1b). One of these pieces can have an extension to be used as a handle for the box (*see* **Note 2**).
4. Drill two holes into two opposite sidewalls and attach two electrode ports on two opposite sides of the box (Fig. 1c).
5. Connect the electrodes to the mini-chamber with platinum wires (Fig. 1c) (*see* **Note 3**).
6. Apply and seal a coverslip using clear silicone to the bottom of the mini-chamber (*see* **Note 4**).
7. A cover made of the same transparent plexiglass material can be designed and used when growing pollen tubes for extended periods of time to reduce evaporation (Fig. 1c).

3.2 Pollen Tube Growth in the Chamber

1. Hydrate the pollen grains and prepare the media (*see* **Notes 5** and **6**).
2. Pollen grains are to be grown at the bottom of the chamber, in close contact with the cover slip for efficient confocal imaging if using short working distance objectives.
3. Pour the agar/agarose containing medium in a Petri dish and let it set. This medium will be used to extract 3-mm-thick media pads on the surface of which pollen tubes will be grown (*see* **Note 7**).
4. Cut an appropriately sized pad by placing the Petri dish on the inverted chamber using a sharp blade for increased precision and to reduce the amount of deformations in the pad (Fig. 2a).
5. Lift the pad with a wide spatula and invert it onto a glass slide to expose the flat bottom surface (Fig. 2b).
6. Wet the edge of a paper with the liquid medium; then gently touch the surface of the hydrated pollen. The pollen will stick to the wet edge of the paper which serves as a stamp to transfer the pollen grains (Fig. 2c).
7. Gently blot the pollen onto the surface of the agarose pad creating a thin line of pollen grains on the surface of the pad. This distribution of pollen grains will induce most pollen tubes to grow perpendicular to the line (Fig. 2d, e).
8. Invert the pad so that the pollen line is in contact with the cover slip and perpendicular to the electrodes (Fig. 3a).
9. Gently apply a small volume of liquid medium to the mini-chamber to keep the surface of the pad wet and ensure electrical conductivity.
10. Let the pollen tubes germinate and then apply the electrical field.
11. 1.5 V/cm DC electrical potential can induce a plasma membrane hyperpolarization response in pollen tubes of 15–20 μm diameter. This field strength can be decreased or increased for narrower or wider tubes, respectively.
12. Start recording time lapse images at the desired magnification and score the time and angle of pollen tube reorientation (*see* **Notes 8** and **9**).
13. Pharmacological agents or dyes can be applied directly to the chamber by sucking the liquid medium off the pad with a pipette and replacing with liquid medium containing the agent.
14. Brightfield or fluorescence (confocal laser scanning) imaging can be done by placing the chamber on an inverted microscope (Fig. 3b).
15. After terminating an experiment, clean the chamber with ethanol followed by a rinse with distilled water and let it dry (*see* **Notes 4** and **10**).

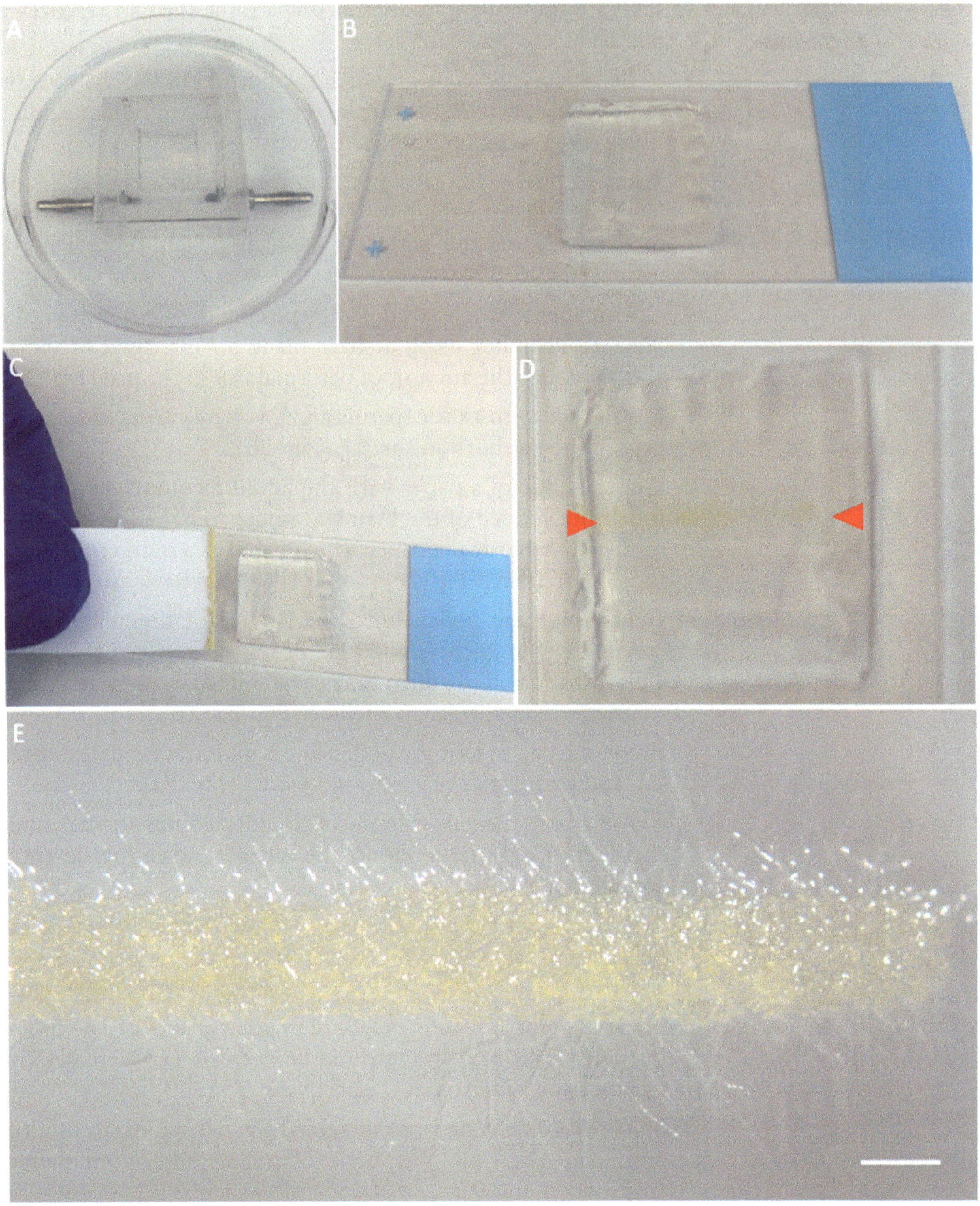

Fig. 2 Application of the pollen to the agar pad. Preparation of agar layer and cutting of suitably sized pad by placing it above the box (**a**). Placing the agar pad on a glass slide (**b**). Transfer of pollen to the edge of the paper (**c**) and blotting to the pad. Red arrow heads indicate line of pollen grains (**d**). Micrograph showing line of germinated pollen with tubes growing primarlly perpendicular to the line (**e**). Bar = 250 μm

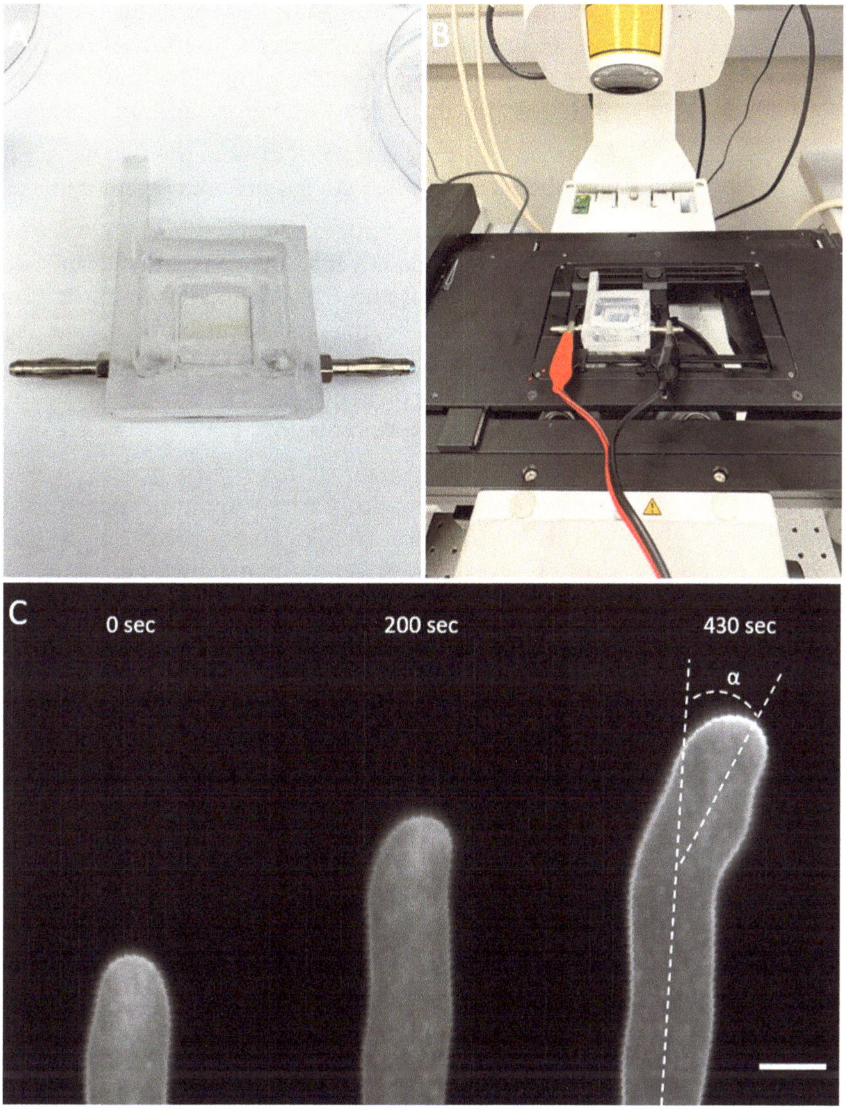

Fig. 3 Preparation of chamber for confocal imaging. (**a**) The pad with pollen is inverted onto the cover slip in the chamber so that the pollen is in contact with the glass. (**b**) After germination, the chamber is attached to the power supply and mounted on the microscope stage. (**c**) Confocal laser scanning micrographs of pollen tube stained with FM4-64. The electric field is oriented perpendicular to the long axis of the cell. The time lapse series shows the tube at the time of current application (0 s), beginning of the tropic response (200 s) and after it responded with a change in growth direction at an angle of 25°

4 Notes

1. Mark the plexiglass pieces to be cut with a marker and use an electric drill or a coping saw to extract the pieces. Remember to wear goggles to protect the eyes from flying plexiglass shreds.

2. Clean the cut pieces of plexiglass with ethanol to remove the marker lines followed by water to remove the plexiglass dust. Glue the pieces together using superglue and allow a minimum of 30 min for the glue to set before drilling in the box.
3. Make sure that the distance between the two wires in the mini-chamber is exactly 1.5 cm for precise current application.
4. The cover slip can easily be replaced if scratched or dirty after use. Make sure the cover slip is constantly clean. Some pollen species (such as lily) have pigments that stain the glass and the cover slip should be changed regularly to reduce background noise and autofluorescence while imaging. Seal the new cover-slip with clear silicone and let it dry for at least one hour. Then wash with distilled water before use.
5. Prepare a humid chamber for pollen hydration by placing a piece of tissue wet with warm water at one side of a petri dish. Cover and leave for 10 min. Apply the pollen to the opposite side of the plate, cover immediately, and let it hydrate for 30 min. Do not let any liquid water touch the pollen.
6. The choice of pollen growth medium depends on the plant species. For efficient use, prepare a ten times concentrated stock (without sucrose), aliquot and store at −20 °C for future use. Once needed, the stock is diluted ten times with the addition of the appropriate sucrose concentration.
7. To prepare solid media, add 1% (w/v) agar/agarose to the liquid media solution, heat in a water bath in the microwave (while avoiding overheating) until the agar is completely dissolved. Wear appropriate protective equipment (goggles and heat-protection gloves). Occasional shaking can help homogenize the media and reduce the heating time. Pour the medium in a petri dish to make a 3-mm-thick layer. In a standard 9 cm Petri dish (8.5 cm base) 17 mL medium is required to make a 3-mm-thick agar layer.
8. To score the time of reorientation, look for changes in the geometry of the tip when growth starts shifting to one side. It works easier to play the time course backward until the change of the tip geometry is detected and recorded (Fig. 3c middle). The angle of deviation can easily be measured using the angle tool in ImageJ.
9. 35% of *Camelliajaponica* pollen tubes respond to a 1.5 V/cm field by reorienting their growth 271 ± 117 s after application of the current. The reorientation angle α (Fig. 3c right) for this species is 20.5 ± 3.5°.
10. Always make sure that the wires are not bent while cleaning the chamber or replacing the cover slip to keep the distance constant between the two electrodes.

References

1. Cheung AY, Wu H-M (2001) Pollen tube guidance—right on target. Science 293:1441–1442
2. Geitmann A, Palanivelu R (2007) Fertilization requires communication: signal generation and perception during pollen tube guidance. Floriculture and Ornamental Biotechnol 1:77–89
3. Cheung AY, Wang H, Wu H-M (1995) A floral transmitting tissue-specific glycoprotein attracts pollen tubes and stimulates their growth. Cell 82:383–393
4. Palanivelu R, Brass L, Edlund AF, Preuss D (2003) Pollen tube growth and guidance is regulated by *POP2*, an *Arabidopsis* gene that controls GABA levels. Cell 114:47–59
5. Higashiyama T, Kuroiwa H, Kuroiwa T (2003) Pollen-tube guidance: beacons from the female gametophyte. Curr Opin Plant Biol 6:36–41
6. Tung C-W, Dwyer KG, Nasrallah ME, Nasrallah JB (2005) Genome-wide identification of genes expressed in *Arabidopsis* pistils specifically along the path of pollen tube growth. Plant Physiol 138:977–989
7. Palanivelu R, Preuss D (2006) Distinct short-range ovule signals attract or repel *Arabidopsis thaliana* pollen tubes *in vitro*. BMC Plant Biol 6:7
8. Prado AM, Porterfield DM, Feijó JA (2004) Nitric oxide is involved in growth regulation and re-orientation of pollen tubes. Development 131:2707–2714
9. Márton ML, Fastner A, Uebler S, Dresselhaus T (2012) Overcoming hybridization barriers by the secretion of the maize pollen tube attractant ZmEA1 from *Arabidopsis* ovules. Curr Biol 22:1194–1198
10. Márton ML, Dresselhaus T (2010) Female gametophyte-controlled pollen tube guidance. Biochem Soc Trans 38:627–630
11. Márton ML, Cordts S, Broadhvest J, Dresselhaus T (2005) Micropylar pollen tube guidance by egg apparatus 1 of maize. Science 307:573–576
12. Okuda S, Tsutsui H, Shiina K, Sprunck S, Takeuchi H, Yui R, Kasahara RD, Hamamura Y, Mizukami A, Susaki D, Kawano N, Sakakibara T, Namiki S, Itoh K, Otsuka K, Matsuzaki M, Nozaki H, Kuroiwa T, Nakano A, Kanaoka MM, Dresselhaus T, Sasaki N, Higashiyama T (2009) Defensin-like polypeptide LUREs are pollen tube attractants secreted from synergid cells. Nature 458:357
13. Takeuchi H, Higashiyama T (2012) A species-specific cluster of defensin-like genes encodes diffusible pollen tube attractants in *Arabidopsis*. PLoS Biol 10:e1001449
14. Higashiyama T, Hamamura Y (2008) Gametophytic pollen tube guidance. Sex Plant Reprod 21:17–26
15. Bou Daher F, Geitmann A (2011) Actin is involved in pollen tube tropism through redefining the spatial targeting of secretory vesicles. Traffic 12:1537–1551
16. Takeuchi H, Higashiyama T (2016) Tip-localized receptors control pollen tube growth and LURE sensing in *Arabidopsis*. Nature 531:245
17. Wang T, Liang L, Xue Y, Jia P-F, Chen W, Zhang M-X, Wang Y-C, Li H-J, Yang W-C (2016) A receptor heteromer mediates the male perception of female attractants in plants. Nature 531:241
18. Zhang X, Liu W, Nagae TT, Takeuchi H, Zhang H, Han Z, Higashiyama T, Chai J (2017) Structural basis for receptor recognition of pollen tube attraction peptides. Nat Commun 8:1331
19. Luo N, Yan A, Liu G, Guo J, Rong D, Kanaoka MM, Xiao Z, Xu G, Higashiyama T, Cui X, Yang Z (2017) Exocytosis-coordinated mechanisms for tip growth underlie pollen tube growth guidance. Nat Commun 8:1687
20. Qu X, Zhang R, Zhang M, Diao M, Xue Y, Huang S (2017) Organizational Innovation of apical actin filaments drives rapid pollen tube growth and turning. Mol Plant 10:930–947
21. Malhó R, Read ND, Pais MS, Trewavas AJ (1994) Role of cytosolic free calcium in the reorientation of pollen tube growth. Plant J 5:331–341
22. Malhó R, Trewavas AJ (1996) Localized apical increases of cytosolic free calcium control pollen tube orientation. Plant Cell 8:1935–1949
23. Sanati Nezhad A, Packirisamy M, Geitmann A (2014) Dynamic, high precision targeting of growth modulating agents is able to trigger pollen tube growth reorientation. Plant J 80:185–195
24. Agudelo CG, Sanati Nezhad A, Ghanbari M, Naghavi M, Packirisamy M, Geitmann A (2013) TipChip: a modular, MEMS-based

platform for experimentation and phenotyping of tip-growing cells. Plant J 73:1057–1068

25. Malhó R, Feijó JA, Pais MSS (1992) Effect of electrical fields and external ionic currents on pollen-tube orientation. Sex Plant Reprod 5:57–63
26. Wang C, Rathore KS, Robinson KR (1989) The responses of pollen to applied electrical fields. Dev Biol 136:405–410
27. Nakamura N, Fukushima A, Iwayama H, Suzuki H (1991) Electrotropism of pollen tubes of camellia and other plants. Sex Plant Reprod 4:138–143
28. Agudelo C, Packirisamy M, Geitmann A (2016) Influence of electric fields and conductivity on pollen tube growth assessed via electrical lab-on-chip. Sci Rep 6:19812
29. Bou Daher F, Chebli Y, Geitmann A (2009) Optimization of conditions for germination of cold-stored *Arabidopsis thaliana* pollen. Plant Cell Rep 28:347–357

Chapter 14

Analyzing Intracellular Gradients in Pollen Tubes

Daniel S. C. Damineli, Maria Teresa Portes, and José A. Feijó

Abstract

Conspicuous intracellular gradients manifest and/or drive intracellular polarity in pollen tubes. However, quantifying these gradients raises multiple technical challenges. Here we present a sensible computational protocol to analyze gradients in growing pollen tubes and to filter nonrepresentative time points. As an example, we use imaging data from pollen tubes expressing a genetically encoded ratiometric Ca^{2+} probe, Yellow CaMeleon 3.6, from which a kymograph is extracted. The tip of the pollen tube is detected with CHUKNORRIS, our previously published methodology, allowing the reconstruction of the intracellular gradient through time. Statistically confounding time points, such as growth arrest where gradients are highly oscillatory, are filtered out and a mean spatial profile is estimated with a local polynomial regression method. Finally, we estimate the gradient slope by the linear portion of the decay in mean fluorescence, offering a quantitative method to detect phenotypes of gradient steepness, location, intensity, and variability. The data manipulation protocol proposed can be achieved in a simple and efficient manner using the statistical programming language R, opening paths to perform high-throughput spatiotemporal phenotyping of intracellular gradients in apically growing cells.

Key words Apical growth, Computational analysis, Intracellular gradient, High-throughput phenotyping, Spatiotemporal analysis, Ratiometric probes, Ion gradients, Ca^{2+}, Tip detection

1 Introduction

Pollen tubes are highly polarized cells with various underlying intracellular gradients that reflect or are causal for cell polarity. Cytosolic ion concentrations match pollen tube polarity and have been reported to change with or even anticipate growth direction, displaying a positive gradient with most cations (Ca^{2+}, H^{+}) and a negative gradient with anions [1–4]. Quantitative spatiotemporal analyses are required to establish reliable and reproducible estimates of these gradients, while a mechanistic understanding or their role is difficult, if not impossible without computational and

Electronic supplementary material: The online version of this chapter (https://doi.org/10.1007/978-1-0716-0672-8_14) contains supplementary material, which is available to authorized users.

Anja Geitmann (ed.), *Pollen and Pollen Tube Biology: Methods and Protocols*, Methods in Molecular Biology, vol. 2160, https://doi.org/10.1007/978-1-0716-0672-8_14,

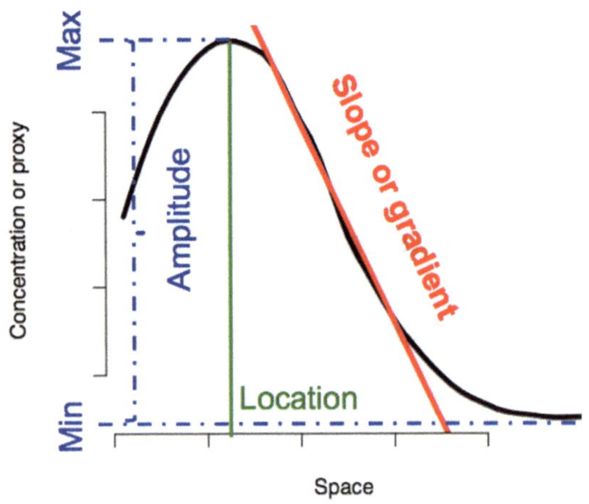

Fig. 1 Gradient landmarks. Basic gradient parameters can be estimated with simple techniques and then used for phenotyping

mathematical approaches, especially in regulatory networks that give rise to oscillations [5].

Three key features can be used to characterize a simple gradient: amplitude, location, and steepness (Fig. 1) [6]. The protocol described herein aims to provide a simple method to estimate these parameters of interest having in mind the usual data produced with pollen tubes, root hairs, and other apically growing cells. We start with a video of a growing cell with any reporter, extracting a kymograph along the midline of the cell. The growing cell apex is detected, allowing to reconstruct the gradient with a series of local polynomial fits. However, when the cell changes growth regime within the same time series the average gradient is no longer statistically representative, as we would be assuming the gradient remains the same independently of growth. In fact, growth arrested tubes show dramatic Ca^{2+} spiking behavior [7, 8], with intracellular gradients oscillating between abnormally high concentrations at the tip and then complete vanishing [7]. Thus, to restrict the gradient analysis to periods when the pollen tube is growing normally, we removed time points with growth arrest based on objective and statistically sound criteria. Finally, the gradient landmarks can be estimated from the average curve and by fitting an expression to the population data in the linear portion of the overall curve. The last step to obtain a gradient with true units is to calibrate the sensor, which is outside the scope of this protocol, but can be easily achieved by interpolating the values on a calibration curve.

2 Materials

2.1 Data

Any data of tip growing cells with fluorescent reporters or dyes can be used. Here we use ratiometric florescence imaging of a genetically encoded Ca^{2+} (Yellow CaMeleon 3.6) reporter in Arabidopsis pollen tubes growing in vitro as an example. This data set is publicly available on Dryad [7]. Arabidopsis pollen tubes were transformed and germinated according to [9].

2.2 Software

All software generated and used is free, open-source, and multiplatform, namely Fiji (ImageJ) [10] and R [11], with "tidyverse" [12] and "mixtools" [13] packages. Tip finding was performed with CHUKNORRIS [7], exemplified here with our web application (https://feijolab.shinyapps.io/CHUK/; under development). Ratiometric kymographs were analyzed within R with basic scripts borrowed from CHUKNORRIS [7] and an original computational protocol to analyze gradients provided here (Data S1).

3 Methods

3.1 Extract Kymograph

3.1.1 Trace Midline

Starting with a video of an apically growing cell with a fluorescence reporter, open the file in Fiji and trace the midline of the cell with the toolbar option "Segmented Line" in the dropdown menu, making sure the trace starts from shank to tip (Fig. 2). In the case of multiple channels, copy the mask to the other channels with the command "control (or command) + e." Beware that specific import option may be necessary to split multiple channels, we used the "Plugins > Bio-Formats > Bio-Formats Importer" for this particular data set to split the two channels acquired in DeltaVision and get its metadata (*see* **Note 1**).

3.1.2 Extract Kymograph for All Channels

Once the mask goes through the midline of all time points, and away from the cell to get some additional background values, the Kymograph for each channel can be extracted with the "Analyze > Multi Kymograph > Multi Kymograph." Here we used a 5 pixels interval, but this interval should be increased if results are too noisy or decreased if too blurry (Fig. 2). Save the image to a text file by exporting with "File > Save As > Text Image...."

3.2 Align Kymograph by Tip Boundary

3.2.1 Find Tip

Given a text file with pixel values of a kymograph, the tip of the cell can be estimated with any method, such as establishing a fluorescence threshold and finding the first pixel that crosses it. The goal is to produce a text file with time in the first column and tip location in the second, being the only required output of this step. Here we exemplify using CHUKNORRIS [7], a method we have previously developed that allows greater resolution, available in R and on the

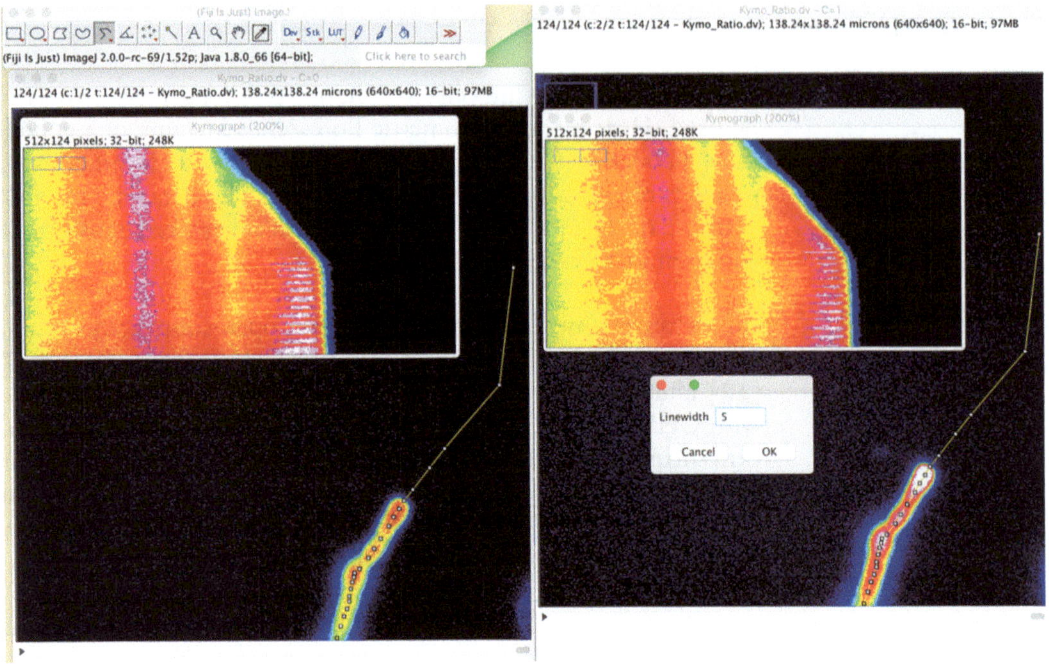

Fig. 2 Extracting Kymographs from multiple channels using Fiji (ImageJ). A segmented line is drawn in the midline of the pollen tube, from shank to tip, including background and a kymograph is extracted. The mask is copied to the other channel yielding another kymograph, allowing ratiometric kymographs to be generated afterwards

web via a Shiny app (feijolab.shinyapps.io/CHUK). Since we are mainly interested in tip detection, we will use the web interface, which requires the kymograph to be a matrix with time as rows and position as columns—with a cell growing toward the right (Fig. 3 top panel). After uploading the kymograph of the strongest channel (here YFP), accept the default parameters, unless the subpixel detection shows errors. The performance of the algorithm can be visualized for any time step (Fig. 3 lower panel), with the major fluorescence drop from left to right corresponding to the tip where CHUKNORRIS fits a linear model and estimated the intercept (red and magenta lines) in the pixel grid. In case parameters of the algorithm need to be tuned to achieve an adequate performance, the pipeline description can be found in [7]. Otherwise the user can proceed to the "Save" tab and download the zip file of all analyses. Here we are only interested in the file "All_time_series.txt" that contains the tip location estimate for each time point (Fig. 3 upper panel magenta points), found in the unzipped analysis folder. Despite the gain in resolution, using the fluorescence signal to estimate tip location and analyze intracellular concentrations entails an interdependency that may be problematic under certain scenarios[7] (*see* **Note 1**).

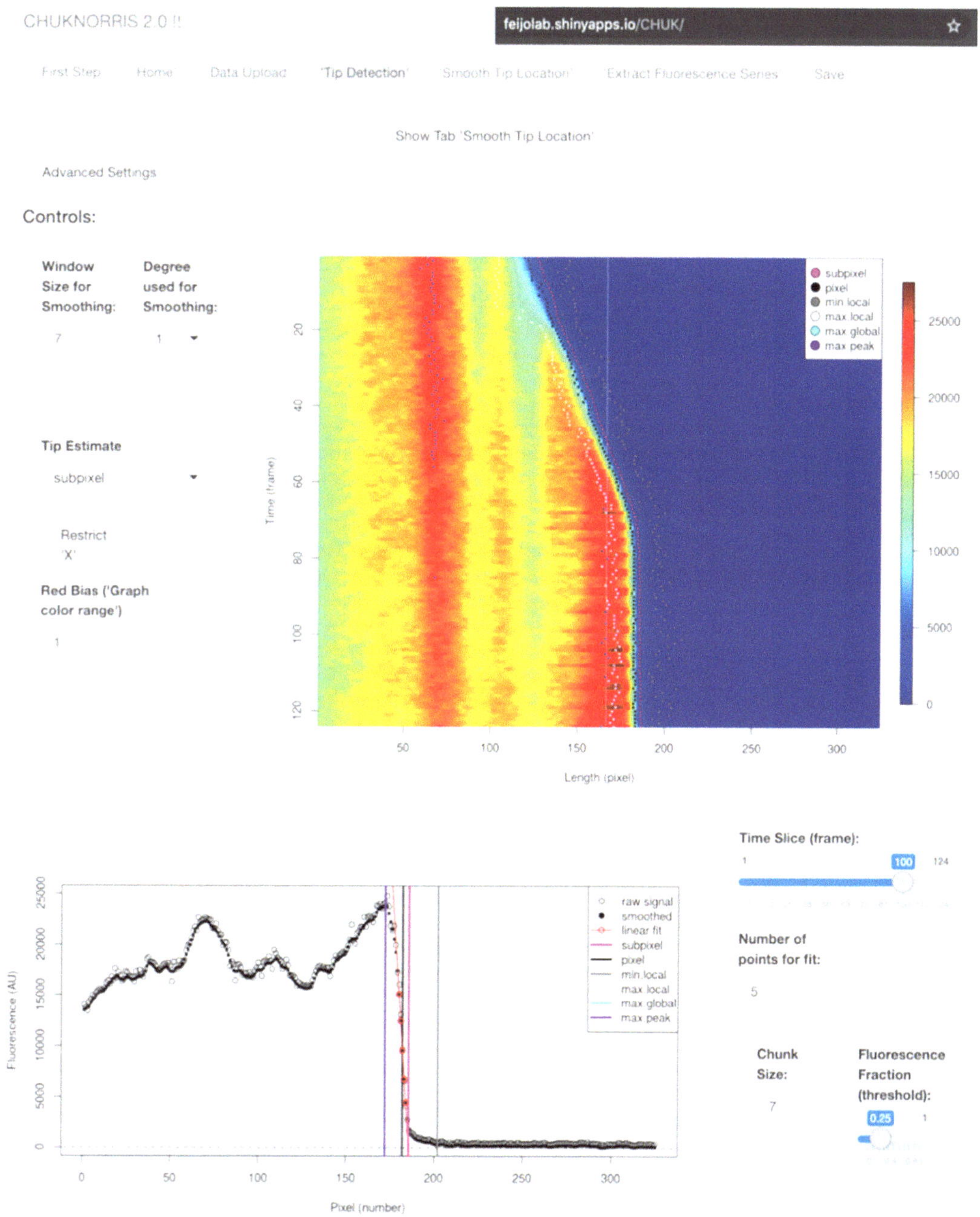

Fig. 3 Tip finding with CHUKNORRIS. Subpixel tip detection is performed with a linear fit of the fluorescence decay (bottom panel), after entering a text file with a kymograph oriented as shown in the top panel. The web app was used (internet address on top right corner; under construction), although one could choose the original CHUKNORRIS R script or any other tip detection method to generate a tip location series, which will be used in the next steps

3.2.2 Invert Columns

The workflow from here on is performed exclusive on the statistical programming language R [11], with RStudio [14] as a suggested interface. All programming details are provided in the accompanying script (Data S1) showing a rationale that can be easily implemented with other languages or analysis software.

Next, the user should open the tip location file (here "All_time_series.txt") file in R [11] with RStudio [14] as a suggested interface, also opening the text files for both kymographs (Data S1). From each kymograph matrix, with growth toward the right (Fig. 2 and 3), invert columns at the tip location coordinates (Fig. 4a, b; Data S1). In the case of subpixel tip detection, pixel values are recalculated using the average of neighboring pixels weighted by the fraction of each pixel necessary to compose a new row, starting at the tip estimate and ending with the first pixel in the shank region (Data S1). The resulting tip aligned kymograph can then be used to compare fluorescence across all time points in the same relative region of the cell (Fig. 4a, b). Here, the ratio of YFP/CFP channels is used to eliminate artifacts due to uneven distribution of reporter proteins within the cell, achieving an adequate depiction of the intracellular Ca^{2+} gradient at each time point (Fig. 4c). Background subtraction and other correction methods could also be used to adjust the ratiometric values [7], however here just the ratio is used for simplicity (*see* **Note 2**).

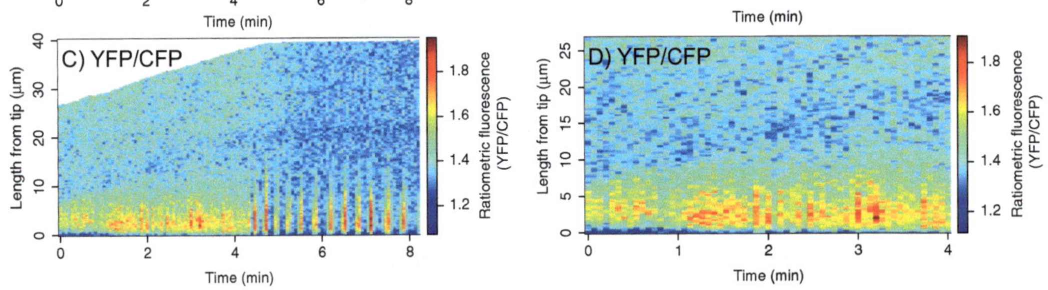

Fig. 4 Tip aligned kymographs. The tip of the pollen tube was detected in the strongest channel (YFP) using CHUKNORRIS and then used to align each kymograph: (**a**) YFP and (**b**) CFP. Then the ratio (**c**) YFP/CFP was performed to correct artifacts yielding, then restricting the analysis to time points where the tube was growing (**d**)

3.3 Filter Time Points

3.3.1 Establish Confounding Factors

Often, not all time points represent the gradient faithfully, whether due to experimental error or changes in growth regime. These time points can either be removed with arbitrary criteria or based on objective statistical estimates. Here we statistically define growth arrest to remove the time points where the gradient is highly oscillatory (Fig. 4c), using the distribution of growth rate in all time points (Fig. 5). We fit a mixture of two Gaussian distributions to the data using the R package "mixtools" [13], estimating the mean, variance, and proportion of the population with slow and fast growth (respectively, red and green curves in Fig. 5).

3.3.2 Determine Cut-Offs

After estimating the distribution of fast and slow growth time points, we find the intercept between the distributions which sets a cut-off that defines growth arrest as points below this threshold value (Fig. 5 magenta line). Removing these points is an adequate choice since during growth arrest the cytosolic Ca^{2+} concentration is either abnormally high at the tip or low and homogeneous throughout the tube (Fig. 4c and [7]). This filtering allows defining a spatiotemporal region of interest in the kymograph with a higher degree of representativity (Fig. 4d). Given that the goal is to average the spatial profile across all time points, pronounced

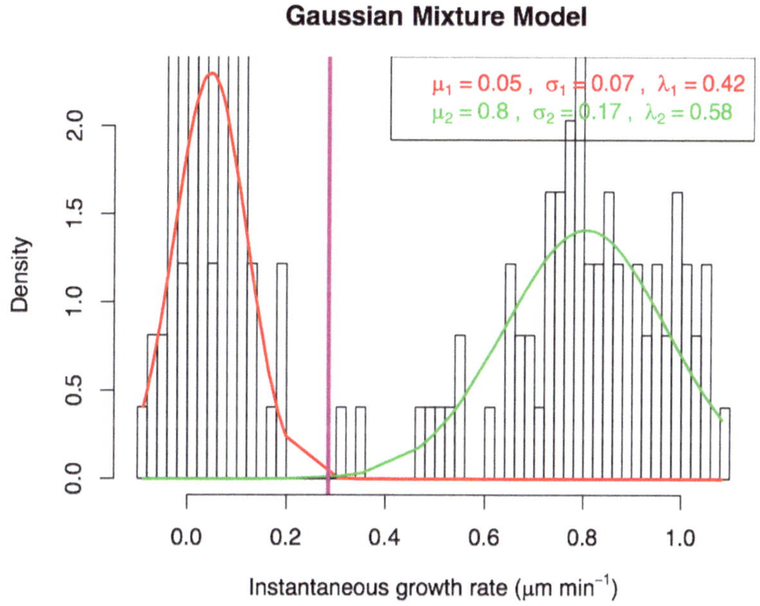

Fig. 5 Gaussian Mixture Model of the instantaneous growth rate across all time points. Pollen tube growth rate determined with CHUKNORRIS can be split into growing (green) and nongrowing (red) regimes by fitting a mixture of two gaussian distributions using the R package "mixtools." The intersection between both distributions (magenta line) yields an objective criterion to filter out time points where there is growth arrest

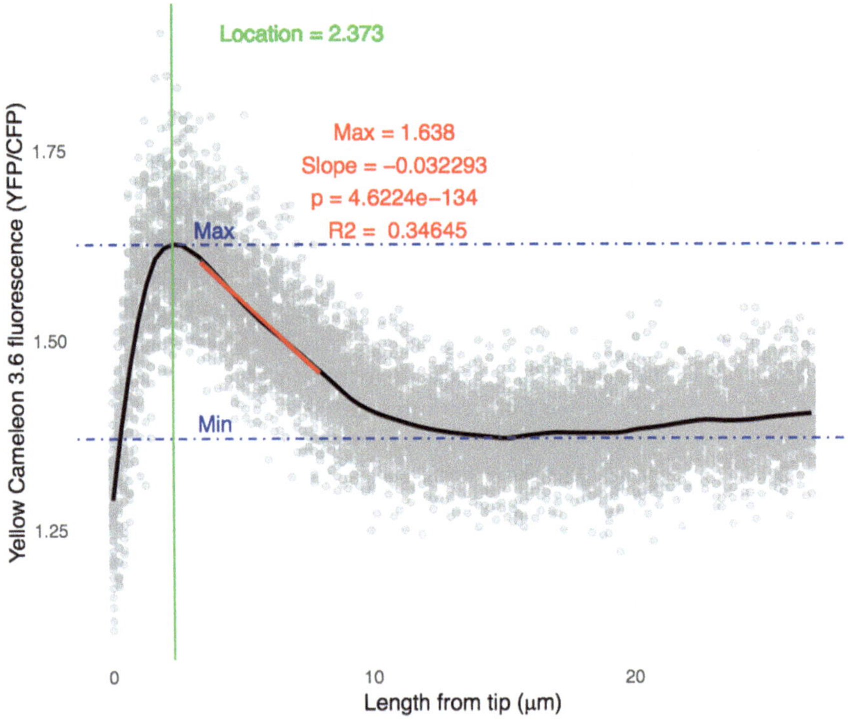

Fig. 6 Average gradient and parameter estimates. Ratiometric fluorescence values of all time points ($n = 65$) of a tip aligned kymograph (gray points) smoothed with a weighted local polynomial fit, "loess" (black line) allows to estimate the average minimum (blue), maximum (blue), and its location (green). The gradient steepness is estimated by fitting a linear model to the linear part of the fluorescence decay (red segment encompassing 15 pixels and text)

oscillations in intracellular gradient should be analyzed separately (*see* **Note 3**).

3.4 Reconstruct Intracellular Gradient

3.4.1 Fit Local Polynomials

Once a sufficiently representative number of time points has been chosen, and the length of the kymograph has been trimmed to the maximum allowing all time points to be compared (Fig. 4d), we can estimate the average gradient. This was done with "loess" (Data S1), a local polynomial fit resembling a moving average but with linear or quadratic fits performed in each window [11]. This smoothing technique allows to approximate the shape of a complex function by weighing a series of local regressions, yielding an estimate of the average gradient (Fig. 6).

3.4.2 Estimate Gradient Parameters

Key parameters of the gradient can readily be obtained by quantifying the basic landmarks (Fig. 1), as the minimum, maximum, and their location (Fig. 6; Data S1). Estimating the slope, or the gradient per se, requires fitting the linear portion of the fluorescence decay. The choice of this region can be done objectively by choosing a number of pixels to fit, (here 15 pixels, with 65 time points in

Fig. 6) and finding which segment has the mean sharpest decay (Fig. 6; Data S1). Next, the real values of position are subtracted by the location of the maximum fluorescence, which will turn the intercept estimated in the linear model into another proxy of the maximum. Finally, the linear regression is performed allowing a quantitative assessment of the gradient steepness (slope in Fig. 6), which only requires calibrated data to correspond to real units. This protocol can be adapted to treat a massive amount of data, where population statistics and phenotyping can be performed with these individual estimates.

4 Notes

1. The estimate of tip location is not independent of the Ca^{2+} signal in this case, as both rely on the same fluorescent reporter. Thus, estimates of gradient location may become unreliable if there are phenotypes with drastic changes Ca^{2+} gradient location. Ideally an independent reporter should be used for tip detection, while bright field tracking is an option that offers greater tracking challenges.
2. Fluorescence data can contain many types of artifacts that compromise the reliability of this gradient analysis. Photobleaching of the fluorophores, for example, is often present to some extent creating a long-term decline in the fluorescence baseline. Although correction can be attempted by fitting simple expressions, or even with "loess," distortions are introduced and may dominate the signal, especially in the ratiometric case.
3. Averaging may become meaningless if there are pronounced oscillations in intracellular gradients, such as observed during growth arrest. However, the data do not have to be discarded as the gradient can be quantified in distinct populations, for example during and after spiking.

Acknowledgments

This work was supported by the National Science Foundation (MCB 1616437/2016 and 1930165/2019) and the University of Maryland.

References

1. Feijó JA, Malho R, Obermeyer G (1995) Ion dynamics and its possible role during in-vitro pollen germination and tube growth. Protoplasma 187:155–167
2. Feijó JA, Sainhas J, Holdaway-Clarke T, Cordeiro MS, Kunkel JG, Hepler PK (2001) Cellular oscillations and the regulation of growth: the pollen tube paradigm. BioEssays 23:86–94

3. Gutermuth T, Lassig R, Portes M, Maierhofer T, Romeis T, Borst J, Konrad KR (2013) Pollen tube growth regulation by free anions depends on the interaction between the anion channel SLAH3 and calcium-dependent protein kinases CPK2 and CPK20. Plant Cell 25:4525–4543
4. Gutermuth T, Herbell S, Lassig R, Brosché M, Romeis T, Feijó JA, Hedrich R, Konrad KR (2018) Tip-localized Ca^{2+} channels control pollen tube growth via kinase-dependent R- and S-type anion channel regulation. New Phytologist 218:1089–1105
5. Damineli DSC, Portes MT, Feijó JA (2017a) One thousand and one oscillators at the pollen tube tip: the quest for a central pacemaker revisited. In: Obermeyer G, Feijó J (eds) Pollen tip growth. Springer, Cham
6. Nelson DL, Lehninger AL, Cox MM (2008) Lehninger principles of biochemistry. W.H. Freeman, New York, NY
7. Damineli DSC, Portes MT, Feijó JA (2017b) Oscillatory signatures underlie growth regimes in Arabidopsis pollen tubes: computational methods to estimate tip location, periodicity, and synchronization in growing cells. J Exp Bot 68(12):3267–3281
8. Iwano M, Entani T, Shiba H, Kakita M, Nagai T, Mizuno H, Miyawaki A, Shoji T, Kubo K, Isogai A (2009) Fine-tuning of the cytoplasmic Ca2+ concentration is essential for pollen tube growth. Plant Physiol 150:1322–1334
9. Wudick MM, Portes MT, Michard E, Rosas-Santiago P, Lizzio MA, Nunes CO, Campos C, Damineli DS, Carvalho JC, Lima PT, Pantoja O, Feijó JA (2018) CORNICHON sorting and regulation of GLR channels underlie pollen tube Ca2+ homeostasis. Science 360 (6388):533–536
10. Schindelin J, Arganda-Carreras I, Frise E, Kaynig V, Longair M, Pietzsch T, Preibisch S, Rueden C, Saalfeld S, Schmid B, Tinevez JY (2012) Fiji: an open-source platform for biological-image analysis. Nat Methods 9 (7):676
11. Core Team R (2019) R: a language and environment for statistical computing. R Foundation for Statistical Computing, Vienna. https://www.R-project.org/
12. Wickham H (2017) tidyverse: easily install and load the 'Tidyverse'. R package version 1.2.1. https://CRAN.R-project.org/package=tidyverse
13. Benaglia T, Chauveau D, Hunter DR, Young D (2009) mixtools: an R package for analyzing finite mixture models. J Stat Softw 32(6):1–29. http://www.jstatsoft.org/v32/i06/
14. RStudio Team (2019) RStudio: integrated development for R. RStudio, Inc., Boston, MA. http://www.rstudio.com/

Chapter 15

Silicone Chambers for Pollen Tube Imaging in Microstructured In Vitro Environments

Hana Bertrand-Rakusová, Youssef Chebli, and Anja Geitmann

Abstract

Live cell imaging at high resolution of pollen tubes growing in vitro requires an experimental setup that maintains the elongated cells in a single optical plane and allows for controlled exchange of growth medium. As a low-cost alternative to lithography-based microfluidics, we developed a silicone-based spacer system that allows introducing spatial features and flexible design. These growth chambers can be cleaned and reused repeatedly.

Key words Live cell imaging, Fluorescence imaging, Micromanipulation, Silicone chambers

1 Introduction

Pollen tube elongation is characterized by unidirectional growth that is spatially confined to the very tip of the cell where delivery of membrane and cell wall precursors by exocytosis and the assembly of cell wall polymers occur [1, 2]. Maintaining cell polarity and producing a directional response to external signals involves multiple signaling processes [3, 4]. These characteristics make the pollen tube an ideal model cell system to study tip growth, endomembrane trafficking, and cell wall assembly. Monitoring intracellular processes and cellular growth behavior of the growing pollen tube is possible in vitro and requires high-resolution live cell imaging. Because of the constantly increasing length of the cell, high-resolution time-lapse imaging of pollen tubes can be challenging as the pollen tube tip does not necessarily stay in a given focal plane. Depending on the size of the pollen grain, the space available in *z*-direction can be confined by sandwiching the cells tightly between slide and coverslip, but this has the potential to stress the cells thus potentially causing growth arrest or bursting. Also, a simple slide observation system is very limited in terms of the design of an experimental assay and does not readily allow exchange of medium

Anja Geitmann (ed.), *Pollen and Pollen Tube Biology: Methods and Protocols*, Methods in Molecular Biology, vol. 2160, https://doi.org/10.1007/978-1-0716-0672-8_15,

without risking the displacement of the cells. Over the past decade, microfluidics and MEMS technology have enabled the design of sophisticated experimental devices that allow for high resolution imaging and experimentation on individual cells in microstructured environments. This has been accomplished both with commercial chambers [5] and in-house produced lithography-based devices [6–17]. The commercial chambers are affordable but do not offer much flexibility in terms of structural design features. In-house produced microfluidics and MEMS devices can be produced en masse once a design is optimized functionally, but each new design is quite costly, time consuming to produce, and the fabrication requires relevant expertise.

To be able to fabricate devices for experiments for rapid screening of multiple pollen tubes as is necessary for screening of mutant lines, we adopted a technology based on a micropatterned silicone spacer. The device is fabricated by cutting out desired features from a silicone sheet and sandwich it between two cover slips. The pattern is cut using a commercial blade cutter which limits the spatial resolution of the features. However, for experiments that do not require features with extremely high spatial resolution this approach is an easy and low-cost alternative to lithography-based microfluidics.

The shape of the chamber is designed in the program provided by the manufacturer of the blade cutter. The silicone sheet easily adheres to the glass or directly to the cover slip and can be used to create a chamber in which pollen tube growth is restricted to a narrow vertical space facilitating time-lapse imaging of growing tubes. The cutting device affords the possibility to adjust speed, force, initial down force, and blade thickness and it is able to cut silicone sheets up to 250 μm thickness. A sheet thickness of 100 μm works well for most pollen species.

With respect to conventional in vitro setups, the silicone chambers offer the same advantage as microfluidic devices: the liquid volume in the experimental chamber is very small, minimizing the consumption of solutions (for instance antibodies used for immunofluorescence label). Also, several independent chambers can be placed on one single microscope slide providing identical experimental conditions for all samples (temperature, humidity, light, etc.). The sandwich structure limits evaporation of the medium while imaging. If staining, washing, or drug treatments are to be performed during imaging, inlets and outlets can be incorporated in the covering layer to allow for access.

In this chapter, we first describe the steps required to design and fabricate the chambers. We explain the considerations made to determine the design features. As an example, we describe a simple method based on the use of a liquid *Camellia* pollen culture.

2 Materials

2.1 Material for Spacer Cutting

1. GRAPHTEC® Cutting plotter CE6000-40.
2. Transparent silicone sheet 0.25 mm ±5%, SiN 400T—Silicone; STERNE®, Silicone performance.
3. CD with the Graphtec Studio program (provided with the cutter). Alternatively, download the software for free from Graphtec website.
4. Polystyrene Petri dish with clear lid, raised ridge, 60 mm × 15 mm.
5. Tweezers.
6. Empty pipette tip box.
7. Kimwipes®.
8. Parafilm®.
9. Coverslips #1.5, thickness (0.17–0.18 mm).

2.2 Pollen Germination Medium for Camellia

1. 10× stock solution: 100 μg/ml H_3BO_3, 300 μg/ml $Ca(NO_3)_2$, 100 μg/ml KNO_3, 200 μg/ml $MgSO_4{\cdot}7H_2O$.
2. Sucrose.
3. Double distilled H_2O.
4. Low-melting agarose (optional).
5. Heating block (if using agarose).

2.3 Imaging

1. Inverted epifluorescence microscope (here: Nikon Eclipse TE 2000).
2. Objectives: 4×, 20×, 40×.

3 Methods

All procedures can be carried out in the laboratory at room temperature.

3.1 Cutting of Spacers and Mounting of Chambers

1. Install the Graphtec Studio software on the computer. Using Graphtec Studio, design the chamber shape and size specific for the intended application (Fig. 1). The toolbars on the left panel allow designing geometrical features by placing and combining shapes. Clicking on the pattern reveals the dimensions. To change length units when working on a PC, choose "Preferences," "Measurements," and select the preferred units. On a Mac, choose the cutters window icon to show the cutters panel. Choose Advanced (CE6000-40) and select the length unit. *See* also **Note 1**.

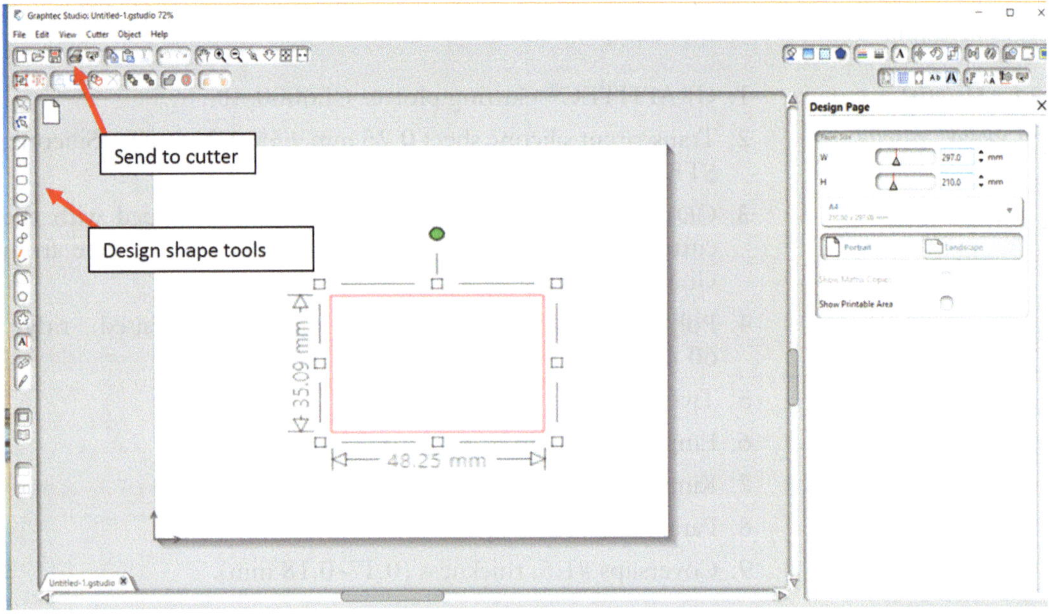

Fig. 1 Graphtec Studio user interface. The dimensions of the designed shape are shown when clicking on the shape

2. The cutting process of the silicone chamber is illustrated in Fig. 2. Start by removing one side of the protective foil from the silicone sheet and insert it in the cutter with the silicone on top. Load the silicone sheet by moving the wheels up using the handle at the back—place the silicone sheet and secure it by putting the wheels back down. *See* also **Note 2**.
3. Position the scalpel above the desired area. On the screen select cutting from the "Current position."
4. To build a chamber, two options exist. Either only the spacer layer is designed and fabricated from silicone and placed between two cover slips (closed chamber) or both a spacer layer and a cover layer with inlets and outlets are cut from silicone. In this case, the cover layer replaces the top cover slip in the final mount (Figs. 3a, b). *See***Notes 2** and **3**.
5. Use tweezers to place the silicone spacer layer on the cover slip. The silicone adheres to the glass and the seal prevents any liquids from escaping the device. Place the silicone spacer delicately on the cover slip to avoid the formation of air bubbles.
6. For experiments requiring longer observation, it is preferable to use the Petri dish with glass bottom instead. Also, place a piece of wet tissue inside the petri dish and seal with Parafilm® to avoid drying (Fig. 3c, d).

Fig. 2 Table top GRAPHTEC Cutting plotter CE6000-40. (**a**) Overview of the device. (**b**) Inserted silicone sheet with the silicone layer on the top. (**c**) Blade (arrow) cutting the designed shape into the silicone layer

3.2 Growth Medium Preparation

1. The 10× concentrated stock solution without sucrose is ideally stored in 1 ml aliquots at −20 °C. Make sure all salts are dissolved before aliquoting (*see***Notes 4** and **5**).
2. Prepare the growth medium from the 10× stock solution by adding doubly distilled water to dilute to 1× and adding 5% sucrose.
3. If solid medium is required, 0.5% or 1% of low melting agarose can be added. Heat the mixture to dissolve (on a heating block or in hot water) and let it cool down to just above gelling temperature before adding the pollen grains. Wear appropriate protective equipment such as goggles and heat-protective gloves as hot agarose can splash.

3.3 Pollen Hydration and Germination

1. Collect, dehydrate, and store *Camellia japonica* pollen at −20 °C for later use (*see* **Note 6** and consult Chapter 13 for details on pollen storage).
2. Prior to experimentation, thaw and rehydrate a few milligrams of pollen in humid atmosphere for 30 min. To do so transfer

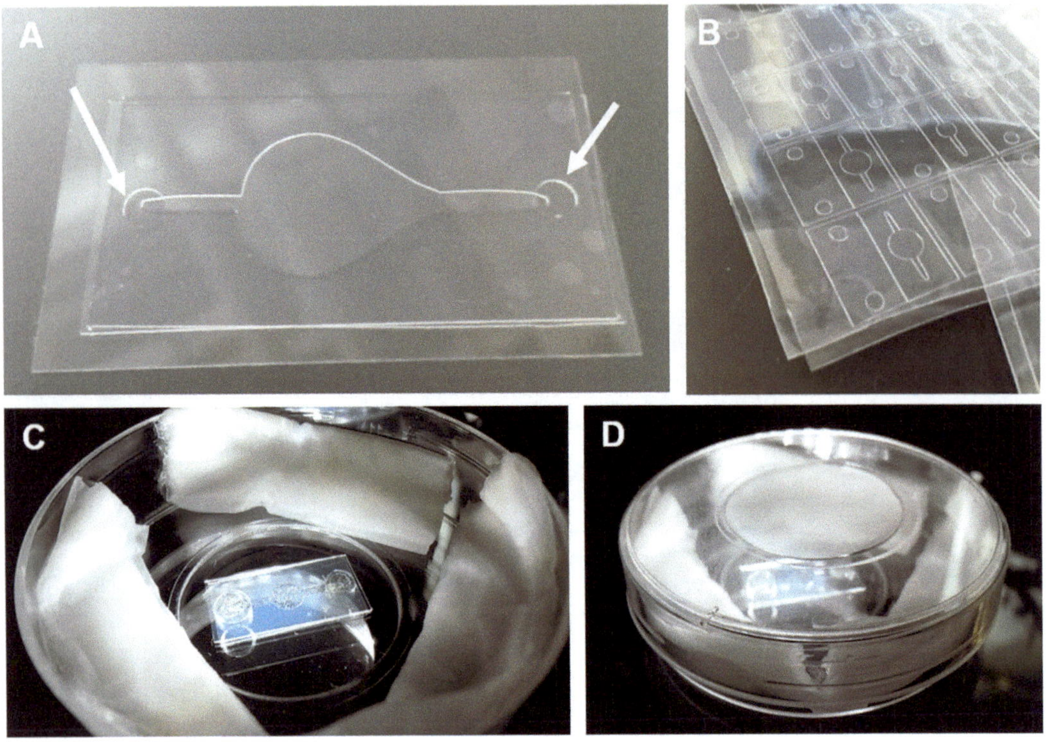

Fig. 3 Mounting of experimental chamber. (**a**) Chamber created using cover slip at the bottom, silicone spacer layer with patterned cut out, and cover silicone layer including inlet and outlet (arrows). (**b**) Patterned spacer and cover layers can be cut in series. (**c**) For experiments requiring longer observation periods, the bottom layer is the glass bottom of the Petri dish and wet Kimwipes® are positioned at the margin of the Petri dish. (**d**) The Petri dish is kept closed and sealed with Parafilm®

pollen on the small Petri dish and place in the upper compartment in the empty pipette tip box. Pour hot tap water in the lower compartment. Keep the box closed to maintain a humid atmosphere. Consult Chapter 13 for alternative ways to hydrate pollen.

3. Suspend the pollen grains in 1 mL liquid growth medium. Pollen density is critical for this type of assay and must be adjusted based on preliminary tests and the desired type of assay. Agitate gently to mix the suspension, pipet the solution in the chamber, and close it with the second layer.
4. To keep the environment humid and prevent drying out the solutions, insert a wet piece of paper tissue next to the chamber, close the lid of the Petri dish, and seal with Parafilm®.
5. Place the microfluidic platform under the microscope to limit mechanical manipulation during pollen germination. While it is possible to let the pollen germinate on the bench, placing the chamber on the microscope immediately minimizes environmental changes and shaking during the transfer which could

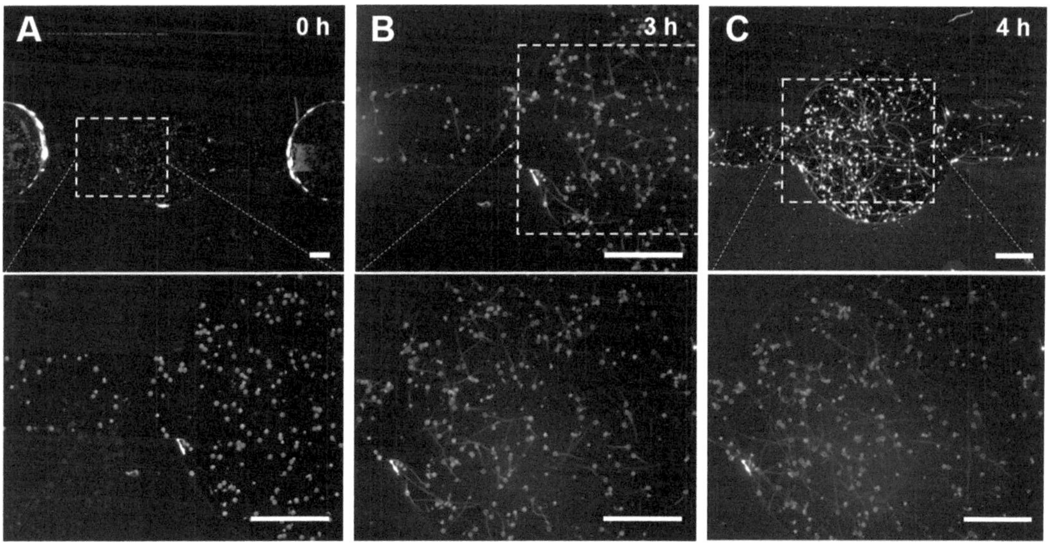

Fig. 4 Fluorescence micrographs of germinating *Camellia* pollen in silicone chamber. Bottom row images show close-ups of the respective images in the upper row. (**a**) Pollen should ideally be distributed evenly in the chamber to create uniform conditions. (**b**) At 3 hours after injection pollen grains have germinated and formed tubes with substantial length. (**c**) At 4 h after injection even very long pollen tubes are kept in horizontal position and in close proximity to the cover slip to facilitate in focus imaging of the entire pollen tube. Scale bars = 1 mm.

cause problems as elongated pollen tubes are sensitive to manipulation (*see* **Notes** 7 and **8**).

6. Starting one hour after incubation, check germination repeatedly every half hour to verify whether the desired pollen tube length has been reached (Fig. 4). *Camellia* pollen usually germinates after 30 min.
7. Perform the desired experimental assay.

3.4 Experimental Assays

Depending on the intended experimental strategy, the chamber dimensions and design features need to be adapted to the pollen species and the application. The following considerations should be made:

- At what length are the tubes to be observed?
  - Depending on the length of the pollen tubes, the density of grains should be adjusted to avoid overcrowding.
- Will the experiment require exchange of medium, addition of solutions?
  - The upper layer of the experimental chamber should be designed with holes (inlet and outlet) to allow exchange of solutions. Two sample designs are provided (Figs. 5 and 6).
- Does a chemical gradient need to be created?

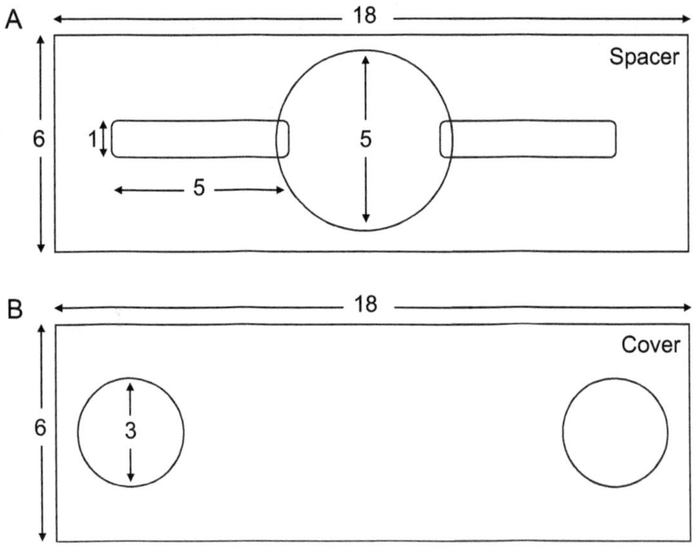

Fig. 5 Schematic design of a simple chamber for the germination of *Camellia* pollen grains as seen in Fig. 4. (**a**) Spacer layer with two channels connected to a central chamber used for the observation of pollen tube growth. (**b**) Cover layer with inlet and outlet. Numbers indicate dimensions in mm

- A static chemical gradient can be created by placing inlets strategically and supplying different solutions. However, this does not allow for precise control of the local concentration or the creation of sharp gradients. Sharp gradients require laminar flow [12].

- Will liquid or solid medium be used?
  - If solid medium is to be used, the hydrated pollen is mixed with warm (38–45 °C) medium including dissolved low-melting agarose. Tight control of the temperature is crucial to ensure satisfactory pollen germination. The injection of the warm solution into the chamber should be done on a heating block to prevent precocious solidification before the medium is evenly distributed in the chamber.

3.5 Solution Exchange

Solutions can be exchanged in chambers designed with openings in order to refresh the medium, to stain the pollen tubes, to apply pharmacological agents, or to chemically fix the cells. Place a tip of a Kimwipes® at one opening to suck the medium and apply the solution of choice at another opening by pipetting slowly. This can be done directly on the inverted microscope.

3.6 Reusing the Chambers

Chambers as well as the Petri dishes can be washed with ethanol and soap and rinsed with double distilled H_2O. After proper drying, they can be used again.

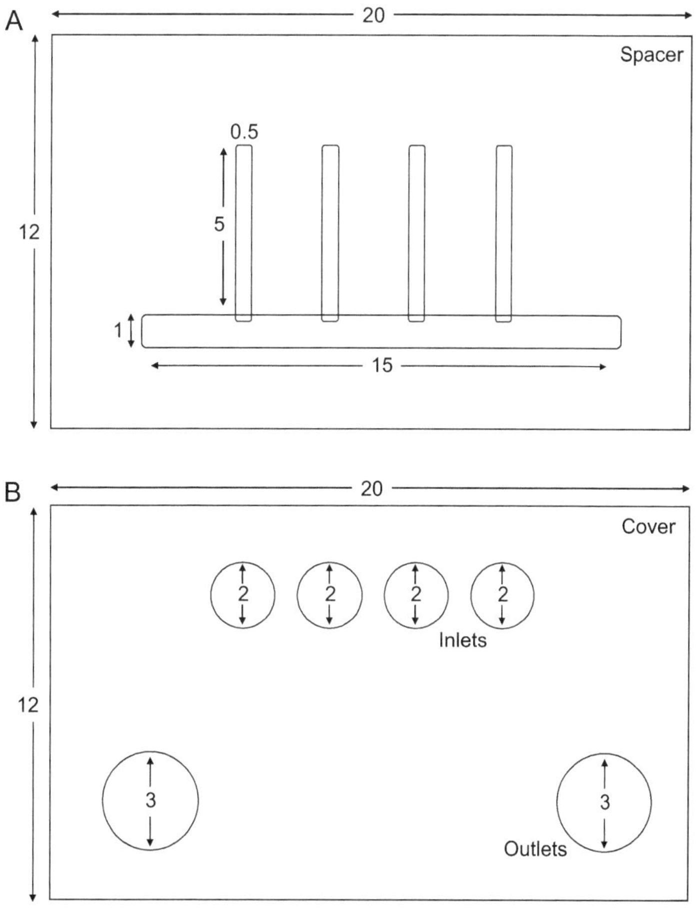

Fig. 6 Second example of a chamber design for germination of *Camellia* pollen grains. Parallel treatments can be applied to pollen separated into four channels. (**a**) Spacer layer with four smaller channels for experimentation and one channel to collect liquid toward outlets. (**b**) Cover layer with four inlets and two outlets. Assembly: (1) Mount the spacer layer (**a**) on a cover slip. (2) Pipet pollen suspension in each experimentation channel. (3) Mount cover layer on spacer carefully aligning inlets and outlets with channels. Each experimentation channel can be used to inject a different solution. Numbers indicate dimensions in mm

4 Notes

1. Be careful to use simple curves, as small dimension arcs are difficult to cut. We have noticed that straight cuts can be smaller than curved designs. The smallest radius that can be cut is around 0.5 mm.
2. The plotter has sensors at the front and the back of the blade. Make sure that the silicone sheets cover both sensors (the error

message indicating that this is not the case says "Lead Media"). If the design comprises holes to create inlets and outlets, they can be cut using the cutter or they can be punched with a punch tool.

3. Once a design is optimized and tested, it can be cut in series to save time and material (Fig. 3b).
4. This medium is suitable for many types of pollen (e.g., *Nicotiana*, *Papaver*, *Petunia*, *Camellia*) but it does not work for monocots such as lily, and it is not optimal for *Arabidopsis thaliana*.
5. Dissolving all the crystals can take a while (even overnight).
6. *Camellia japonica* pollen was harvested once a year from a plant at the Botanical Garden of Montreal.
7. The sample has to be handled with care. Pollen tubes are sensitive to mechanical stress and can burst if the microscope slide hits the surface.
8. Pollen tubes can be imaged using an inverted or upright microscope. An inverted microscope offers the possibility to apply solutions or interfere in other ways during the experiment more easily than on an upright microscope.

References

1. Rakusova H, Geitmann A (2017) Control of cellular morphogenesis through intracellular trafficking. In: Obermeyer G, Feijó J (eds) Pollen tube tip growth: from biophysical aspects to systems biology. Springer, New York, NY, pp 129–148
2. Chebli Y, Kroeger J, Geitmann A (2013) Transport logistics in pollen tubes. Mol Plant 6:1037–1052
3. Franklin-Tong VE (1999) Signaling and the modulation of pollen tube growth. Plant Cell 11:727–738
4. Li HJ, Meng JG, Yang WC (2018) Multilayered signaling pathways for pollen tube growth and guidance. Plant Reprod 31:31–41
5. Chebli Y, Geitmann A (2015) Live cell and immuno-labeling techniques to study gravitational effects on single plant cells. In: Blancaflor E (ed) Plant gravitropism, Methods in molecular biology. Humana, New York, NY, pp 209–226
6. Agudelo CG, Packirisamy M, Geitmann A (2016) Influence of electric fields and conductivity on pollen tube growth assessed via electrical lab-on-chip. Sci Rep 6:19812
7. Agudelo CG, Sanati Nezhad A, Ghanbari M, Naghavi M, Packirisamy M, Geitmann A (2013) TipChip – a modular, MEMS (microelectromechanical systems)-based platform for experimentation and phenotyping of tip growing cells. Plant J 73:1057–1068
8. Agudelo CG, Packirisamy M, Geitmann A (2013) Lab-on-a-chip for studying growing pollen tubes. In: Žárský V, Cvrčková F (eds) Plant cell morphogenesis: methods and protocols, Methods in molecular biology. Springer, New York, NY, pp 237–248
9. Agudelo CG, Sanati Nezhad A, Ghanbari M, Packirisamy M, Geitmann A (2012) A microfluidic platform for the investigation of elongation growth in pollen tubes. J Micromech Microeng 22:115009
10. Geitmann A (2017) Microfluidics and MEMS (microelectromechanical systems)-based platforms for experimental analysis of pollen tube growth behavior and quantification of cell mechanical properties. In: Obermeyer G, Feijó J (eds) Pollen tube tip growth: from biophysical aspects to systems biology. Springer, New York, NY, pp 87–103
11. Ghanbari M, Sanati Nezhad A, Agudelo CG, Packirisamy M, Geitmann A (2014) Microfluidic positioning of pollen grains in lab-on-a-chip for single cell analysis. J Biosci Bioeng 117:504–511

12. Sanati Nezhad A, Packirisamy M, Geitmann A (2014) Dynamic, high precision targeting of growth modulating agents is able to trigger pollen tube growth reorientation. Plant J 80:185–195
13. Sanati Nezhad A, Ghanbari M, Agudelo CG, Naghavi M, Packirisamy M, Bhat R, Geitmann A (2014) Optimization of flow assisted entrapment of pollen grains in a microfluidic platform for tip growth analysis. Biomed Microdevices 16:23–33
14. Sanati Nezhad A, Naghavi M, Packirisamy M, Bhat R, Geitmann A (2013) Quantification of cellular penetrative forces using lab-on-a-chip technology and finite element modeling. PNAS 110:8093–8098
15. Sanati Nezhad A, Naghavi M, Packirisamy M, Bhat R, Geitmann A (2013) Quantification of the Young's modulus of the primary plant cell wall using Bending-Lab-On-Chip (BLOC). Lab Chip 13:2599–2608
16. Sanati Nezhad A, Ghanbari M, Agudelo CG, Packirisamy M, Bhat RB, Geitmann A (2012) PDMS microcantilever-based flow sensor integration for lab-on-a-chip. IEEE Sensors J 13:601–609
17. Yanagisawa N, Higashiyama T (2018) Quantitative assessment of chemotropism in pollen tubes using microslit channel filters. Biomicrofluidics 12:024113

Chapter 16

Analyzing the Impact of Protein Overexpression on Ca^{2+} Dynamics and Development in Tobacco Pollen Tubes

Chunxia Zhang, Leonie Steinhorst, and Jörg Kudla

Abstract

Overexpression of RFP-tagged proteins in growing tobacco pollen tubes together with the genetically encoded Ca^{2+} sensor YC3.6 allows to analyze localization and dynamics of the protein of interest, as well as the impact of its overexpression on Ca^{2+} dynamics and pollen tube growth. Here, we describe a step-by-step instruction for transient transformation of *N. tabacum* pollen and subsequent in vitro germination and Ca^{2+} imaging.

Key words Tobacco pollen tube, Gold particle bombardment, Ca^{2+} indicator YC3.6, Ca^{2+} dynamics, FRET Ca^{2+} imaging

1 Introduction

The male gametophyte of plants represents a highly specialized cell type. Pollen tubes growing out of a pollen grain exhibit tip growth, which depends on the establishment and maintenance of an oscillating tip-focused Ca^{2+} gradient [1]. Many processes, such as organization of the cytoskeleton, vesicle dynamics, and ion fluxes across membranes, have to be tightly controlled to enable pollen germination, directed tube growth, and eventually successful fertilization. They are all controlled and integrated by signalling networks that rely on second messengers, and among them, Ca^{2+} has important functions in pollen tube growth and development [2]. Heterologous expression of fluorophore-tagged proteins of interest in growing tobacco pollen tubes is a useful tool to analyze their function in this specialized cell type. Simultaneous Ca^{2+} imaging can additionally reveal if overexpression of these fluorophore-tagged proteins affects pollen tube Ca^{2+} dynamics.

Electronic supplementary material: The online version of this chapter (https://doi.org/10.1007/978-1-0716-0672-8_16) contains supplementary material, which is available to authorized users.

Anja Geitmann (ed.), *Pollen and Pollen Tube Biology: Methods and Protocols*, Methods in Molecular Biology, vol. 2160, https://doi.org/10.1007/978-1-0716-0672-8_16, © Springer Science+Business Media, LLC, part of Springer Nature 2020

Here, we describe a hands-on approach for the analysis of the impact of protein overexpression on the pollen tube's Ca^{2+} dynamics and polar tip growth, as performed in Steinhorst et al. [3]. For this, the protein of interest is fused to a red fluorophore—such as mCherry, and co-expressed with Ca^{2+} indicator YC3.6 in growing tobacco pollen tubes. Transient transformation of the pollen is performed via bombardment with plasmid-coated gold particles [4–6]. After bombardment the transformed pollen are cultivated for germination in a setup that allows for high-quality imaging of the Ca^{2+} dynamics [7–9].

2 Materials

2.1 Gold Particles and Plasmid DNA Preparation

1. Gold particles: Wash 50 mg gold particles with 1 mL absolute ethanol by vortexing for 60 sec and centrifugation at maximum speed (13,400 × *g*) in a benchtop centrifuge for 10 s. Repeat wash step twice. Then wash gold particles with 1 mL sterilized double-distilled H_2O (ddH_2O) by vortexing 60 s and centrifugation at maximum speed for 10 s. Resuspend gold particles in 1 mL ddH_2O and divide into 25 μL aliquots. Store at 4 °C (*see* **Note 1**).
2. DNA constructs: Expression of the YC3.6 calcium sensor is driven by the LAT52 pollen tube specific promoter [10]. Expression of mCherry alone as control or fused with the protein of interest is also driven by the LAT52 promoter.
3. Plasmid DNA: Purify plasmid DNA by a plasmid midi prep kit with a final concentration of approximately 1 μg/μL (*see* **Note 2**).
4. 0.1 M spermidine: A 1 M spermidine stock solution should be prepared first and divided into aliquots of 100 μL each, store at −80 °C. 0.1 M spermidine can be prepared by diluting 1 M spermidine stock solution with ddH_2O. Prepare 50 μL aliquots and store them at −20 °C (*see* **Note 3**).
5. 2.5 M $CaCl_2$: Weigh 0.2775 g $CaCl_2$ and dissolve it in 1 mL ddH_2O. Make aliquots 100 μL each and store at 4 °C (*see* **Note 4**).

2.2 Pollen Germination Medium

Pollen germination medium recipe with final concentrations [8, 11]: 1 mM MES, 0.2 mM $CaCl_2$, 1.6 mM H_3BO_3, adjust pH to 5.8 with Tris, 130 g D (+)-sucrose per 1 L solution (13%). All ingredients are dissolved in Milli-Q water (*see* **Note 5**).

2.3 Pollen Grain Harvest and Cultivation

1. 50 mL falcon tube.
2. Medical scissors.
3. Gauze.

4. Cellulose acetate filters.
5. Whatman filter paper.
6. Petri dish (90 × 15 mm).
7. Vacuum pump system with vacuum flask.
8. Glass-bottom dish (29 mm, with 14 mm micro-well).
9. Cover slip.
10. Kitchen paper.
11. Styrofoam box.
12. Low melt point agarose.

2.4 Bombardment Transformation

1. Helium-driven particle accelerator (PDS-1000/He; Bio-Rad).
2. 1350 psi rupture discs (Bio-Rad).
3. Macrocarrier.
4. Stopping screens.
5. Clean bench.
6. Isopropanol.

2.5 Calcium Imaging

1. Epifluorescence microscope, for example an inverted ZEISS Axio 239 observer (Carl Zeiss Microimaging GmbH, Goettingen, Germany), equipped with an 240 emission filter wheel (LUDL Electronic Products, Hawthorne, NY, USA) and a 241 Photometrics cool SNAPHQ2 CCD camera (Tucson, Arizona). 485 nm and 535 nm emission filters are used for CFP and YFP, respectively. 561 nm is used for exciting mCherry.
2. 40×/0.75 Dry objective, 440 nm excitation supplied by a xenon short arc reflector lamp (Hamamatsu, Japan),
3. Metafluor software (Meta Imaging series 7.7 from molecular devices, Dowingtown, PA).
4. ImageJ software.

3 Methods

3.1 Coating Gold Particles with Plasmid DNA

1. Vortex 1.25 mg of gold particles (in 25 μL ddH_2O) extensively for 3 min.
2. Add 10 μL of 0.1 M spermidine to the gold particles and vortex for 1 min.
3. While vortexing, add 3–5 μg of each plasmid to gold particles. Vortex for 1 min (*see* **Note 6**).
4. Slowly add 25 μL of 2.5 M $CaCl_2$ while constantly vortexing. Vortex for 3 min.

5. Precipitate gold particles by centrifugation at maximum speed (13,400 × *g*) for 10 s and remove the supernatant.
6. Wash gold particles with 200 μL of absolute ethanol and precipitate by centrifugation at maximum speed (13,400 × *g*) for 10 s.
7. Remove the supernatant and resuspend the gold particles in 18 μL of absolute ethanol (*see* **Note 7**).
8. The gold particles can be either immediately used, or stored at 4 °C overnight (*see* **Note 8**).

3.2 Collection, Transformation, and Cultivation of Pollen Grains

1. Harvest mature pollen grains from 8–12 flowers of 6- to 8-week-old tobacco (*Nicotiana tabacum*) plants by cutting the anthers and transferring them into 2 mL of pollen germination medium in a 50 mL falcon tube (*see* **Note 9**).
2. Vortex vigorously for 1 min to release the pollen grains into the medium. Remove anthers by filtering with gauze (Fig. 1).
3. Filtering of pollen grains onto cellulose acetate filters by vacuum filtration:
   (a) Connect the running water tap with the filtering flask by rubber pipes (Fig. 2A).
   (b) Mount one cellulose acetate filter paper and turn on the vacuum system. Pre-wet the cellulose acetate filter paper with 2 mL pollen germination medium (Fig. 2B, C) (*see* **Note 10**).

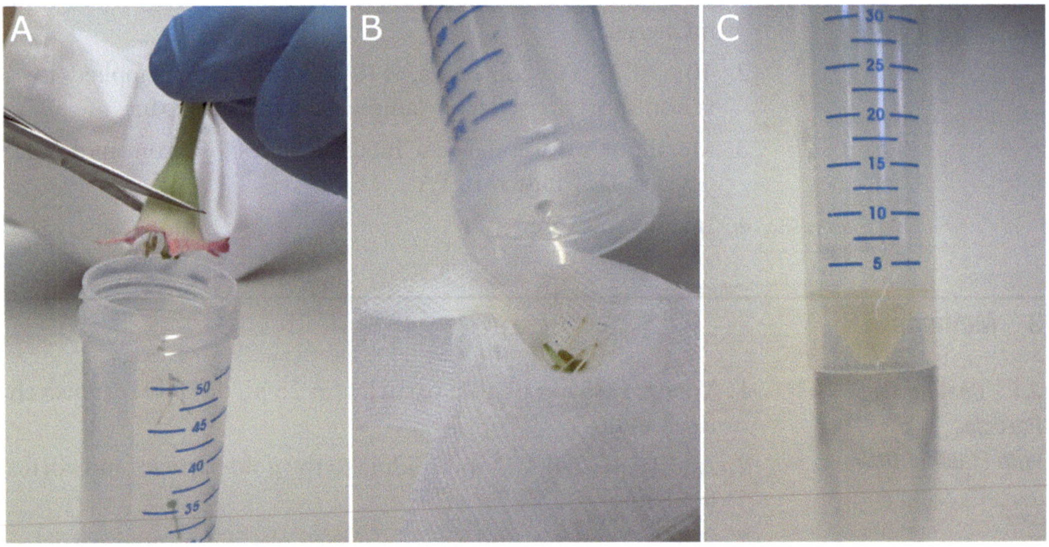

Fig. 1 Tobacco pollen grain collection. (**A**) Cutting the anthers from 6- to 8-week-old tobacco flowers. (**B**) Removing anthers by filtering with gauze. (**C**) Pollen grains were released into the medium

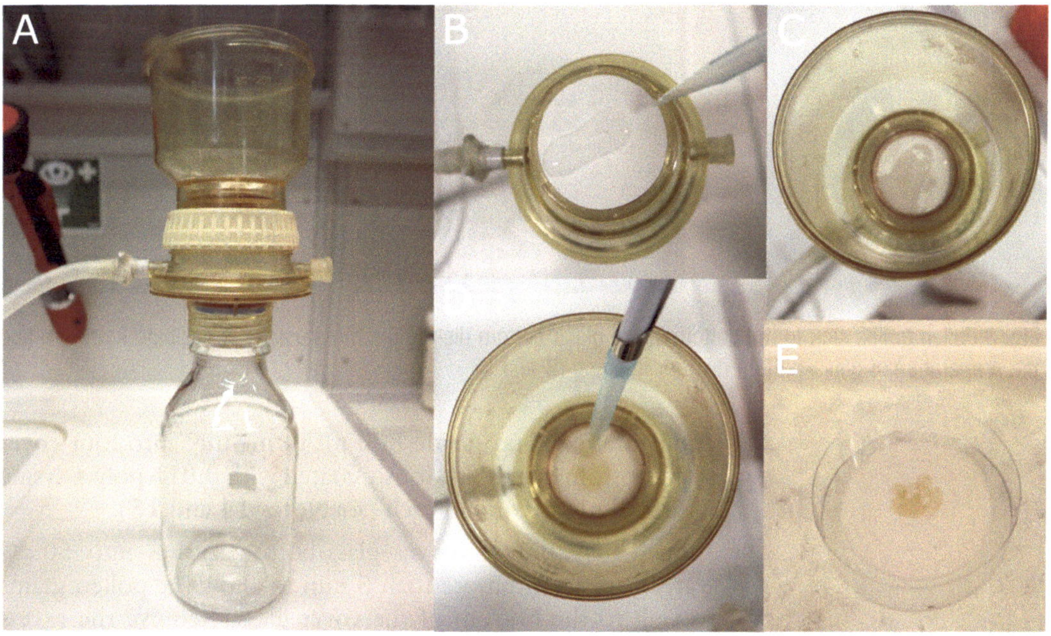

Fig. 2 Filtering pollen grains onto the cellulose acetate filter by vacuum filtration. (**A**) Vacuum pump system with vacuum flask connected by rubber pipes to the running water tap. (**B**) Pre-wetting the mounted cellulose acetate filter paper with 2 mL pollen germination medium. (**C**) Turning on the vacuum system. (**D**) Pipietting the pollen grain suspension onto the pre-wet filter paper. (**E**) Transferring the cellulose acetate filter paper with pollen grains facing up onto the 90 mm petri dish with two layers of whatman filter papers pre-wet with pollen germination medium and keep harvested pollen on ice

(c) Drop pollen grain suspension onto the pre-wet filter paper and the pollen grains are collected and evenly distributed onto the filter paper (Fig. 2D).

(d) Transfer the cellulose acetate filter paper with pollen grains facing up onto the 90 mm petri dish with two layers of Whatman filter papers pre-wet with pollen germination medium (Fig. 2E) (*see* **Note 11**).

4. Set up the PSD-1000/He particle delivery system as follows: 1350 psi, 28 mmHg vacuum. (*see* **Note 12**).
5. Dip the rupture disc, stopping screen and the rupture disc retaining cap into isopropanol for few seconds and air dry in clean bench.
6. Load-coated gold particles (6 μL) onto the macrocarrier and air dry (*see* **Note 13**).
7. After bombardment, collect the pollen grains from bombarded areas, which are easily visible on the cellulose acetate filter paper as brown color of the gold particles, using a small spatula or spoon. Transfer the bombarded pollen grains into a 1.5 mL Eppendorf tube with 800 μL 40 °C warm pollen germination medium supplemented with 1% low melt point agarose (final

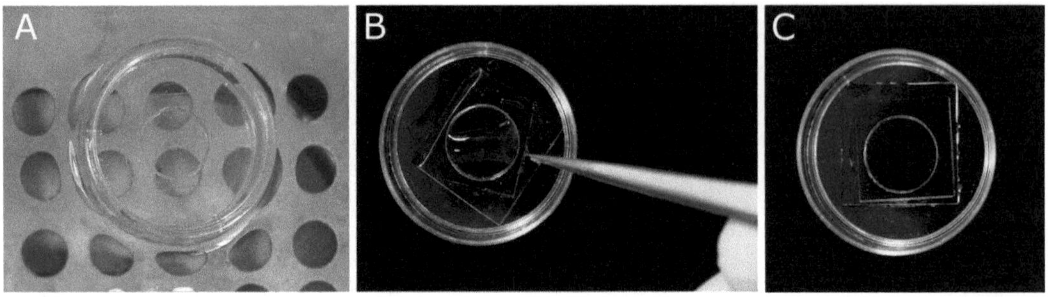

Fig. 3 Pollen grains sink to the bottom of the glass-bottom dish in the thermostatic water bath kettle (**A**) and the preparation of glass-bottom dish for imaging (**B** and **C**)

concentration) and divide 800 μL of mixture into four cover glass-bottom dishes previously coated with 0.01% poly-L-lysine with 200 μL mixture in each (*see* **Notes 14** and **15**).

8. Immediately transfer cover glass-bottom dishes into 50 °C water bath for no more than 5 min to allow the pollen grains to sink to the bottom of the cover glass. Remove the excess medium. The remaining medium is allowed to solidify at room temperature. In this way, pollen tubes will grow on the surface of the bottom cover glass, which is suitable for imaging (Fig. 3A) (*see* **Note 16**).
9. Transfer glass-bottom dishes with pollen grains into the styrofoam box with two layers of kitchen paper soaked with tap water to keep a high humidity during pollen germination.
10. Keep the styrofoam box at room temperature and let pollen tubes grow for 4 h (*see* **Note 17**).

3.3 Monitoring Ca^{2+} Dynamic Changes During Pollen Tube Growth

1. For imaging, use a 40×/0.75 dry objective, and excite the CFP of the YC3.6 at a wavelength of 440 nm.
2. Drop approximately 100 μL Milli-Q water into the well of the cover glass-bottom dish and cover the well with another cover slip to prevent drying of the pollen germination medium (Fig. 3B, C).
3. Mount the cover glass-bottom dish onto the microscope.
4. Set the acquisition time at a 6 s interval and 18 min in total.
5. Detect the $CFP_{Emission}$ (485 nm) and the $YFP_{Emission}$ (535 nm) to measure FRET (fluorescence resonance energy transfer) between CFP and YFP during pollen tube growth.
6. Save the series of images of $YFP_{Emission}$ and $CFP_{Emission}$ [7].
7. Post-acquisition analysis using ImageJ to generate a movie file to show the Ca^{2+} dynamics in pollen tubes.
   (a) By performing File → Import → image sequence to import the series of images of YFP and CFP into ImageJ.

(b) Perform Plugins → Ratio plus, then go to Image → Look up table → 16 colors.

(c) Perform Image → Adjust → Brightness/Contrast to improve contrast.

(d) Save the movie as AVI format file (Supplemental Movies S1 and S2).

3.4 Study Impact of Protein Function on Ca^{2+} Dynamics and Development of Pollen Tubes

1. Fuse protein X with mCherry fluorophore to generate X-mCherry construct.
2. Co-transform X-mCherry and YC3.6 constructs into pollen grains (refer to Subheadings 3.1 and 3.2). Here we take mCherry + YC3.6 as control and CBL2-mCherry + YC3.6 as an example (Fig. 4) (*see* **Note 18**).
3. Check the expression of the mCherry fusion protein (excite with 561 nm and detect emission at 582–638 nm).
4. Monitoring Ca^{2+} dynamics (refer to Subheading 3.3).
5. Perform post-acquisition analysis with ImageJ software.

   (a) Create a movie with series of CFP and YFP images captured from the experiments (refer to **step 7** in Subheading 3.3).

   (b) Load the movie into ImageJ, then go to Image → Stacks → Z-project (choose maximum intensity).

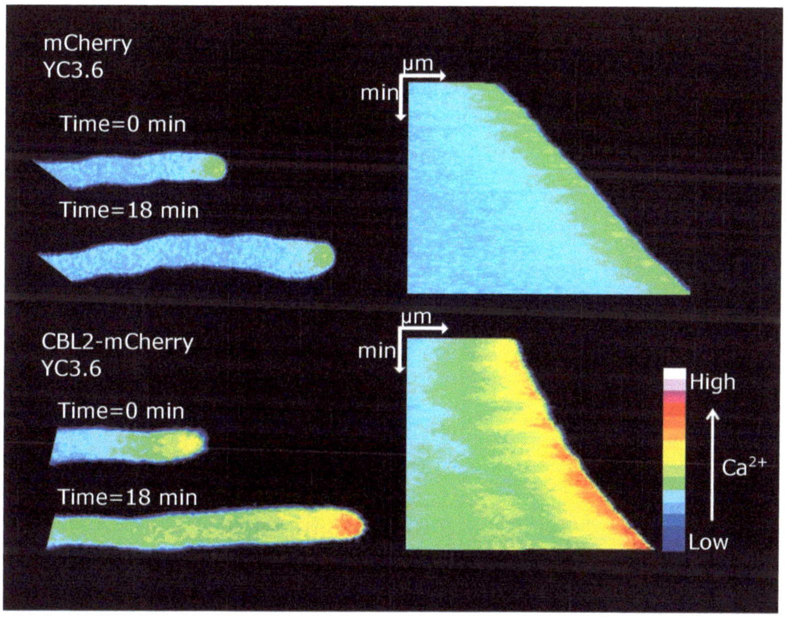

Fig. 4 Kymograph of Ca^{2+} dynamics measured with the Ca^{2+} sensor YC3.6 co-expressed with mCherry (upper panel) or CBL2-mCherry (lower panel), respectively in tobacco pollen tubes

(c) Draw a segmented line in the middle of the pollen tube.

(d) Switch to the original stack file and transfer the drawn line by performing Edit → Selection → Restore Selection.

(e) Generate kymograph pictures by performing Image → Stacks → Reslice (Fig. 4).

4 Notes

1. Gold particles (1.0 μm; Bio-Rad, cat. no. 165–2263).
2. Elute plasmid DNA with sterilized double distilled H_2O.
3. The 1 M spermidine stock solution should be sterilized by filtration through a 0.22 μm syringe filter and store at −80 °C. 0.1 M spermidine stores at −20 °C for up to 1 year.
4. If the experiments need to be sterile in the whole procedure, 2.5 M $CaCl_2$ solution should be sterilized by filtration through a 0.22 μm syringe filter before dividing into aliquots.
5. The pollen germination medium should be freshly prepared before use.
6. If you need to co-express two plasmids, use 5 μg plasmid DNA for each.
7. 18 μL gold particles can be used for three bombardments, 6 μL for each.
8. Preparing the gold particles coated with plasmids freshly optimizes transformation efficiency.
9. To get higher pollen germination rates, the use of fresh pollen grains is recommended. The 50 mL falcon tube with pollen germination medium should be always kept on ice to prevent premature pollen germination.
10. Make sure that no air bubbles are under the cellulose acetate filter paper.
11. Keep the petri dish with the filtered pollen grains on ice, to prevent germination.
12. Wear safety glasses during the operation of the PSD-1000/He particle delivery system.
13. To achieve good transformation efficiency, ensure that gold particles are evenly distributed in the center of the macrocarrier.
14. The cover glass-bottom dishes should be coated with 0.01% poly-L-lysine 12 h before use.
15. To prepare 1% low melt point agarose pollen growth medium mixture, first, dissolve 2% low melt point agarose with Milli-Q water in a microwave oven and then mix with 2× pollen

growth medium to get 1% final concentration of low melt point agarose in pollen growth medium.

16. The temperature of the water in water bath kettle should be a little higher than 40 °C (50–55 °C works well), so that the cover glass-bottom dish reaches 40 °C when it is floating on the water in bath kettle.
17. Because pollen tubes are sensitive to temperature and mechanical stress, the pollen grains should be germinated in the same place where the microscope is located to protect pollen tubes from the changes of temperature and the mechanical stress caused by movement immediately before the measurement. The temperature of the room should be 24–27 °C.
18. CBL2 is calcineurin B-like protein 2 (At5g55990) which functions in pollen tube growth in Arabidopsis [3].

Acknowledgement

This work was supported by grants from DFG and NSFC.

References

1. Pierson ES, Miller DD, Callaham DA, van Aken J, Hackett G, Hepler PK (1996) Tip-localized calcium entry fluctuates during pollen tube growth. Dev Biol 174:160–173
2. Steinhorst L, Kudla J (2013) Calcium-a central regulator of pollen germination and tube growth Biochim. Biophys Acta 1833:1573–1581
3. Steinhorst L, Mähs A, Ischebeck T, Zhang C, Zhang X, Arendt S, Schültke S, Heilmann I, Kudla J (2015) Vacuolar CBL-CIPK12 Ca^{2+}-sensor-kinase complexes are required for polarized pollen tube growth. Curr Biol 25:1475–1482
4. Kost B, Spielhofer P, Chua NH (1998) A GFP-mouse talin fusion protein labels plant actin filaments in vivo and visualizes the actin cytoskeleton in growing pollen tubes. Plant J 16:393–401
5. Ischebeck T, Stenzel I, Heilmann I (2008) Type B phosphatidylinositol-4-phosphate 5-kinases mediate Arabidopsis and *Nicotiana tabacum* pollen tube growth by regulating apical pectin secretion. Plant Cell 20:3312–3330
6. Mähs A, Steinhorst L, Han J, Shen L, Wang Y, Kudla J (2013) The Calcineurin B-like Ca^{2+} sensors CBL1 and CBL9 function in pollen germination and pollen tube growth in Arabidopsis. Mol Plant 6:1149–1162
7. Behera S, Kudla J (2013) High-resolution imaging of cytoplasmic Ca^{2+} dynamics in Arabidopsis roots. Cold Spring Harb Protoc. https://doi.org/10.1101/pdb.prot073023
8. Gutermuth T, Lassig R, Portes MT, Maierhofer T, Romeis T, Borst JW, Hedrich R, Feijó JA, Konrad KR (2013) pollen tube growth regulation by free anions depends on the interaction between the anion channel SLAH3 and calcium-dependent protein kinases CPK2 and CPK20. Plant Cell 25:4525–4543
9. Wang H, Jiang L (2011) Transient expression and analysis of fluorescent reporter proteins in plant pollen tubes. Nat Protoc 6:419–426
10. Muschietti J, Dircks L, Vancanneyt G, McCormick S (1994) LAT52 protein is essential for tomato pollen development: pollen expressing antisense LAT52 RNA hydrates and germinates abnormally and cannot achieve fertilization. Plant J 6:321–338
11. Read SM, Clarke AE, Bacic A (1993) Stimulation of growth of cultured *Nicotiana tabacum* W 38 pollen tubes by poly (ethylene glycol) and $Cu_{(II)}$ salts. Protoplasma 177:1–14

Chapter 17

Imaging and Analysis of the Content of Callose, Pectin, and Cellulose in the Cell Wall of Arabidopsis Pollen Tubes Grown In Vitro

Ana R. Sede, Diego L. Wengier, Cecilia Borassi, José M. Estevez, and Jorge P. Muschietti

Abstract

To achieve fertilization, pollen tubes have to protect and properly deliver sperm cells through the pistil to the ovules. Pollen tube growth is a representative example of polarized growth where new components of the cell wall and plasma membrane are continuously deposited at the tip of the growing cell. The integrity of the cell wall is of fundamental importance to maintain apical growth. For this reason, pollen tube growth has become an excellent model to study the role of polysaccharides and structural cell wall proteins involved in polar cell expansion. However, quantification of structural polysaccharides at the pollen tube cell wall has been challenging due to technical complexity and the difficulty of finding specific dyes. Here, we propose simple methods for imaging and quantification of callose, pectin, and cellulose using specific dyes such as Aniline Blue, Propidium Iodide, and Pontamine Fast Scarlet 4B.

Key words Pollen tubes, Imaging, Callose, Cellulose, Pectin, Cell wall

1 Introduction

Spatial distribution of different polysaccharides and the role of structural proteins such as hydroxyproline-rich glycoproteins (HRGPs) are crucial to maintain the integrity of the pollen tube cell wall. Generally, in angiosperms, the apical region of pollen tubes is composed of a single, flexible, and dynamic wall layer which is constantly remodeling at the tip. In distal regions where expansion has ceased, the cell wall gets stiffened through enzymatic action promoting polymer linkages, for example the gelation of pectin, and through deposition of stiffer materials such as cellulose and callose [1]. This stiffening locks in place the width of the tube and maintains the constant diameter of the tubular shank of the cell [2]. In contrast to other plant cells, the cell wall of the pollen tube

Anja Geitmann (ed.), *Pollen and Pollen Tube Biology: Methods and Protocols*, Methods in Molecular Biology, vol. 2160, https://doi.org/10.1007/978-1-0716-0672-8_17,

contains a large amount of callose and pectin and only relatively low amounts of cellulose.

Callose is a (1 → 3)-β-D-glucan synthetized at the plasma membrane by callose synthase complexes. It is deposited along the pollen tube under the fibrous pectic layer in the cylindrical portion of the tube and is absent at the tip [3]. The first reports describing the detection of callose with Aniline Blue staining date back to 1949 (reported in Currier [4]). Aniline Blue has been widely used to stain callose in pollen tubes both in vitro and in vivo, and to detect callose associated to plasmodesmata between neighboring cells and phloem sieve elements, and plant cell responses to pathogen attack [4, 5].

Pectin is a complex polymer mainly composed of homogalacturonan (HG) and rhamnogalacturonan I and II (RGI and RGII) [6] that is deposited at the apical region of the pollen tube in a highly methylesterified state. The degree of pectin esterification decreases toward the distal regions of the pollen tube where the enzyme pectin methylesterase remains active [3]. In distal regions, free pectin carboxyl groups interact with Ca^{2+} ions, forming a gel that stiffens the region preventing any further lateral expansion [3]. Propidium iodide (phenanthridinium, 3,8-diamino-5-[3-(diethylmethylammonio)propyl]-6-phenyl-,diiodide) is a red-fluorescent dye which competes with Ca^{2+} for binding to carboxyl residues of HGs and has been used as an indicator of the degree of crosslinking of pectic HGs in pollen tube cell walls [7].

Cellulose, a polysaccharide with a lower relative abundance in pollen than in most other plant cells, is a (1 → 4)-β-D-glucan. Cellulose is synthetized at the plasma membrane by multimeric cellulose synthase (CESA) complexes [8]. In pollen tube cell walls, cellulose is organized as crystalline microfibrils that form a tight network with pectin and callose [3]. Calcofluor white has been used for cellulose staining, although this dye is not specific since it has also affinity for callose and for (1 → 4)-β-*N*-acetyl-D-glucosamine (known as chitin). Alternatively, Pontamine Fast Scarlet 4B (S4B) (also known as Direct Red 23 and Levacell Scarlet E-3B) has been shown to be selective for polysaccharides containing β-D-(1 → 4) bonds and fluoresces more intensely in the presence of cellulose than with any other plant polysaccharide [9]. S4B has been used to stain cellulose microfibrils in the primary cell wall of *Arabidopsis thaliana* roots [9] and recently, in pollen tubes [10], and cotyledons [11].

The synthesis and assembly of the plant cell wall components in *Arabidopsis thaliana* are controlled by a myriad of proteins with different expression patterns, degrees of functional redundancy, and contribution to the overall polymer production. A typical approach to study the role of genes involved in plant cell wall synthesis and assembly implies obtaining and characterizing single and multiple mutants for the genes of interest. While the quantification of gene expression in different mutant backgrounds is relatively straightforward, determining the final effect on the cell wall secretion and assembly can be challenging. For example, localized production and deposition of polymers (as in the growing pollen tube tip), requires sensitive and unambiguous detection and quantification methods that allow direct comparison and statistical analysis. Here, we present protocols for the detection, quantification, and statistical analysis of the abundance of callose, pectin, and cellulose in growing pollen tubes of Arabidopsis.

2 Materials

2.1 Plant Material

Arabidopsis thaliana var. Col-0 (wild type) seeds are obtained from the Arabidopsis Biological Resource Center (ABRC) or the Nottingham Arabidopsis Stock Centre (NASC). To perform the assays, select one-day open flowers (Fig. 1); at this stage, flowers are fully opened and mature pollen grains are released from the anthers (*see* **Note 1**).

2.2 Pollen Germination Media

Two alternative pollen germination media (PGM) can be used: *semi*-solid and *semi*-liquid. Prepare stock solutions for PGM according to Boavida and McCormick (2007) [12]. Final concentrations are 0.01% H_3BO_3, 5 mM KCl, 5 mM $CaCl_2$, 1 mM $MgSO_4$, and 10% sucrose. Set the pH to 7.5–8 with 0.1 M KOH.

- To prepare *semi*-solid medium, reach the final volume with MilliQ H_2O, add low-melting agarose to a final concentration of 1% and melt it at 95 °C for 5 min using a block heater (*see* **Note 2**). This medium makes a gel resistant to compression.
- To prepare *semi*-liquid medium, reach the final volume using 0.15% agarose pre-melted solution prepared in MilliQ H_2O (*see* **Note 3**). Pollen tubes grow on the solution surface and do not sink.

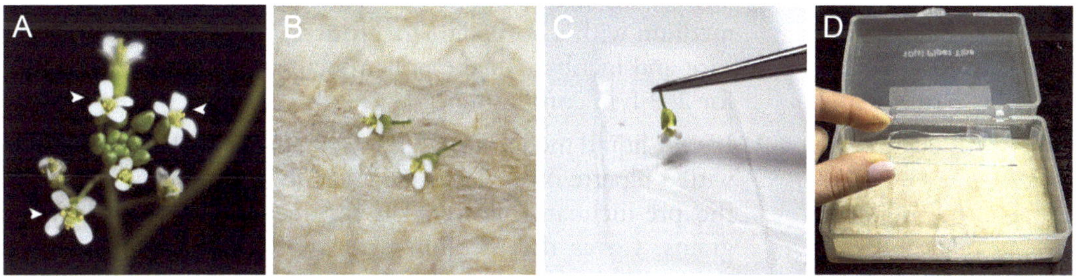

Fig. 1 Step-by-step pollen germination method on *semi*-solid medium. One-day open *Arabidopsis thaliana* flowers (**a**, white arrowheads) are positioned over wet paper towels (**b**) and incubated for 30 min at room temperature in a humidity chamber (similar to **d**). Pollen grains from pre-incubated flowers are transferred to the medium using forceps (**c**). Finally, a coverslip is placed on top of the medium and slides are incubated in a humidity chamber at 22 °C for 2–3 h (**d**)

2.3 Dyes

All dye stock solutions must be prepared in MilliQ H_2O and stored at 4 °C protected from light. To avoid osmotic shock, use liquid PGM (reach the final volume by using MiliQ H_2O without agarose addition) to prepare fresh working solutions from all dye stocks before performing the assays.

1. Propidium iodide (PI) stock solution: 1 mg/ml. To prepare a PI working solution, make a fresh 1/5 dilution.
2. Aniline blue (AB) stock solution: 0.1% (w/v) in 108 mM K_3PO_4 pH ~11 [13]. Store the solution overnight at 4 °C and then filter it. Finally, add glycerol to a final concentration of 2% (v/v). To prepare an AB working solution, make a fresh 1/2 dilution.
3. Pontamine Fast Scarlet 4B (S4B) stock solution: 1% (w/v). To prepare a S4B working solution, make a fresh 1/100 dilution.

2.4 Equipment

1. Pollen incubation chamber: A tip rack can be used to build a humidity chamber, placing a wet paper towel at the bottom of the rack (Fig. 1).
2. Block heater.
3. pH meter.
4. Confocal microscope.
5. ImageJ/Fiji Software.

3 Methods

3.1 Pollen Germination

Collect at least two one-day open flowers and pre-incubate them directly on wet paper towels for 30 min at room temperature in a pollen incubation chamber (Fig. 1).

1. If *semi*-solid medium is used, melt the agarose and quickly place a thin layer of medium with a pipette on a glass slide. Once PGM has gelled, use forceps to gently brush the pre-incubated flowers on the medium to spread pollen grains. Cover the medium with a coverslip (*see* **Note 4**) by applying some pressure and incubate in the pollen incubation chamber at 22 °C for 2–3 h in continuous light (Fig. 1).
2. If *semi*-liquid medium is used, place a drop (~20 μl) of medium with a pipette on a glass slide and use forceps to gently brush the pre-incubated flowers on the medium to spread pollen grains. Cover the medium with a coverslip and incubate in the pollen incubation chamber at 22 °C for 2–3 h under continuous light (Fig. 1).

3.2 Pectin Staining

1. Carefully lift the coverslip, add 5 μl of the PI working solution in the *semi*-solid or *semi*-liquid PGM and cover it again.

2. Incubate 5 min before imaging (*see* **Note 5**). PI has an excitation maximum at 535 nm and an emission peak at 617 nm. PI can be alternatively excited with a 488 or a 532 nm laser line.

3.3 Callose Staining

1. Carefully lift the coverslip, add 5 μl of the AB working solution in the *semi*-solid or *semi*-liquid PGM and cover it again.
2. Proceed with the imaging. AB has an excitation maximum at 320–380 nm and an emission peak at 450 nm.

3.4 Cellulose Staining

1. Carefully lift the coverslip, add 5 μl of S4B working solution in the *semi*-solid or *semi*-liquid PGM and cover it again.
2. Proceed with the imaging. S4B has an excitation maximum at 405 nm and an emission peak at 570 nm.

4 Image Acquisition

Confocal scanning settings must be optimized according to the microscope in use and its hardware configuration (*see* **Note 6**). The images shown here were obtained with a Leica SPE confocal laser scanning microscope, using a 1024 × 1024 format, bidirectional scanning at a speed of 600 Hz, an APO/ACS 63×/1.30 oil objective and a 3× digital zoom (Fig. 2). Pinhole was set to 1 Airy unit. The transmitted light was used to focus and analyze pollen tube tips (*see* **Note 7**). Laser lines were selected depending on the excitation/emission spectra of the fluorophores.

4.1 ImageJ Analysis

To quantify the fluorescence intensity at the pollen tube margins, make measurements along the perimeter of the pollen tubes, including their apical and subapical regions (Fig. 3).

1. Open the ImageJ/Fiji application. In "File" go to "Open" and find the location of the file.
2. Click the "Line tool" from the toolbar (fifth option from the left), set "Line Width" in 15 by double clicking and activate

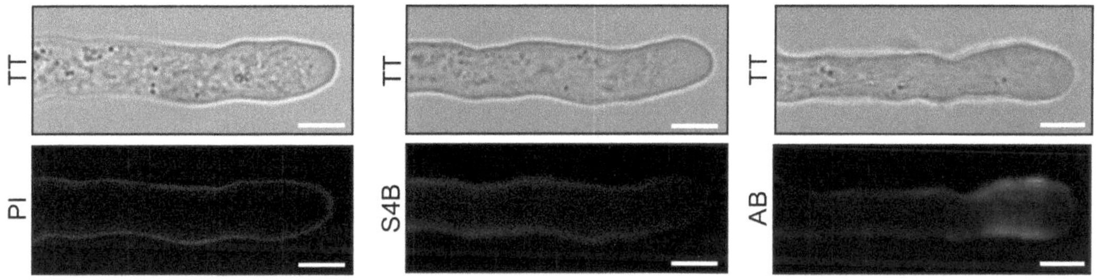

Fig. 2 Representative confocal images of Col-0 pollen tubes stained with propidium iodide (PI) for pectin (left panel), Pontamine Fast Scarlet 4B (S4B) for cellulose (middle panel), and aniline blue (AB) for callose (right panel). *TT* transmitted light. Scale bar = 5 μm

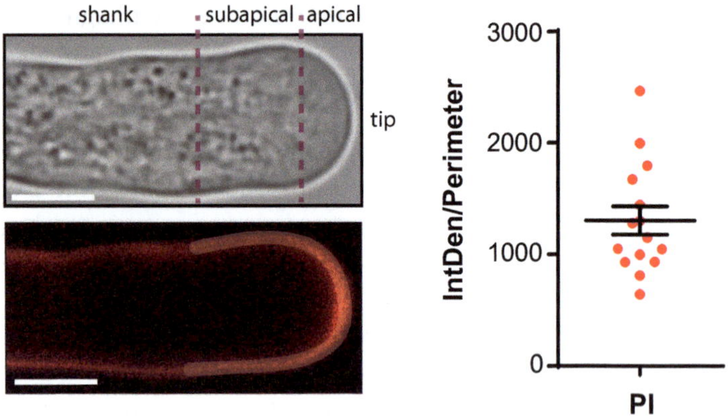

Fig. 3 Pollen tube spatial organization: tip, apical region, subapical region, and shank are indicated (left, upper panel). Representative image of a Col-0 pollen tube stained with PI (left, bottom panel). Measurement of PI fluorescence intensity at the margins of Col-0 pollen tubes; the fluorescent signal was normalized to the perimeter of the measured region. Recorded fluorescence of $n = 15$ pollen tubes from one independent experiment is shown. Scale bar = 5 μm

"Spline fit" option (*see* **Note 8**). Then, by pressing the right button of the mouse, select the "Segmented line" option.

3. Draw the perimeter of the pollen tube, including the apical and subapical region, and finish the drawing by double-clicking on the last traced point.
4. To record the fluorescence, go to "Analyze" and press "Measure" (or Ctrl+M). A new window will appear with the data.
5. Copy and paste the data into the appropriate data analysis software for further analysis.

To quantify the fluorescence intensity of the cytoplasm and the margins of the pollen tube, trace a longitudinal or a transversal line (Fig. 4a, b):

1. Open the Image J/Fiji application. In "File" go to "Open" and find the location of the file.
2. Click the "Line tool" from the toolbar (fifth option from the left) and set "Line width" in 30 by double clicking (*see* **Note 9**). Then select the "Straight Line" option.
3. Trace the longitudinal or the transversal line, go to "Analyze" and press "Plot profile" (or Ctrl+K). A two-dimensional graph with the intensity recorded along the line drawn will appear in a new window.
4. To export the data, select "More" at the bottom of the window, then press "Copy All Data" and paste it into the appropriate data analysis software for further analysis.

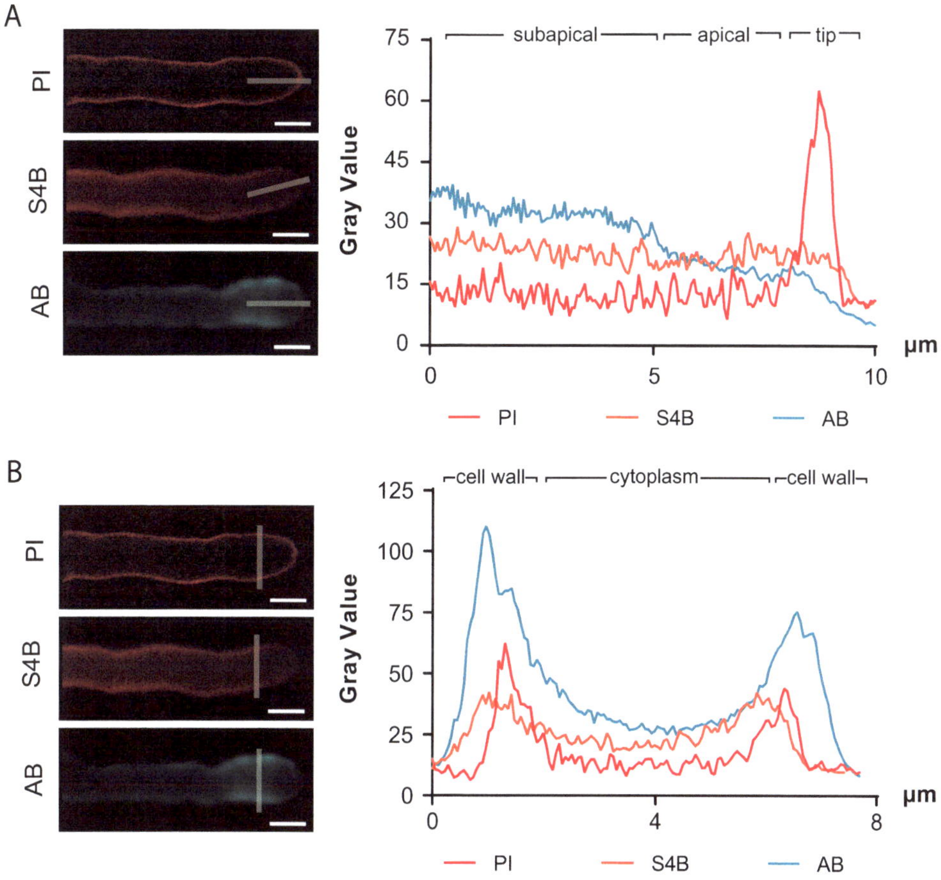

Fig. 4 Representative images and intensity measurement of cell wall components of Col-0 pollen tubes stained with propidium iodide (PI), Pontamine Fast Scarlet 4B (S4B), and aniline blue (AB). (**a**) Longitudinal fluorescence measurement (upper panel). (**b**) Transversal fluorescence measurement (bottom panel). Scale bar = 5 µm

4.2 Statistical Analysis of Data

Common errors when generating and analyzing pollen tube data are: grouping data from independent experiments (performed on different days), using data from experiments with low pollen germination rates, comparing genotypes that were not analyzed simultaneously, comparing raw data without prior normalization, and using an erroneous N in statistical analysis.

Unlike other species, reproducible germination rates for Arabidopsis pollen are quite difficult to obtain in vitro. Pollen germination after 3 h of incubation can vary from 0% to 60% or more in any given experiment, even when the same PGM is used and the assay is performed in the same growth chamber. In fact, several articles have focused on improving pollen germination and pollen tube growth [12, 14, 15]. In terms of biological variability, pollen tube populations may be different on different days. In terms of sample representability, we assume that the higher the germination rate, the more adequate the sample is to infer population parameters. Inversely, experiments with low germination rates could produce estimates of biased population parameters because they only consider the most successful pollen tubes grown in suboptimal conditions. For this reason, we suggest that only experiments in which wild-type pollen germination rates are at least 30–40%, after 3 h of incubation, should be considered.

Another source of variation arises from the experiment *per se*. In this case, staining will fluctuate with dye preparations and fluorescence detection will be different between different confocal microscopes or imaging sessions even when identical settings are used. If germination rates are reproducible through experiments and minimal variation is introduced during imaging, sample estimates of population parameters could be obtained and compared. However, if that is not the case, it is necessary to implement normalization strategies to compare data between experiments (*see* **Notes 10** and **11**).

5 Notes

1. For best results, avoid using the first flowers of the primary bolt since most pollen is infertile.
2. H_3BO_3 can be added to the PGM after the agarose melting.
3. Agarose serves as a support and at the same time improves pollen germination and pollen tube growth.
4. Germination rates can be improved when pollen grains are grown on the interface between the coverslip and the PGM.
5. PI is toxic to pollen tubes at high concentrations and after long incubation periods. When the pollen tube is dead, the PI enters the cell and fluorescence is observed throughout the cytoplasm.
6. An optimal microscope configuration must be set for each fluorophore and must be kept constant during imaging sessions and between experiments.
7. Only live pollen tubes should be included for analysis. To ensure that a pollen tube is alive, use transmitted light and verify that the tubes show the typical cytoplasm streaming. The proximity of the big vacuole to the apical region indicates the impending death of the pollen tube.
8. The line width (in pixels) is set according to the image resolution and the zoom chosen. Select a line width thick enough to only encompass the cell wall. Here, the line width for a 1024×1024 image captured with a digital zoom $3\times$ was calculated.

9. In this case, a thicker line width is selected to record and integrate the fluorescence of a larger number of pixels and decrease the measurement error.
10. The first strategy to minimize imaging variations is to subtract the background signal and normalize data. For example, when determining the fluorescence intensity in transects, the background signal (detected in the same image) is subtracted from each value in the series of linear data. Thus, normalization is done by dividing each value to the highest value of the series and data are scaled to the same reference system (between 0 and 1). In each experiment, the observations for each genotype (i.e., Col-0) will be averaged to obtain a sample estimate of the population parameter, in the same way that the sample mean provides an estimate of the population mean. In each case, the number of observations required will be determined (in our experience, 15 observations per genotype per experiment are adequate). Because the purpose of each experiment is to obtain sample estimates for the population parameters, N corresponds to the number of replications per genotype and not to the total number of observations within each replicate or all samples in the experiments. For example, 15 pollen tubes for Col-0 in three independent replicates correspond to an $n = 3$, and not $n = 45$. If data for each genotype are comparable between experiments after this normalization (that is, the standard errors are 10% of the average of the observations), no further normalization is required.
11. A second normalization strategy is to define the variable to be measured as a ratio between the sample estimates of the population parameters for the genotypes or treatments of interest. For example, a mutant versus a wild-type genotype or a treated versus an untreated sample. This strategy is suitable for highly variable experiments. As before, sample estimates can be calculated based on an appropriate number of observations (for example, a number that produces a standard error of 10% of the average). In this case, N corresponds to the number of independent ratios generated.

Acknowledgments

This work was supported by PICT2014-0423, PICT205-0078 and PICT2017-0076 to J.M. and ICGEB CRP/ARG16-03 and PICT2016-0132 and PICT2017-0066 to J.M.E. In addition, this research is funded by Instituto Milenio iBio – Iniciativa CientÚfica Milenio, MINECON to J.M.E.

References

1. Geitmann A, Steer M (2006) The architecture and properties of the pollen tube cell wall. Plant Cell Monogr 3:177–200
2. Fayant P, Girlanda O, Chebli Y, Aubin CE, Villemure I, Geitmann A (2010) Finite element model of polar growth in pollen tubes. Plant Cell 8:2579–2593
3. Chebli Y, Kaneda M, Zerzour R, Geitmann A (2012) The cell wall of the Arabidopsis pollen tube-spatial distribution, recycling, and network formation of polysaccharides. Plant Physiol 160:1940–1955
4. Currier HB (1957) Callose substance in plant cells. Am J Bot 44:478–488
5. Eggert D, Naumann M, Reimer R, Voigt CA (2014) Nanoscale glucan polymer network causes pathogen resistance. Sci Rep 4:4159. https://doi.org/10.1038/srep04159
6. Harholt J, Suttangkakul A, Scheller HV (2010) Biosynthesis of pectin. Plant Physiol 153:384–395
7. Rounds CM, Lubeck E, Hepler PK, Winship LJ (2011) Propidium iodide competes with Ca^{2+} to label pectin in pollen tubes and Arabidopsis root hairs. Plant Physiol 157:175–187
8. Lampugnani ER, Khan GA, Somssich M, Persson S (2018) Building a plant cell wall at a glance. J Cell Sci 131:jcs207373
9. Anderson CT, Carroll A, Akhmetova L, Somerville C (2010) Real-time imaging of cellulose reorientation during cell wall expansion in Arabidopsis roots. Plant Physiol 152:787–796
10. Sede AR, Borassi C, Wengier DL, Mecchia MA, Estevez JM, Muschietti JP (2018) Arabidopsis pollen extensins LRX are required for cell wall integrity during pollen tube growth. FEBS Lett 592:233–243
11. Altartouri B, Bidhendi AJ, Tani T, Conrad C, Chebli Y, Liu N, Karunakaran C, Scarcelli G, Geitmann A (2019) Pectin chemistry and cellulose crystallinity govern pavement cell morphogenesis in a multi-step mechanism. Plant Physiol 181:127–141
12. Boavida LC, Mccormick S (2007) Temperature as a determinant factor for increased and reproducible in vitro pollen germination in *Arabidopsis thaliana*. Plant J 52:570–582
13. Mori T, Kuroiwa H, Higashiyama T, Kuroiwa T (2006) GENERATIVE CELL SPECIFIC 1 is essential for angiosperm fertilization. Nat Cell Biol 8:64–71
14. Fan L, Wang Y, Wang H, Wu W (2001) In vitro Arabidopsis pollen germination and characterization of the inward potassium currents in Arabidopsis pollen grain protoplasts. J Exp Bot 52:1603–1614
15. Rodriguez-Enriquez MJ, Mehdi S, Dickinson HG, Grant-Downton R (2013) A novel method for efficient in vitro germination and tube growth of *Arabidopsis thaliana* pollen. New Phytol 197:668–679

Chapter 18

Characterization of Growth Behavior and the Resulting Forces Applied by Pollen Tubes in a 3D Matrix

Ronny Reimann and Delf Kah

Abstract

The question of how pollen tubes orient themselves on their way to the egg cell is a major focus of plant reproduction research. The role of physical guidance through the tissues of the pistil in relation to the mechanical perception and growth adaptation of the pollen tubes has not been sufficiently investigated. In order to advance research on the mechanical perception of pollen tubes and their force application during invasive growth, we present simple methods for the observation and mechanical characterization of pollen tubes in vitro, which can be established with little effort in any biological laboratory with standard equipment. Pollen grains are germinated in a hydrogel containing agarose and their growth is recorded in 3D using brightfield microscopy. Using suitable analysis software, parameters such as growth rate and pollen tube diameter can then be determined to estimate the exerted penetration force.

Key words Pollen tube, Polar growth, Force measurement, ClickPoints, Mechanical guidance, Single cell, Hydrogel, Agarose, Indentation, Microindenter, 3D matrix

1 Introduction

In order to reach the egg cells, pollen tubes must penetrate or interact with at least seven different pistil cell types and tissues: the stigma, the transmission tracts of the style and the ovary, as well as the septum epidermis, the funiculus, the micropyle, and the synergids [1]. Along their growth path pollen tubes are mainly guided by chemotropic cues, but probably not exclusively. Several experimental results strongly suggest an additional mechanical guidance by the tissues and cells of the pistil [2–4]. To better understand the processes of mechanical guidance, it is necessary to reproduce the physical architecture of the pistil in experimental setups. So far, however, only few studies have worked with obstacles or combined areas with different resistance of the growth medium, to simulate the tissues of the pistil [2–5]. Furthermore, nearly all pollen tube cultivation protocols use 2D systems and ignore the 3D nature of the pistil.

Anja Geitmann (ed.), *Pollen and Pollen Tube Biology: Methods and Protocols*, Methods in Molecular Biology, vol. 2160, https://doi.org/10.1007/978-1-0716-0672-8_18,

Pollen tubes are not only gently guided through the pistil, they must apply force to grow through tissues. Especially the transmission tract of plants with a solid style represents an obstacle. It is filled with tissue and the pollen tubes must grow through cell layers embedded in the extracellular matrix [6, 7]. The most important parameter for characterizing the invasive growth of pollen tubes is the force they exert when penetrating obstacles. It just became feasible to measure the penetration force of single pollen tubes in recent years with the advance of microfluidic systems and their combination with methods for single cell force measurement [2, 3, 5]. A current drawback of these methods is that only a small number of pollen tubes can be investigated at once. Individual pollen tubes often show large variations in their growth parameters compared to each other. Therefore, a larger number of pollen tubes should be analyzed in order to make well-founded statements about pollen tube behavior based on extensive statistical analyses. Furthermore, working with microfluidic devices and single cell force measurement techniques requires special equipment and expertise. Methods that can be carried out in any laboratory without special equipment would accelerate pollen tube research.

To answer these problems in the characterization of the growth behavior of pollen tubes, we present a combination of different methods, which can be established without much effort in any biological laboratory with standard equipment. To simultaneously characterize the invasive growth of high numbers of pollen tubes, we combined automated brightfield microscopy, image processing, and semi-automated image analysis. Pollen tubes are grown in 3D matrices, solidified with low melting agarose. In contrast to classical pollen tube growth systems, these hydrogels allow the growth of pollen tubes in all directions and thus better resemble the in vivo environment of the pistil. Furthermore, the matrix represents a mechanical resistance adjustable by the agarose concentration. Experimental setups with defined agarose concentrations or with zones of different agarose concentrations can be used to characterize the invasive pollen tube growth behavior under different mechanical resistances. To determine the penetration forces of pollen tubes, we here suggest an indenter-based method. Pollen tube-shaped objects with different diameters are driven into the 3D matrix. Thereby the speed and the necessary force are recorded and used to derive a general formula, which relates the penetration force to the speed, the diameter, and the medium strength. This formula can then be used to determine the penetration force of pollen tubes.

We also present a specialized sub-protocol to determine the maximal penetration force of pollen tubes. The respective experimental setup forces pollen tubes to grow into a 3D matrix with high strength. Pollen tube growth can then be analyzed with the developed image analysis pipelines and the indenter method is used to estimate the corresponding force application.

2 Materials

2.1 Plant Material

1. *Nicotiana tabacum* L. cv. Petit Havana SR1 plants cultivated at 22 °C and 60% relative humidity under long day conditions (16 h light, 8 h dark).
2. Depending on the setup of the experiment also pollen of other plant species can be used (*see* **Note 1**).

2.2 Cultivation of Pollen Tubes in Artificial 3D Matrices

1. Paper towels.
2. Sealable box.
3. Forceps.
4. Scissors.
5. Two sample heaters.
6. Pollen tube culture medium (depending on the species), e.g., Brewbaker-Kwak (BK) medium for *N. tabacum* (1 mM $CaCl_2$, 1 mM $Ca(NO_3)_2$, 1 mM $MgSO_4$, 1.62 mM H_3BO_3, 10% (w/v) sucrose, pH 7.5) [8].
7. Suitable vessels for the prepared growth media (e.g., 50 mL tubes).
8. pH meter.
9. Low melting agarose.
10. Microwave.
11. 6-Well plates (*see* **Note 2**).
12. 2-mL tubes with lids.
13. 10-μL pipette tips (only for the maximal force measurement).
14. Petri dish in a format suitable for your experimental setup (only for the maximal force measurement).
15. Adhesive laboratory tape (only for the maximal force measurement).

2.3 Brightfield Microscopy

1. Light microscope with 5× or 10× objective and a motorized microscope stage (moveable in *y*, *x*, and *z* direction).
2. Software to acquire images with relevant metadata (i.e., magnification, positions in *x*-, *y*-, and *z*-plane) and time stamps for later readout by the analysis software.

2.4 Image Analysis

1. Computer with Python3 installation (free graphical installers for Windows, Mac, and Linux can be downloaded at www.anaconda.com/distribution).

2. Image analysis software ClickPoints (installer, instructions, and documentation are available at www.clickpoints.readthedocs.io) [9].

2.5 Indentation Experiments

1. Micromanipulator with an open application program interface (API), so that it can be controlled by self-written Python scripts (e.g., Eppendorf InjectMan NI 2).
2. Computer with Python3 installation (free graphical installers for Windows, Mac, and Linux can be downloaded at www.anaconda.com/distribution).
3. Digital microscale with a resolution <0.1 mg and an open API, so that it can be controlled by self-written Python scripts.
4. If necessary, connection cable (serial to USB) for micromanipulator and scale.
5. Acupuncture needles with different diameters.
6. Attachment for needles to the micromanipulator.
7. Sandpaper in different grit sizes (ca. 200–2000).
8. Hot glue gun and sticks.
9. Combination pliers or another suitable tool for cutting needles.
10. Pollen tube culture medium solidified with different concentrations of low melting agarose (*see* Subheading 3.1).
11. Sample heater.
12. Single wells (ca. 250 μL volume) from 8 × microplate stripes (dimensions of single 96-wells).
13. Phosphate-buffered saline (PBS).

3 Methods

3.1 Growth Medium Preparation

The selection of the pollen tube growth medium depends on the experimental question and the species studied. Simple media compositions such as BK-medium are best suited for agarose, as components of more complex media such as polyethylene glycol (PEG) may precipitate with agarose.

1. Prepare the pollen tube growth medium required for your experimental setup.
2. Add the desired concentration of low melting agarose.
3. Boil the suspension shortly and only once in the microwave. Do not let the medium boil over.
4. Split the growth medium into 2-mL aliquots (in 2-mL tubes) or another volume that fits your needs (*see* **Note 3**).
5. Store aliquots at −20 °C until use.

3.2 Pollen Isolation and Sample Preparation

1. Carefully remove the anthers from the flowers using forceps (*see* **Note 4**) (*see* also Chapter 22 by Fritz and Kost).
2. Hydrate the entire anthers including the pollen grains in a sealable box with moist cloths for at least 1 h. Take care not to submerge the anthers in water.
3. For each agarose concentration to be investigated in your experimental setup, liquefy 5.5–6 mL of the agarose medium (*see* **Note 5**) at 75 °C for 20–25 min in a sample heater.
4. Fill 1.5 mL agarose medium of every agarose concentration in one cavity of a 6-well plate. Distribute the medium evenly at the bottom and let the medium solidify at room temperature for 15 min (*see* **Note 6**).
5. Cool down the rest of the agarose medium to 37 °C in the sample heater (*see* **Note 7**) and keep it at this temperature (*see* **Note 8**).
6. Transfer the anthers containing the hydrated pollen grains into a small volume (100–200 μL) of liquid medium without agarose and vortex intensively (*see* **Note 9**).
7. For each agarose concentration to be investigated, add 10–20 μL of the prepared pollen grain suspension to 2 mL medium (37 °C) with the respective agarose concentration. Vortex in regular intervals to avoid pollen grain sedimentation.
8. Fill the resulting suspensions as quickly as possible into the cavities with the same agarose concentration and let the agarose solidify at room temperature for 10 min.
9. Add 1.5 mL agarose medium of every concentration in the corresponding wells and let the agarose solidify at room temperature for 5 min (*see* **Note 10**). You should end with one layer of pollen grain containing agarose medium between two layers without pollen grains in each well.

3.3 Bright Field Image Acquisition

1. Transfer the 6-well plate carefully to the motorized microscope stage.
2. Search for areas with a suitable pollen grain density (*see* **Note 11**).
3. Save each position with *y*, *x*, and *z* coordinates.
4. Take image stacks at these positions, with 100 images in 10 μm distance to each other in regular intervals of 10 min, or adjust these parameters to your needs.

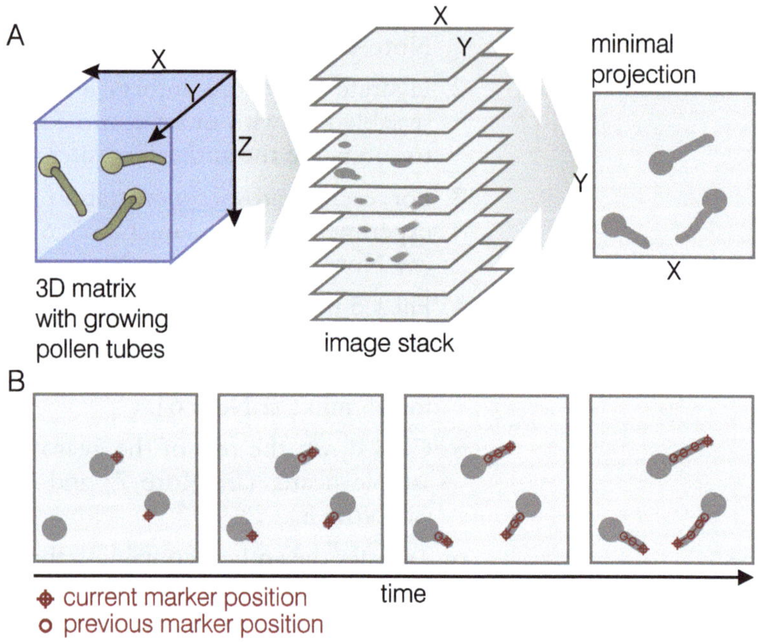

Fig. 1 Acquisition and processing of image data for the analysis of pollen tube growth in a 3D matrix. (**a**) Schematic representation of the image stack acquisition and the subsequent processing to minimal projections. Such image stacks are obtained at regular intervals to obtain time series. (**b**) The time series of the minimal projections are used to track pollen tubes with the image analysis software ClickPoints by setting markers to the pollen tube tips in every time frame

5. Save the image stacks sorted by position or ensure later sorting based on the image information.

3.4 Image Analysis of Bright Field Images

The following steps require a computer program that is capable of transforming image stacks into minimal projections (*see* Fig. 1a and **Note 12**). For simplicity we recommend using the image processing program ImageJ for an individual time series of z-stacks as described below. Alternatively, you can use an individual script that can automatically create minimum projections for multiple image series.

1. Sort your image stacks according to the position of image acquisition in subfolders, if not already done. Each folder should contain a time series consisting of z-stacks with a constant number of images per time step.
2. Open the time series in ImageJ with "File → Import → Image Sequence..." by selecting the first image.
3. Select the image sequence and create a series of minimum projections with "Image → Stacks → Tools → Grouped ZProjection." Choose the option "Min Intensity" in the "Projection

method" spinbox and set the number of images per stack in the "Group size" textbox.

4. Save the sequence of minimum projections in a new folder with "File → Save As → Image Sequence." Choose the original file format in the "Format" spinbox.
5. Load the obtained minimal projection time series into ClickPoints.
6. Press F2 to open the hidden menu bars.
7. Right click on marker 1 in the upper corner on the left side.
8. Change the type of marker 1 to "TYPE_track".
9. Left click on the tip of pollen tube of interest and hold.
10. Switch (while keeping the left mouse button pressed) to the next time step with the right arrow key on your keyboard.
11. Repeat **steps 9** and **10** until the pollen tube has been completely measured (Fig. 1b).
12. Measure as many pollen tubes as possible or required.
13. Save the measurement. It will be saved as .cdb-file (database).
14. Extract datasets from the database with a suitable Pythons script (*see* **Note 13**).

3.5 Indentation Experiments

To estimate the invasive forces that pollen tubes exert in vitro, experiments with a microindenter can be conducted. The aim of such experiments is to determine the force required to drive a pollen tube-shaped object at a controlled speed deep into agarose-based hydrogels that serve as the 3D matrix. By varying the penetration velocity, the indenter radius, and the agarose concentration of the hydrogel, an empirical expression for the invasive force can be found. The following steps require a computer script capable of controlling the micromanipulator and simultaneously reading and storing the measured data from the scale. We recommend the open programming language *Python*.

1. Cut an acupuncture needle with combination pliers at a point quite far from the tip where it has the specified radius.
2. Move the blunt needle tip in circular movements over sandpaper to grind it into a hemispherical shape. It is recommended to use sandpaper with several grain sizes (first coarse, then fine).
3. Check the tip shape with a brightfield microscope and repeat **step 2** if necessary.
4. Attach the needle as indenter to the micromanipulator using the hot glue gun.
5. Thaw the agarose sample for 20 min at 75 °C. For samples >3% agarose use higher temperatures.

6. Pipette approx. 150 μL liquid agarose gel into each 250 μL well from the microplate stripes and allow the sample to cool for 20 min. Avoid the formation of air bubbles and pipette them out of the sample if necessary (*see* **Note 14**).
7. Fill the remaining 96-well with PBS to avoid drying out the agarose sample (*see* **Note 15**).
8. Put the sample on the microscale and align the needle parallel to the wall of the 250 μL well (*see* **Note 16**).
9. Approach the agarose hydrogel manually so that there is just enough space at the tip of the indenter to see it by eye and drive the needle with the micromanipulator into the PBS layer.
10. Zero the scale (*see* **Note 17**).
11. Start the measurement. The following steps should be implemented in a software which is written in such a way that it queries the microscale with the maximum possible frequency and can adapt the micromanipulator in response in real time.
12. Move the indenter vertically toward the sample surface at a moderate speed of 20 μm/s.
13. When the microscale measures more than 5 μg (i.e., when the sample surface is reached), start the actual measurement protocol, which moves the indenter at a fixed speed over a certain distance (Fig. 2a). A distance of 4 mm is suitable for half-filled 250 μL wells, so you have many measuring points in the sample bulk and do not reach the bottom of the well.
14. During the measurement the program should save the time points and weights in an array and save them afterwards as a readable .txt-file (saving during the measurement may cost too much performance, which is needed for the microscale query).
15. After the measurement, the program should drive the indenter up exactly in the opened channel, in order to avoid unnecessary breaking of further material.

3.6 Force Estimation from Indentation Experiments

The following section briefly describes how the force required to break up the agarose matrix can be deduced from the measurement data of the indentation experiment described above.

1. Load the measurement data so that penetration depth and weight are available for each time step in a list or array.
2. Determine the penetration depth at which the corresponding weight first exceeds a threshold of 5 mg (or another meaningful value that ensures that the indenter has reached the agarose surface for the first time at this point). Normalize the penetration depths by subtracting the numerical value of this position from all penetration depths (Fig. 2b).

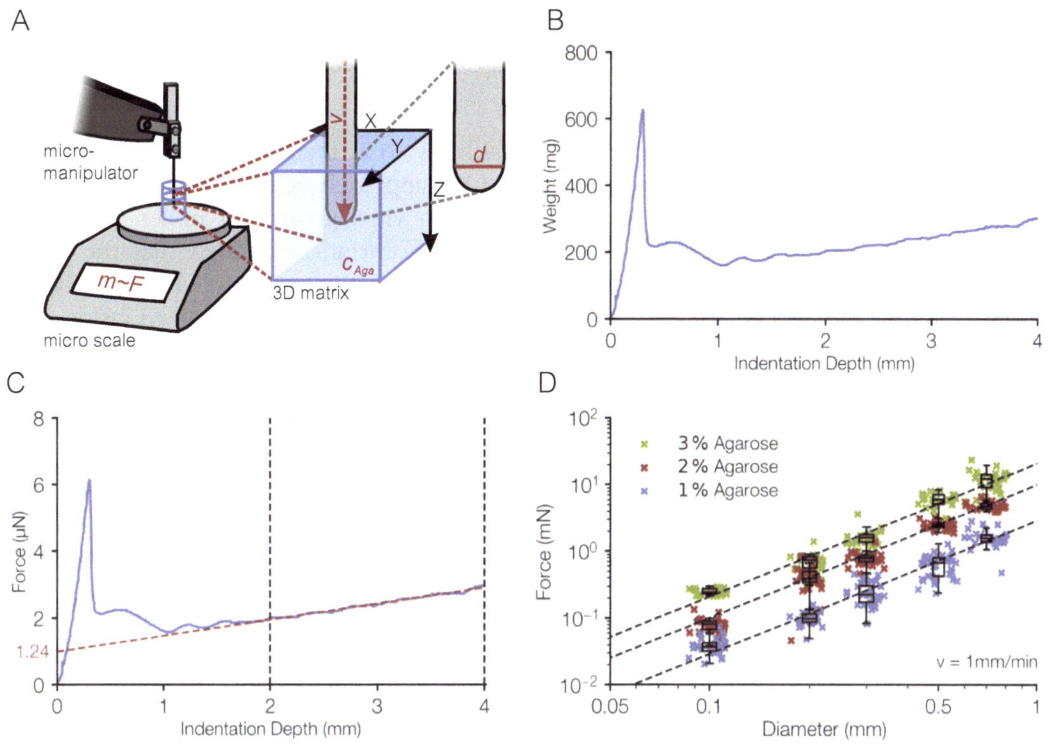

Fig. 2 Microindenter-based method to estimate the invasive force of pollen tube shaped objects. (**a**) Setup for indentation experiments. A blunt acupuncture needle with hemispherical tip is mounted on a micromanipulator and driven into an agarose hydrogel sample. A microscale is used to monitor changes in weight and hence in force. (**b**) Two phases of matrix indentation. Firstly, a lot of force is necessary to overcome the gel's surface tension [10]. Then, the force measured increases due to friction. (**c**) A friction-independent invasion force can be determined by interpolating the force at *0 depth*. This force is found to be dependent on the indentation velocity and indenter diameter, as well as on the agarose concentration (**d**)

3. Convert the measured weights into forces by multiplying each weight by the nominal gravitational constant $g = 9.81\ m/s^2$.
4. Create a mask that includes all measured values for penetration depths between 2 and 4 mm (indicated through dotted lines in Fig. 2c) (*see* **Note 18**).
5. Perform a linear fit of force (measured data that meet mask criteria) over penetration depth.
6. The *Y*-axis section corresponds to the friction corrected penetration force in the bulk material (*see* **Note 19**).
7. Measure the indentation force for as many combinations of agarose concentration, indenter diameter, and speed as possible.
8. Fit the measurement data to an expression of the form $\text{Force} = a \times \text{diameter}^b \times c_{\text{agarose}}{}^c \times \text{velovity}^d$ (indicated

through dotted lines in Fig. 2d) using the Generalized reduced gradient method (e.g., with the solver tool in Microsoft Excel) (*see* **Note 20**).

3.7 Maximum Force Estimation of Pollen Tubes

Here, we present a possibility to estimate the ***maximum*** invasive force of pollen tubes. Our approach is to perform the already presented method for the in vitro growth of pollen tubes in a hydrogel with an agarose concentration so high that the pollen tubes are no longer able to penetrate the gel. Pollen tubes of *N. tabacum*, for instance, can grow ca. 500 μm deep inside a gel with a concentration of 12% agarose before coming to a stop (and sometimes bursting). In such a gel *N. tabacum* can thus apply the maximum penetration force which can be determined from the relationship between force, diameter, agarose concentration, and velocity taken from the indenter experiments as described in Subheading 3.6. However, it is not possible to seed pollen grains in an agarose gel of more than ca. 3% because of the increasing melting point and faster solidification (which complicates handling massively). Therefore, we present a method in which the pollen grains are first germinated in a less concentrated agarose gel inside a pipette tip and then grow into a gradually stiffening external medium.

1. Prepare agarose medium with a high agarose concentration and keep it at 100 °C.
2. Cut off tips from 10 μL pipette tips (ca. 1 cm); one for every agarose concentration you want to investigate.
3. Prepare plant material and 1% agarose medium (*see* **Note 21**) as described in Subheadings 3.1 and 3.2.
4. Keep agarose medium with pollen grains at 37 °C.
5. Carefully fill 10 μL agarose medium with pollen grains with a pipette in the cut-off tips (from the cut side towards the tip opening.). Avoid air bubbles (*see* **Note 22**).
6. Fixate 2–4 of the cut-off tips in parallel with adhesive tape on the bottom of a Petri dish with a format that is suitable for microscopy.
7. Pipette a drop (ca. 500 μL) of hot highly concentrated agarose gel close to the tip openings.
8. Quickly place a small cover slip on the drop and the tip openings and push it slightly in the direction of the tips to ensure the openings are sealed with the high concentrated agarose gel (*see* **Note 23**).
9. Repeat the **steps 5–8** until you have a sufficient number of samples. Cover the Petri dish between the steps to avoid evaporation.

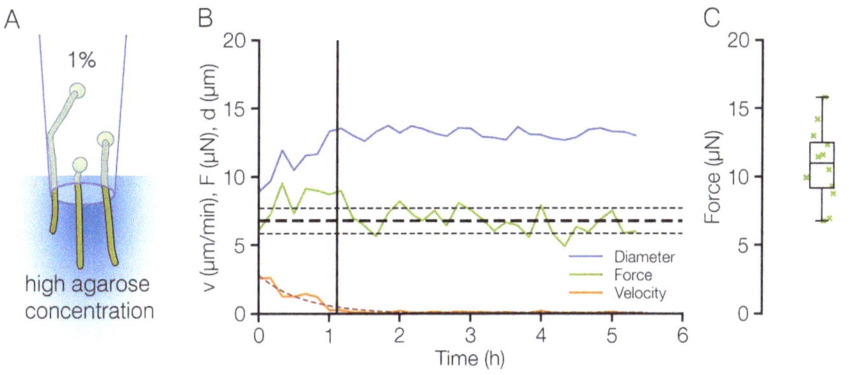

Fig. 3 Maximum force estimation. (**a**) Experimental setup: Pollen grains are germinated in low concentrated agarose gel inside a pipette tip and are guided to a certain extent by the tip into an outside matrix with a high agarose concentration. (**b**) Evaluated diameter, velocity, and force of exemplary pollen tubes of *N. tabacum* growing in 12% agarose. While pollen tubes of *N. tabacum* decrease their velocity over time (and ultimately stop), they simultaneously increase in diameter, resulting in a constant invasive force under a threshold velocity of 0.5 µm/min. (**c**) The resulting mean forces for *N. tabacum* are 10.9 ± 2.8 µN ($n = 12$)

10. Fill the Petri dish with liquid medium until everything is covered (*see* **Note 24**).
11. Image the area of the pipette tip opening for ca. 12 h using brightfield microscopy as described in Subheading 3.3 (*see* schematic illustration in Fig. 3a).
12. Repeat the experiment with varying agarose concentrations in the outer gel until the pollen tubes grow out of the pipette tip but slowdown in the solid medium and finally stop (*see* **Note 25**).
13. Analyze pollen tube diameter and velocity as described in Subheading 3.4 (exemplary data for *N. tabacum* pollen tubes are shown in Fig. 3b).
14. Estimate the penetration force for every time step using the empirically derived expression described in Subheading 3.6. (exemplary data for *N. tabacum* pollen tubes are shown in Fig. 3c).

4 Notes

1. The methods have also been successfully tested on pollen from a wide range of other species, such as *Arabidopsis thaliana*, *Eschscholzia californica*, *Capsicum chinense*, *Solanum chacoense*, *Campanula trachelium*, and *Hemerocallis fulva*.
2. For smaller pollen tubes such as pollen tubes of *A. thaliana* 12-well plates are sufficient.

3. The 2-mL tubes are well suited because they can be heated quickly and safely in heating blocks.
4. Pollen grains tend to adhere to surfaces. Therefore, we suggest keeping the pollen grains in the anthers during the first steps of our protocol if possible. But of course, pollen grains can also be hydrated directly and used further.
5. It is better to prepare more agarose medium because there are always residues of medium in the experimental vessels.
6. It is easier to distribute the medium evenly at the cavity bottom at 75 °C. Cover the 6-well plate to prevent evaporation.
7. If a second heating block is available, it can be set to 37 °C to save time.
8. This temperature was chosen to prevent pollen grains from damage. Some species might also tolerate higher temperatures.
9. Especially in media with higher agarose concentrations, pollen grains cannot be distributed evenly without suspending them in liquid medium before.
10. The third and last layer ensures that the analysis does not include pollen tubes that grow on the surface of the solid medium. Of course, the experiment can also be carried out with the purpose to analyze the growth of pollen tubes on the surface. For this purpose pollen grains can be spread directly on the solid medium prior to imaging.
11. There should be enough pollen grains so that a sufficient amount of pollen tubes can be analyzed. At the same time a too high density of pollen grains would prevent a good and uninterrupted tracking of the pollen tubes.
12. For minimal projections the first image of a stack is saved and only the pixels that are darker in the following pictures are overwritten. Thus only the darkest pixels along the *z*-axis of an image stack are kept at each position (*see* Fig. 1a).
13. See https://clickpoints.readthedocs.io/en/latest/examples.html for some examples on how to extract data from ClickPoints databases using Python.
14. Air bubbles in the agarose sample significantly falsify the results of the indenter measurement and should be avoided as much as possible. If the indenter pushes through an air bubble during a measurement, this is indicated in the raw data by a sudden positive peak followed by a negative peak in the measured force. First, the penetrator acts briefly against the surface tension of the air bubble and then requires less invasive force within the air bubble than in the agarose hydrogel.
15. By storage at room temperature the moisture evaporates from the agarose samples over time, which has an influence on the

measured forces. Even in the presence of the PBS layer, sufficient moisture can evaporate after ca. 2 h to cause a measured increase in force in the order of the otherwise measured forces at narrow needle diameters and low speeds due to the resulting higher stiffness.

16. A small metal angle is useful for the proper alignment of indenter and 250 μL well.
17. Many digital microscales have an automatic zero smoothing function that compensates for weight changes below a certain threshold around the zero point. When measuring samples with a low agarose concentration or a small indenter diameter, it is recommended to disable this function if possible.
18. A solidified agarose matrix in the 250 μL wells typically has a surface area of approx. 1 μm thickness (*see* Fig. 2b). This differs considerably from the bulk material by its inhomogeneity and the additional surface tension before puncture. This area must therefore be excluded for the subsequent fit of the force over penetration depth.
19. With a higher penetration depth there is also more surface of the indenter in the agarose gel, whereby at constant speed the measured force increases linearly by friction (*see* Fig. 2c). To correct this, it is necessary to extrapolate to the point of 0 penetration depth.
20. We performed such experiments for agarose concentrations of 1, 2, and 3%, indenter diameters of 0.1, 0.2, 0.3, 0.5, and 0.7 mm, and velocities of 0.2, 0.5, 1, 2, and 5 mm/min. We were able to fit all data to the expression $\text{Force} = 7.29 \times 10^{-4}\mu\text{N} \times \text{diameter}^2 \times c_{\text{agarose}}{}^{0.2} \times \text{velocity}^{1.8}$ with a correlation between fitted and measured forces of $R^2 = 0.9918$. Note, that without restrictions the optimal fit parameter for the diameter dependency turned out to be 1.78 and was then set to be 2 in order to represent an intuitively expected relation between indenter cross-section and material breakage in the agarose gel.
21. In addition, you can distribute fluorescent beads in the low concentrated agarose gel to later image the transition to the higher concentrated gel.
22. The cut-off pipette tip canalizes the growth of the pollen tubes in one direction (*see* Fig. 3a).
23. The hot agarose will melt the 1% agarose at the contact area and will create a narrow region with a density gradient. The pollen tube will therefore not face a sudden stiff border in the medium. Also at concentrations above 12% agarose it is very difficult to pipette agarose medium. It is possible to place the

agarose on the tip with the help of two small spatulas that are preheated in the heating block.

24. We have found that this results in much better pollen growth. Perhaps the liquid column compensates evaporation or buffers other unknown effects.
25. It is not a big problem if the pollen tubes burst as long as they show a time range of relatively constant penetration force before bursting. In some species, this force can also fluctuate strongly around an average value.

References

1. Palanivelu R, Tsukamoto T (2012) Pathfinding in angiosperm reproduction: pollen tube guidance by pistils ensures successful double fertilization. Wiley Interdiscip Rev Dev Biol 1:96–113. https://doi.org/10.1002/wdev.6
2. Sanati Nezhad A, Naghavi M, Packirisamy M et al (2013) Quantification of cellular penetrative forces using lab-on-a-chip technology and finite element modeling. Proc Natl Acad Sci 110:8093. https://doi.org/10.1073/pnas.1221677110
3. Burri JT, Vogler H, Läubli NF et al (2018) Feeling the force: how pollen tubes deal with obstacles. New Phytol 220:187–195. https://doi.org/10.1111/nph.15260
4. Gossot O, Geitmann A (2007) Pollen tube growth: coping with mechanical obstacles involves the cytoskeleton. Planta 226:405–416. https://doi.org/10.1007/s00425-007-0491-5
5. Ghanbari M, Packirisamy M, Geitmann A (2018) Measuring the growth force of invasive plant cells using Flexure integrated Lab-on-a-Chip (FiLoC). Technology 06:101–109. https://doi.org/10.1142/S2339547818500061
6. Cheung AY, Wu H-M, di Stilio V et al (2000) Pollen-Pistil interactions in *Nicotiana tabacum*. Ann Bot 85:29–37. https://doi.org/10.1006/anbo.1999.1016
7. Lennon KA, Roy S, Hepler PK, Lord EM (1998) The structure of the transmitting tissue of *Arabidopsis thaliana* (L.) and the path of pollen tube growth. Sex Plant Reprod 11:49–59. https://doi.org/10.1007/s004970050120
8. Brewbaker JL, Kwack BH (1963) The essential role of calcium in pollen germination and pollen tube growth. Am J Bot 50:859–865. https://doi.org/10.1002/j.1537-2197.1963.tb06564.x
9. Gerum RC, Richter S, Fabry B, Zitterbart DP (2017) ClickPoints: an expandable toolbox for scientific image annotation and analysis. Methods Ecol Evol 8:750–756. https://doi.org/10.1111/2041-210X.12702
10. Fakhouri S, Hutchens SB, Crosby AJ (2015) Puncture mechanics of soft solids. Soft Matter 11:4723–4730. https://doi.org/10.1039/C5SM00230C

Chapter 19

Flow Chamber Assay to Image the Response of FRET-Based Nanosensors in Pollen Tubes to Changes in Medium Composition

Theresa Maria Reimann

Abstract

Pollen tubes growing in the transmitting tract are presented with an extracellular matrix rich in a variety of substances. The expression of a multitude of genes for transport proteins in the pollen tube indicates that pollen tubes take up at least some of the components provided by the transmitting tract, for example nutrients, ions, or signaling molecules. FRET (Förster resonance energy transfer)-based nanosensors are perfectly suited to study the uptake of these molecules into pollen tubes. They are genetically encoded and can easily be expressed in Arabidopsis pollen tubes. Furthermore, the method is noninvasive and nanosensors for a wide range of substances are available. This chapter will describe the design of plasmids required to generate stable Arabidopsis lines with a pollen tube-specific expression of nanosensor constructs. We also present a method to germinate Arabidopsis pollen tubes in a flow chamber slide that allows the perfusion of the pollen tubes with liquid medium supplemented with the substrate of the nanosensor. Simultaneous evaluation of the FRET efficiency of the nanosensor by confocal microscopy reveals whether the substance is taken up by the pollen tubes. Together with the great number of available nanosensors this method can generate a detailed picture of the substances that are taken up during pollen tubes growth.

Key words FRET, Nanosensors, Pollen tubes, Arabidopsis, Pollen germination, Confocal microscopy, Perfusion chamber

1 Introduction

Soon after pollen germination the emerging pollen tubes enter the transmitting tract, a specialized tissue connecting stigma, style, and ovary. In plants of different species, the transmitting tract can be shaped quite variably. However, a common feature of all transmitting tracts is their high content of extracellular matrix that consists of a variety of substances including lipoproteins, glycolipids, glycoproteins, polysaccharides, amino acids, and free sugars [1]. The components of the transmitting tissue promote pollen tube growth by providing physical support as well as nutrients and directional cues [2]. Both, nutrients and some molecules for pollen tube

Anja Geitmann (ed.), *Pollen and Pollen Tube Biology: Methods and Protocols*, Methods in Molecular Biology, vol. 2160,
https://doi.org/10.1007/978-1-0716-0672-8_19,

guidance that are sensed intracellularly, are probably taken up into the pollen tube across the plasma membrane via transport proteins. In *Arabidopsis thaliana*, transcriptome analysis revealed the expression of at least 459 genes for putative transport proteins in pollen and pollen tubes [3–5]. However, the presence of a transport protein does not necessarily correlate with the uptake of its potential substrate, for example due to regulatory mechanisms. Therefore, it is very interesting to analyze which substances are actively taken up into pollen tubes in vivo. Genetically encoded, FRET (Förster resonance energy transfer)-based nanosensors provide a perfect tool for such uptake analyses. The common feature of these nanosensors is a protein that specifically binds the substrate of interest with two fluorophores capable of FRET fused to the *N*- and *C*-terminus of this binding protein (Fig. 2a). The conformational change induced by substrate binding to the substrate binding protein leads to a rearrangement of the two fluorophores. As FRET is distance and orientation dependent the conformational change resulting from substrate binding can be monitored by a change in FRET efficiency (Fig. 2a). Besides FRET-based nanosensors, there are also sensors with only a single fluorophore, the emission (intensity and/or wavelength) of which is modulated by ligand binding to the binding protein. In the last years genetically encoded nanosensors for a variety of substances have been developed, for example for ions like Ca^{2+} [6], Zn^{2+} [7], Mg^{2+} [8], Cl^{-} [9, 10], Hg^{2+} [11], Cu^{2+} [12], Cd^{2+} [13], and PO_4^{3-} [14], amino acids like leucine [15], methionine [16], lysine [17], glutamate [18], and histidine [19, 20], sugars like glucose [18, 21], sucrose [22], maltose [23], arabinose [23], and ribose [24], phytohormones like abscisic acid [25, 26] and gibberellic acid [27], as well as various metabolites like ATP [28, 29], cAMP [30], NADH [31], NO [32], diacylglycerol [33], phosphatidylinositol-3,4,5-trisphosphate [34], vitamin B12 [35], lactate [36], citrate [37], and ethanol [38]. Most of these sensors have been developed for the application in animal, yeast, or bacterial cells, but some have also been used to detect substrate fluxes and uptake rates in plants [21, 25–27, 39–42]. However, only few nanosensors have been exploited to characterize ion fluxes or uptake of molecules in pollen tubes [43, 44]. Most of these experiments were carried out with pollen tubes of *Nicotiana tabacum* as these pollen tubes are easy to cultivate and can be transformed transiently by particle bombardment. However, transient transformation of pollen often results in a poor transformation rate which complicates imaging. Therefore, stable transgenic lines expressing the desired nanosensor construct would be preferable. Compared to *N. tabacum* stable transgenic lines of *Arabidopsisthaliana* can be generated much faster and easier. Additionally, an enormous database of mutant lines is available for Arabidopsis, which is a further point in favor of using this model plant for nanosensor analyses. However, there are also some disadvantages when using Arabidopsis pollen tubes for nanosensor experiments: Pollen of the commonly used ecotype Col-0 have a very low in vitro germination rate in liquid medium. Furthermore, the exchange of medium, which is necessary to apply different concentrations of the substance of interest, leads to a movement of the quite small pollen tubes of Arabidopsis compared to other species, making imaging almost impossible. Here a method is presented that overcomes these drawbacks and allows the imaging of Arabidopsis pollen tubes with nanosensors under constant perfusion with liquid medium that can be supplemented with the respective ligand. The pollen tubes of plants stably expressing a nanosensor construct are grown semi-in vivo through stigmata which ensures a high pollen germination rate. The pollinated stigmata are positioned in a chamber slide filled with solid germination medium. Only after the pollen germinated and the tubes emerged from the stigmata the chamber is filled up with liquid medium and attached to two peristaltic pumps. The stigma and some pollen tubes that penetrated the solid germination medium hold the semi-in vivo pollen tubes in place even when the chamber is perfused with medium. To test whether the ligand of the nanosensor is taken up by the pollen tubes the ligand is added to the medium and the response of the nanosensor is monitored by confocal microscopy. A change of FRET efficiency after the ligand reached the pollen tubes shows that the ligand is taken up into the cytosol of the pollen tube and can bind to the nanosensor. For some nanosensors variants with different affinities for the ligand are available. The repetition of the experiment with these variants can be used to additionally estimate the steady-state concentration of the ligand in the cytosol of the pollen tube. In the following, we describe the design of constructs for the expression of nanosensor constructs in pollen tubes of *Arabidopsis thaliana*, the germination of Col-0 pollen tubes in a flow chamber and the imaging and evaluation of FRET-based nanosensors in response to their substrate.

2 Materials

2.1 Plant Material

1. *Arabidopsis thaliana* expressing a FRET-based nanosensor construct under the control of a pollen tube-specific promotor (*see* Subheading 3.1) grown for 3 weeks under a short-day regime (8 h of light/16 h of dark) and afterwards under a long-day regime (16 h of light/8 h of dark) until flowering (*see* **Note 1**). Furthermore, plants expressing the appropriate control constructs have to be grown in parallel.

2.2 Pollen Tube Flow Chamber

1. Commercially available channel slide or custom-made microscope slide according to Fig. 1a, c (*see* **Note 2**), that can be attached to two peristaltic pumps and has a channel/chamber depth of at least 4 mm.

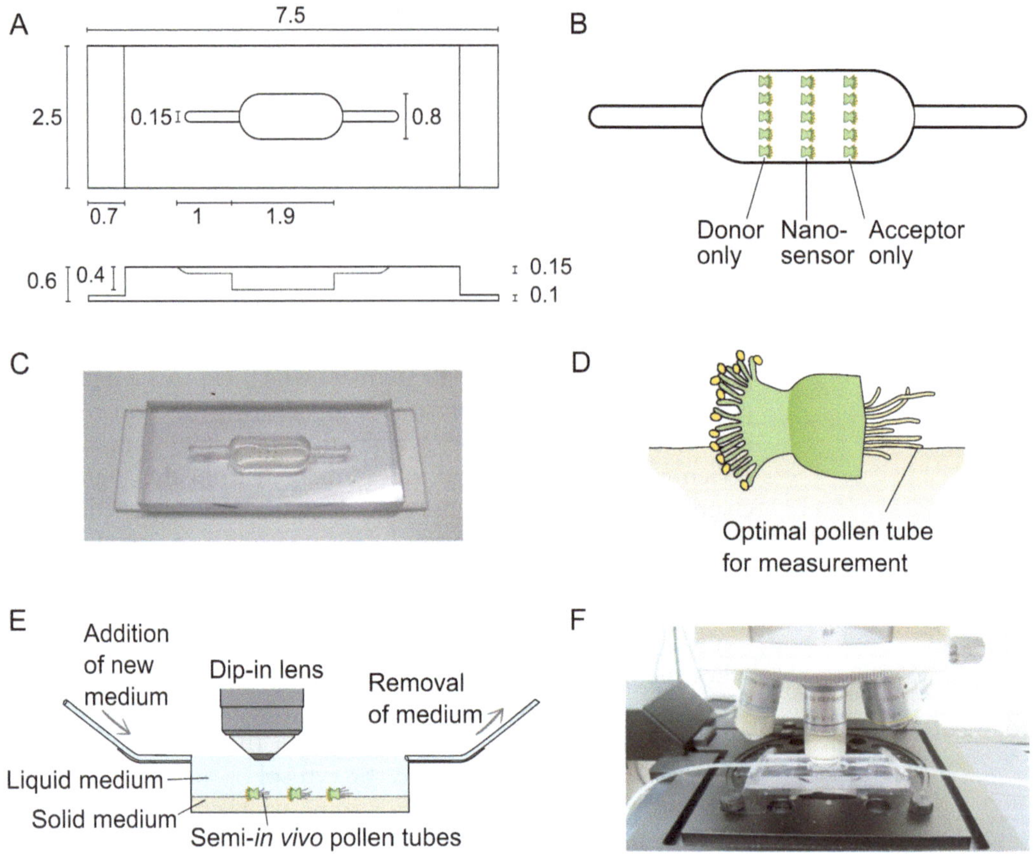

Fig. 1 Technical setup of the pollen tube flow chamber experiment. (**a**) Dimensions (in cm) of the custom-made pollen tube flow chamber. The chamber was made from a piece of acrylic glass. Silicon tubing can be pressed into the indentations to the left and right of the perfusion chamber to connect the chamber with two peristaltic pumps. (**b**) Positioning of the stigmata pollinated with pollen of nanosensor, donor only, and acceptor only control plant lines in the perfusion chamber. (**c**) Photograph of the custom-made pollen tube flow chamber. (**d**) Schematic drawing of pollen tubes grown semi-in vivo through a stigma. Some pollen tubes grow directly on top of the solid medium. These pollen tubes are especially suitable for the measurements as they barely move even during the flow caused by perfusion. (**e**) Schematic overview of the operational pollen flow chamber. The perfusion chamber is half-filled with solid germination medium, on which the pollinated stigmata are positioned. As soon as the pollen tubes emerge from the cut-surface the rest of the complete perfusion chamber is filled with liquid medium. Imaging is performed with a 25× objective dipping directly into the liquid medium. (**f**) Photograph of the pollen tube flow chamber under the microscope

2.3 Pollen Germination

1. 1 M $CaCl_2$ stock solution.
2. 1 M $Ca(NO_3)_2$ stock solution.
3. 1 M KCl stock solution.
4. 15% (w/v) casein enzymatic hydrolysate stock solution.
5. 10% (w/v) myo-inositol stock solution.
6. 10% (w/v) ferric ammonium citrate stock solution.

7. 1 M γ-amino butyric acid (GABA) stock solution.
8. 0.1 M spermidine stock solution.
9. 5% (w/v) boric acid stock solution.
10. Liquid and solid pollen germination medium according to Rodriguez-Enriquez et al. [45]: 250 mM sucrose, 1 mM $CaCl_2$, 1 mM $Ca(NO_3)_2$, 1 mM KCl, 0.03% (w/v) casein enzymatic hydrolysate, 0.01% (w/v) myo-inositol, 0.01% (w/v) ferric ammonium citrate, 10 mM GABA, 0.1 mM spermidine, 0.01% (w/v) boric acid, 0.5% (w/v) low-melt agarose (for solid medium only).
11. Microwave oven or water bath.
12. 1.5-mL tubes.
13. Thermal shaker for 1.5-mL tubes.
14. Stereomicroscope with light (gooseneck lamp).
15. Tweezers.
16. Razor blades.
17. Humid chamber (small box with wet paper towels).

2.4 Imaging and Analysis

1. Silicone tubing 1.5/1 mm (outside/inside diameter).
2. Two peristaltic pumps.
3. Upright confocal microscope with a 25× dip-in objective and lasers for excitation of the fluorophores of the selected nanosensors. For the FLIPglu sensors the donor (eCFP) was excited at 458 nm and detected between 465 and 482 nm. Acceptor (eYFP) fluorescence was excited at 514 nm, and the detection window ranged from 520 to 550 nm.
4. Software for fluorescence ratio imaging, e.g., Leica FRET Sensitized Emission Wizard.

3 Methods

3.1 Plasmid Design and Plant Transformation

In recent years, FRET-based nanosensors for a variety of substances have been established. Respective plasmids can for example be obtained from Addgene. To generate plants expressing these constructs in pollen tubes, plant transformation vectors have to be constructed that contain a strong pollen tube-specific promotor, the coding sequence for the nanosensor and a resistance marker for plant selection (Fig. 2b). In principle, every plant transformation vector can be used that meets these requirements.

1. A useful vector backbone for cloning of such constructs is for example pTR37 [44]. The coding sequence of the nanosensor can be inserted by LR reaction downstream of the strong pollen tube-specific promotor p*LAT52*.

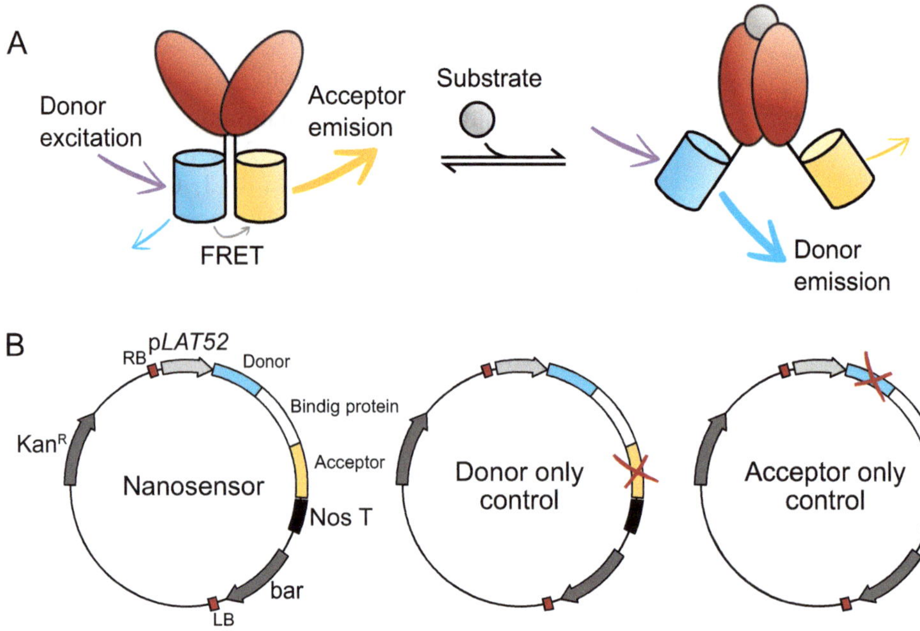

Fig. 2 Principal function of FRET-based nanosensors and plasmids for stable expression of nanosensor constructs in plants. (**a**) Scheme of a FRET-based nanosensor. A donor fluorophore (blue) and an acceptor fluorophore (yellow) are linked via a specific substrate binding protein (red). As long as donor and acceptor are in close vicinity FRET is possible and the excited donor can transfer its energy to the acceptor resulting in the emission of acceptor fluorescence. Binding of a substrate molecule to the substrate binding protein leads to a conformational change resulting in a displacement of the attached fluorophores. In the scheme this movement leads to a spatial separation of donor and acceptor and thus a reduced FRET efficiency. Note that there are other types of nanosenors, where the binding of a substrate leads to an increase of FRET efficiency. (**b**) Possible vectors for the stable expression of nanosensor and control constructs in *Arabidopsis thaliana*. Schematic diagram of plasmids for the expression of the coding sequence (CDS) for a nanosensor consisting of donor, substrate binding protein, and acceptor under the control of the strong pollen-specific promotor p*LAT52*. Donor only and acceptor only control constructs can be obtained by point mutating the coding sequence of the other fluorophore. *Bar* BASTA resistance for selection of transformed plants, *Kan^R* kanamycin resistance for bacterial cloning steps, *Nos T* nos terminator, *RB, LB* right border and left border for *Agrobacterium*-mediated plant transformation

2. Besides the nanosensor construct itself "donor only" and "acceptor only" control plasmids containing only the functional sequence for one of the fluorophores have to be generated (Fig. 2b). If these are not available, use site-directed mutagenesis with mismatch primers to destroy either the acceptor or the donor fluorophore sequence in the nanosensor construct, respectively. For nanosensors with the FRET pair eYFP/eCFP mutate either tyrosine at position 67 of eYFP or tryptophan at position 67 of eCFP to glycine. Clone the control constructs into the same expression vector as the nanosensor construct.

3. Transform the three constructs into *Agrobacterium tumefaciens* and use them to generate stable Arabidopsis lines by floral dip [46].
4. Select transgenic plants with a suitable herbicide/antibiotic.
5. Check for nanosensor fluorescence by analyzing pollen of resistant plants of the T1 generation under an epifluorescence microscope by simply dabbing an open flower onto a microscope slide.
6. Select the plants with the strongest fluorescence of all three lines (nanosensor, donor only, and acceptor only) for the flow chamber experiments.

3.2 Preparation of the Pollen Germination Medium

1. Weigh the needed amount of sucrose into a beaker and add the other components from the stock solutions (*see* **Notes 3** and **4**). Dissolve in $H_2O_{bidest.}$ and adjust the pH value with diluted KOH to 8.0.
2. Add the substance to be tested, e.g., glucose, to an aliquot of the liquid germination medium (*see* **Note 4**).
3. You can store the liquid medium at 4 °C for several days. For long-term storage either filter sterilize or store at −20 °C in 50-mL plastic tubes.
4. For the solid germination medium add 0.5% (w/v) low-melt agarose and dissolve it in the medium by warming the solution in the water bath or a microwave oven (*see* **Note 4**).
5. Pipet 1-mL aliquots of the hot medium into 1.5-mL tubes and store them at −20 °C.

3.3 Semi-In Vivo Pollen Germination in the Flow Chamber

1. Shake an aliquot of solid germination medium in a thermal shaker at ~65 °C until fluid.
2. Pipet ~250 μL of medium into the perfusion chamber of the chamber slide until the bottom is completely covered with a thin film of medium. Put the microscope slide into the humid chamber.
3. Under the stereomicroscope select flowers of the nanosensor plants in floral stage 12–13 [47], where the stigma appears between the sepals prior to the complete opening of the flower (*see* **Notes 5** and **6**).
4. Remove sepals, petals and anthers with fine tweezers. Insert the closed tweezers between two sepals and open them so that sepals and petals are spread apart. Then you can easily grab sepals, petals and anthers with the tweezers and remove them.
5. Pick an open flower and use it to pollinate the prepared stigma.

6. Wait for approximately 15 min for the pollen to adhere to the stigma. During this time, you can already pollinate more stigmata (*see* **Note 7**).
7. Pick the pollinated pistil with fine tweezers and lay it on a glass slide. With a razor blade cut off the stigma and transfer it to the pollen germination medium in the humid chamber (Fig. 1d). If possible, adjust the stigma in a way that the cut surface points away from the inflow connection (Fig. 1e).
8. Prepare at least five stigmata pollinated with pollen from the nanosensor plant.
9. Furthermore, prepare the same number of stigmata with pollen of the donor only and acceptor only control plants.
10. Place the stigmata of each plant line in one row and label them with a colored dot or so on the side of the slide (Fig. 1b).
11. Overlay the solid medium with a thin film of liquid pollen germination medium (*see* **Note 8**).
12. Incubate the pollen in the humid chamber at 22 °C.
13. As soon as the pollen tubes emerge from the cut surface (after ~2–3 h) you can fill the whole perfusion chamber of the slide with liquid germination medium. Let the pollen tubes grow for 2 more hours up to overnight.

3.4 Imaging

Here the imaging of an eCFP/eYFP-based nanosensor using an upright Leica SP5 confocal microscope (*see* **Note 9**) is described. However, the principal workflow can be transferred to other FRET pairs and imaging applications.

1. Connect the chamber slide to the two peristaltic pumps with silicone tubing. Set up the pumps so that one delivers new medium into the chamber and the second pump removes medium from the chamber (*see* **Notes 10** and **11**).
2. Transfer the sample to the confocal microscope. A schematic drawing and a photograph of the imaging setup are presented in Fig. 1e, f.
3. Switch on the microscope and start the Leica FRET SE Wizard.
4. Activate the 458-nm laser for excitation of eCFP and the 514-nm laser for the excitation of eYFP (*see* **Note 12**) and set the detection windows to 465–482 nm and to 520–550 nm, respectively.
5. Use the 25× dip-in objective for imaging. The objective can be directly immersed into the liquid pollen germination medium.
6. Select a pollen tube of the nanosensor line (both FRET fluorophores intact) and zoom in.

7. With both lasers active in the live mode set the gain of the detection windows for donor and acceptor. If necessary, you can also adjust the laser dose.
8. Reduce the 514-nm laser (≙direct excitation of the acceptor) to 0%.
9. Increase the gain of the acceptor detection window (520–550 nm) until the acceptor signal is slightly below saturation (*see* **Note 13**).
10. In the workflow panel, go to the step "define the acceptor settings."
11. Reduce the donor excitation (458-nm laser) to 0%.
12. In the live scan mode slowly increase the laser for acceptor excitation (514 nm) until the acceptor detection window is slightly below saturation (*see* **Note 14**).
13. Go to the step "Corrections" in the workflow panel. This step acquires images of the donor and acceptor only controls that are later necessary to subtract bleed-through and cross-excitation effects (Fig. 3a).
14. Select "FRET" as this specimen is already in focus. Click on "Capture image."
15. Select "Donor only" and "search specimen." You can now move your slide and change the zoom to get a donor only control pollen tube in focus. Exit the "Search specimen" mode. The software will now automatically restore the imaging setup used for the FRET sample (image resolution, zoom factor). You can adjust the position of the donor only pollen tube in the live mode. Then click on "Capture image."
16. Select "Acceptor only" and repeat **step 15** with a pollen tube of the acceptor only control. *See* Fig. 3a for an overview of all correction images required.
17. Proceed to the step "Correction factors."
18. Select "Donor only—Signal." Make sure the respective image is selected in the image index panel on the left. Draw a ROI in the image viewer within the region of the pollen tube of the donor only control and click on "Accept."
19. Select "Acceptor only—Signal" and select the corresponding image from the index panel. Draw a new ROI in the pollen tube and click on "Accept" (*see* **Note 15**).
20. Click "Calculate factors" to get the correction factors and save them.
21. Go to the step "Evaluation" to start your experiment.

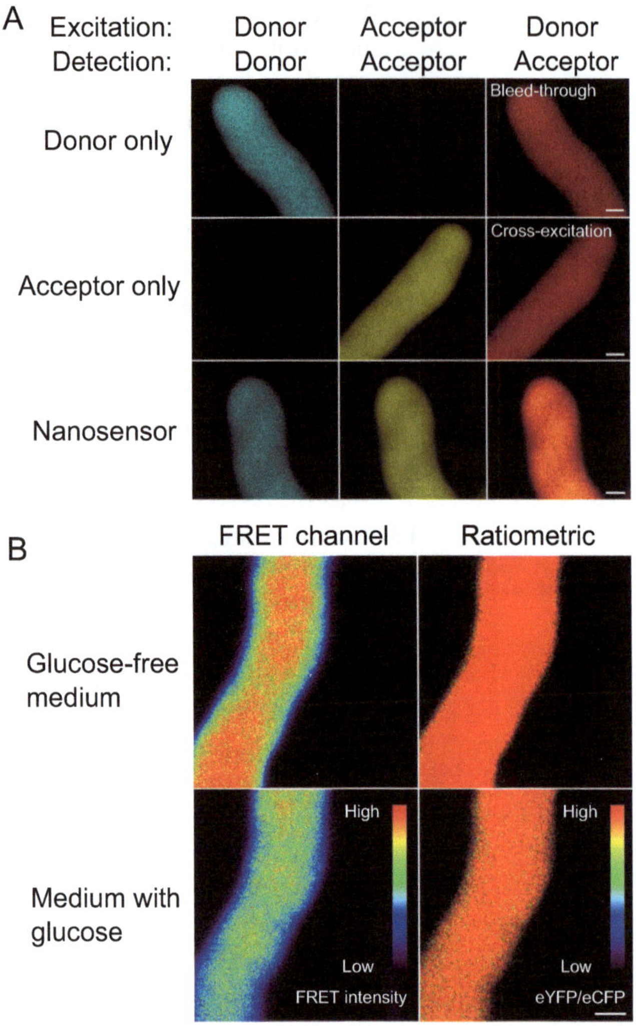

Fig. 3 FRET correction images and FRET signal of nanosensors. (**a**) Correction images of FLIPglu nanosensors in Arabidopsis pollen tubes. Constructs for FLIPglu nanosensors for the detection of glucose were expressed in Arabidopsis pollen tubes under the control of the p*LAT52* promotor [44]. In the donor only control pollen tubes only eCFP is functional, eYFP is deactivated by a point mutation (Y67G). Inversely, the acceptor only pollen tubes contain a functional eYPF whereas eCFP is deactivated by the point mutation W67G. Donor excitation was 458 nm, acceptor excitation was 514 nm. Detection windows ranged from 465 to 482 nm and from 520 to 550 nm for donor and acceptor, respectively. Fluorescence excited with 458 nm but detected in the acceptor window was classified as FRET. However, fluorescence in the FRET channel was also detected with the donor only and acceptor only controls as a result of bleed-through and cross-excitation, respectively. These effects are corrected by the calculation of the FRET efficiency. (**b**) FRET channel and ratiometric images of the same pollen tube in the absence and in the presence of glucose in the medium. The addition of glucose to the medium leads to a reduction of FRET and accordingly to a reduced ratio of eYFP/eCFP fluorescence. Scale bars: 2 μm

22. Select "Method 1" (with correction factors) or "Method 3" (without correction factors) for FRET evaluation (*see* **Note 16**) and switch to "Acquisition mode."
23. Start the peristaltic pumps and perfuse the pollen tubes with your standard medium.
24. Focus the nanosensor samples in the live mode and select a pollen tube which is not moving (*see* **Note 17**).
25. Set up a time-lapse experiment by selecting the Acquisition mode "xyzt." Define your time-lapse settings in the pop-up panel "t." A total imaging time of 1 h with images taken in 10–20 s intervals has proven useful (*see* **Note 18**).
26. Click "Run experiment."
27. After 1 min change the standard liquid medium in the supply pump to the medium supplemented with the substance to be tested.
28. Wait for 5 min, then change to standard medium again.
29. After 5 min you can supply the supplemented medium again.
30. Repeat **steps 28** and **29** until the end of your experiment (*see* **Note 19**).
31. As a control you should also perform the experiment with a substance that is not recognized by the nanosensor.

3.5 Evaluation

1. During the experiment you can sometimes already see the response of the nanosensor molecules to the added substance as intensity changes in the FRET channel and in the FRET SE visualization (Fig. 3b).
2. During or after the experiment you can switch to the "Graph" tab. Draw one or multiple ROIs in the area of the pollen tube. The automatically generated graphs show the FRET efficiency in every ROI over the time (Fig. 4).
3. Export the graph as .jpg or .tif file or save the values as an excel sheet.
4. Correlate the changes in FRET efficiency at certain time points with the changes of the medium composition (*see* **Note 20**). A change in FRET efficiency (*see* **Note 21**) indicates that the pollen tubes took up the substance of interest and it was able to bind to the nanosensor molecules in the cytosol. A return of the FRET efficiency to approximately the baseline value upon the return to perfusion with standard medium indicates that the substance is metabolized or transferred to subcellular compartments of the pollen tube.
5. The addition of medium with a control substance that is not recognized by the nanosensor should not lead to changes in FRET efficiency.

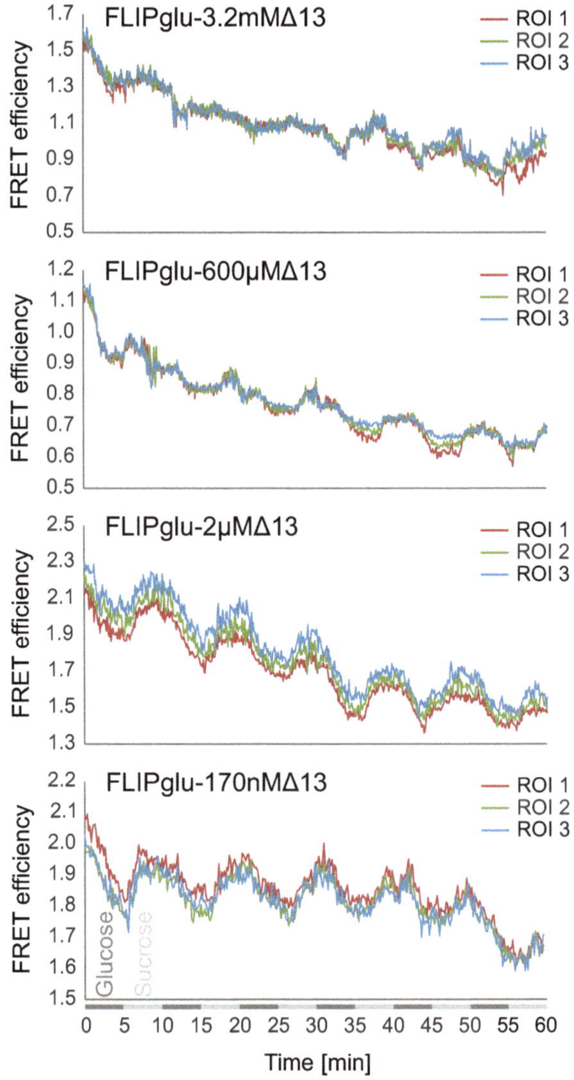

Fig. 4 Exemplary evaluation of a nanosensor experiment. Pollen tubes of *Arabidopsis thaliana* with FLIPglu nanosensors were perfused with liquid pollen germination medium. The sugar concentration in the medium was changed between 250 mM sucrose and 200 mM sucrose +50 mM glucose in intervals of 5 min. FRET was detected in three different ROIs of each pollen tube and evaluated with the Leica FRET SE wizard. The experiment was repeated with pollen tubes of four plant lines expressing variants of the FLIPglu nanosensor with different affinities for glucose [21]. The changes in glucose concentration in the cytosol in response to the changes in the glucose concentration in the medium can only be monitored with the high-affinity sensors FLIPglu-170nMΔ13 and FLIPglu-2μMΔ13 but not with the sensor variants with lower affinities (FLIPglu-600μMΔ13 and FLIPglu-3.2mMΔ13). This indicates that the steady-state glucose concentration in pollen tubes lies within the detection range of the high-affinity nanosensors, i.e., in the low micromolar range

6. For some nanosensors variants with different affinities for the substance of interest are available. If you are using one of those nanosensors you can repeat the experiment with all variants (Fig. 4). This will not only allow to answer a yes-no question on whether a substance is taken up into the pollen tube but also to make a statement on steady-state levels of the substance in the tube. If the concentration is low too few ligands will bind to a low affinity nanosensor, so no changes in FRET efficiency will occur and concentration changes cannot be observed. By contrast, very high concentrations of the ligand will always keep nanosensors with a very high affinity in a saturated state and therefore these will also not be able to detect concentration changes. The steady-state concentration of the substance therefore can be assumed to be similar to the affinity of the nanosensor variant that shows the greatest changes in FRET efficiency.

4 Notes

1. If possible, cultivate the plants in a climate chamber and not in the greenhouse. We have observed that these plants have more flowers in the perfect state for semi-in vivo pollen germination.
2. Most commercially available channel slides do not provide enough height to contain solid medium, liquid medium, and stigmata. Therefore, we constructed an optimized chamber from a piece of acrylic glass.
3. If you frequently prepare pollen germination medium you can also mix all stock solutions in the correct ratios and freeze them as an "ion mastermix" that you just have to add to your pollen germination medium once.
4. For one experiment you need approximately 1 mL of solid and 50 mL of liquid medium with and without the ligand to be tested.
5. For the preparation of the stigmata best lay the whole plant flat and position the inflorescence under the stereomicroscope. In this way you can keep the flower in position by gently holding down the inflorescence stem and/or the pedicel with your hand.
6. You can also use the transgenic plants only as pollen donors and prepare the stigmata from wild-type plants. This is especially recommended, if your transgenic lines have only few flowers.
7. To remember the temporal order of pollination, start with a flower of the main stem and proceed to flowers on the side stems from top to bottom. You can also label pollinated ovaries with a piece of thin sewing thread around the flower stalk.

8. Make sure to add only so much medium that the majority of the pollen grains on the stigmata is not submerged. Otherwise they will not germinate well. However, the thin film of liquid medium is necessary to accustom the emerging pollen tubes to the liquid medium. Otherwise the later addition of liquid medium leads to the immediate burst of the pollen tubes.
9. A detailed protocol for the Leica FRET SE Wizard can be found online at https://webcdn.leica-microsystems.com/fileadmin/academy/Appl_Let_20_FRET_SE_Sep06.pdf.
10. Adjust the flow rate of both pumps prior to your experiment to keep the fill level of the perfusion chamber constant.
11. Fill both silicon tubes with liquid medium prior to attaching them to the chamber to avoid bubbles.
12. Some lasers might need to be started at least 20 min prior to imaging. Otherwise laser intensity and thus fluorescence intensities will vary over time.
13. To check the saturation, you can activate the QLUT mode. Make sure you see no blue regions in the pollen tube as this color indicates overexposure.
14. Adjust the direct excitation of the acceptor only by changing the laser power, not by altering the gain. The latter will also affect the detection of the acceptor signal in the FRET channel.
15. Background subtraction is usually not necessary. If you want to include it, select "Donor only—background" and draw a ROI in the appropriate image outside the region of the pollen tube. Repeat this step for "Acceptor only—background."
16. Method 1 calculates the FRET efficiency E with the formula $E = (B - A \times \beta - C \times \gamma)/C$. A, B, and C correspond to the intensities of the three signals (donor, FRET, and acceptor, respectively). β and γ are the calibration factors generated by acceptor only and donor only references [48]. Method 3 calculates the FRET efficiency E with the formula $E = B/A$, i.e., FRET signal/donor signal. This method is acceptable for nanosensors with an 1:1 stoichiometry of donor:acceptor.
17. Especially pollen tubes growing directly on top of the solid germination medium are perfect for the measurements as they barely move despite the flow caused by the perfusion (Fig. 2d).
18. Depending on the expression level of your nanosensor construct, the flow rate of the peristaltic pumps, and the reaction time of the nanosensor you will have to adjust the time-lapse parameters.
19. It can happen that a pollen tube moves out of focus during the experiment. You can either readjust the focus manually or you always take a stack of images for every time point.

20. Note that it will take some time after you changed the medium in the supply pump until the new medium reaches the pollen tubes in the flow chamber. You can determine this time shift by adding a dye to the medium in a test run. Furthermore, you can also perform a test experiment exchanging the medium for H_2O_{bidest}. As soon as the water reaches the pollen tubes they burst immediately.
21. Depending on the structure of the nanosensor, binding of the substance of interest will either lead to a decrease or an increase of FRET efficiency.

References

1. Cheung AY, Wang H, Wu H (1995) A floral transmitting tissue-specific glycoprotein attracts pollen tubes and stimulates their growth. Cell 82:383–393
2. Johnson MA, Preuss D (2002) Plotting a course: Multiple signals guide pollen tubes to their targets. Dev Cell 2:273–281
3. Pina C, Pinto F, Feijó JA et al (2005) Gene family analysis of the Arabidopsis pollen transcriptome reveals biological implications for cell growth, division control, and gene expression regulation. Plant Physiol 138:744–756
4. Wang Y, Zhang W-Z, Song L-F et al (2008) Transcriptome analyses show changes in gene expression to accompany pollen germination and tube growth in Arabidopsis. Plant Physiol 148:1201–1211
5. Tavares B, Domingos P, Dias PN et al (2011) The essential role of anionic transport in plant cells: the pollen tube as a case study. J Exp Bot 62:2273–2298
6. Miyawaki A, Llopis J, Heim R et al (1997) Fluorescent indicators for Ca^{2+} based on green fluorescent proteins and calmodulin. Nature 388:882–887
7. Dittmer PJ, Miranda JG, Gorski JA et al (2009) Genetically encoded sensors to elucidate spatial distribution of cellular zinc. J Biol Chem 284:16289–16297
8. Lindenburg LH, Vinkenborg JL, Oortwijn J et al (2013) MagFRET: the first genetically encoded fluorescent Mg^{2+} sensor. PLoS One 8:e82009
9. Markova O, Mukhtarov M, Real E et al (2008) Genetically encoded chloride indicator with improved sensitivity. J Neurosci Methods 170:67–76
10. Kuner T, Augustine GJ (2000) A genetically encoded ratiometric indicator for chloride: capturing chloride transients in cultured hippocampal neurons. Neuron 27:447–459
11. Gu Z, Zhao M, Sheng Y et al (2011) Detection of mercury ion by infrared fluorescent protein and its hydrogel-based paper assay. Anal Chem 83:2324–2329
12. Wegner SV, Sun F, Hernandez N et al (2011) The tightly regulated copper window in yeast. Chem Commun 47:2571–2573
13. Vinkenborg JL, van Duijnhoven SMJ, Merkx M (2011) Reengineering of a fluorescent zinc sensor protein yields the first genetically encoded cadmium probe. Chem Commun 47:11879–11881
14. Gu H, Lalonde S, Okumoto S et al (2006) A novel analytical method for in vivo phosphate tracking. FEBS Lett 580:5885–5893
15. Mohsin M, Abdin MZ, Nischal L et al (2013) Genetically encoded FRET-based nanosensor for in vivo measurement of leucine. Biosens Bioelectron 50:72–77
16. Mohsin M, Ahmad A (2014) Genetically-encoded nanosensor for quantitative monitoring of methionine in bacterial and yeast cells. Biosens Bioelectron 59:358–364
17. Ameen S, Ahmad M, Mohsin M et al (2016) Designing, construction and characterization of genetically encoded FRET-based nanosensor for real time monitoring of lysine flux in living cells. J Nanobiotechnol 14:49
18. Deuschle K, Okumoto S, Fehr M et al (2005) Construction and optimization of a family of genetically encoded metabolite sensors by semirational protein engineering. Protein Sci 14:2304–2314

19. Okada S, Ota K, Ito T (2009) Circular permutation of ligand-binding module improves dynamic range of genetically encoded FRET-based nanosensor. Protein Sci 18:2518–2527
20. Hu H, Gu Y, Xu L et al (2017) A genetically encoded toolkit for tracking live-cell histidine dynamics in space and time. Sci Rep 7:43479
21. Deuschle K, Chaudhuri B, Okumoto S et al (2006) Rapid metabolism of glucose detected with FRET glucose nanosensors in epidermal cells and intact roots of Arabidopsis RNA-silencing mutants. Plant Cell 18:2314–2325
22. Lager I, Looger LL, Hilpert M et al (2006) Conversion of a putative Agrobacterium sugar-binding protein into a FRET sensor with high selectivity for sucrose. J Biol Chem 281:30875–30883
23. Kaper T, Lager I, Looger LL et al (2008) Fluorescence resonance energy transfer sensors for quantitative monitoring of pentose and disaccharide accumulation in bacteria. Biotechnol Biofuels 1:11
24. Lager I, Fehr M, Frommer WB et al (2003) Development of a fluorescent nanosensor for ribose. FEBS Lett 553:85–89
25. Waadt R, Hitomi K, Nishimura N et al (2014) FRET-based reporters for the direct visualization of abscisic acid concentration changes and distribution in Arabidopsis. elife 3:e01739
26. Jones AM, Danielson JÅ, ManojKumar SN et al (2014) Abscisic acid dynamics in roots detected with genetically encoded FRET sensors. elife 3:e01741
27. Rizza A, Walia A, Lanquar V et al (2017) In vivo gibberellin gradients visualized in rapidly elongating tissues. Nat Plant 3:803–813
28. Imamura H, Huynh Nhat KP, Togawa H et al (2009) Visualization of ATP levels inside single living cells with fluorescence resonance energy transfer-based genetically encoded indicators. Proc Natl Acad Sci 106:15651–15656
29. Lerchundi R, Kafitz KW, Winkler U et al (2018) FRET-based imaging of intracellular ATP in organotypic brain slices. J Neurosci Res 97:933–945
30. Nikolaev VO, Bünemann M, Hein L et al (2004) Novel single chain cAMP sensors for receptor-induced signal propagation. J Biol Chem 279:37215–37218
31. Hung YP, Albeck JG, Tantama M et al (2011) Imaging cytosolic NADH-NAD+ redox state with a genetically encoded fluorescent biosensor. Cell Metab 14:545–554
32. Pearce LL, Gandley RE, Han W et al (2000) Role of metallothionein in nitric oxide signaling as revealed by a green fluorescent fusion protein. Proc Natl Acad Sci 97:477–482
33. Nishioka T, Aoki K, Hikake K et al (2008) Rapid turnover rate of phosphoinositides at the front of migrating MDCK cells. Mol Biol Cell 19:4213–4223
34. Yoshizaki H, Aoki K, Nakamura T et al (2006) Regulation of RalA GTPase by phosphatidylinositol 3-kinase as visualized by FRET probes. Biochem Soc Trans 34:851–854
35. Ahmad M, Mohsin M, Iqrar S et al (2018) Live cell imaging of vitamin B12 dynamics by genetically encoded fluorescent nanosensor. Sensors Actuators B Chem 257:866–874
36. San Martín A, Ceballo S, Ruminot I et al (2013) A genetically encoded FRET lactate sensor and its use to detect the Warburg effect in single cancer cells. PLoS One 8:e57712
37. Ewald JC, Reich S, Baumann S et al (2011) Engineering genetically encoded nanosensors for real-time in vivo measurements of citrate concentrations. PLoS One 6:e28245
38. Soleja N, Manzoor O, Nandal P et al (2019) FRET-based nanosensors for monitoring and quantification of alcohols in living cells. Org Biomol Chem 17:2413–2422
39. Chaudhuri B, Hörmann F, Lalonde S et al (2008) Protonophore- and pH-insensitive glucose and sucrose accumulation detected by FRET nanosensors in Arabidopsis root tips. Plant J 56:948–962
40. Lanquar V, Grossmann G, Vinkenborg JL et al (2014) Dynamic imaging of cytosolic zinc in Arabidopsis roots combining FRET sensors and RootChip technology. New Phytol 202:198–208
41. Allen GJ, Kwak JM, Chu SP et al (1999) Cameleon calcium indicator reports cytoplasmic calcium dynamics in Arabidopsis guard cells. Plant J 19:735–747
42. Mukherjee P, Banerjee S, Wheeler A et al (2015) Live imaging of inorganic phosphate in plants with cellular and subcellular resolution. Plant Physiol 167:628–638
43. Michard E, Dias P, Feijó JA (2008) Tobacco pollen tubes as cellular models for ion dynamics: improved spatial and temporal resolution of extracellular flux and free cytosolic concentration of calcium and protons using pHluorin and YC3.1 CaMeleon. Sex Plant Reprod 21:169–181
44. Rottmann T, Fritz C, Sauer N et al (2018) Glucose uptake via STP transporters inhibits in vitro pollen tube growth in a HEXOKINASE1-dependent manner in *Arabidopsis thaliana*. Plant Cell 30:2057–2081

45. Rodriguez-Enriquez MJ, Mehdi S, Dickinson HG et al (2013) A novel method for efficient in vitro germination and tube growth of *Arabidopsis thaliana* pollen. New Phytol 197:1–12

46. Clough SJ, Bent AF (1999) Floral dip : a simplified method for Agrobacterium-mediated transformation of *Arabidopsis thaliana*. Plant J 16:735–743

47. Smyth DR, Bowman JL, Meyerowitz EM (1990) Early flower development in Arabidopsis. Plant Cell 2:755–767

48. Wouters F, Verveer P, Bastiaens P (2001) Imaging biochemistry inside cells. Trends Cell Biol 11:203–211

Chapter 20

Quantification of Mechanical Forces and Physiological Processes Involved in Pollen Tube Growth Using Microfluidics and Microrobotics

Jan T. Burri, Gautam Munglani, Bradley J. Nelson, Ueli Grossniklaus, and Hannes Vogler

Abstract

Pollen tubes face many obstacles on their way to the ovule. They have to decide whether to navigate around cells or penetrate the cell wall and grow through it or even within it. Besides chemical sensing, which directs the pollen tubes on their path to the ovule, this involves mechanosensing to determine the optimal strategy in specific situations. Mechanical cues then need to be translated into physiological signals, which eventually lead to changes in the growth behavior of the pollen tube. To study these events, we have developed a system to directly quantify the forces involved in pollen tube navigation. We combined a lab-on-a-chip device with a microelectromechanical systems-based force sensor to mimic the pollen tube's journey from stigma to ovary in vitro. A force-sensing plate creates a mechanical obstacle for the pollen tube to either circumvent or attempt to penetrate while measuring the involved forces in real time. The change of growth behavior and intracellular signaling activities can be observed with a fluorescence microscope.

Key words Pollen tube (PT), Growth, Force sensor, Perceptive force, Penetrative force, Lab-on-a-chip (LOC), Microelectromechanical system (MEMS), Calcium (Ca^{2+}), Imaging, Fluorescence

1 Introduction

Pollen tubes (PTs) have to navigate a labyrinth of obstacles on their way to the ovule where they burst and release the sperm cells they carry to effect double fertilization. This requires a continuous adaptation of growth parameters, such as growth rate and orientation, which is accompanied by changes in physiological parameters, cytoskeletal organization, and cell wall properties (reviewed in: [1]). For example, a tip-focused calcium (Ca^{2+}) gradient is necessary for PT growth [2–4] and could be involved in growth reorientation [5, 6]. A growing number of molecular sensors, mostly genetically encoded fluorescent proteins, have been developed to visualize and quantify many of the intracellular factors involved [7].

Anja Geitmann (ed.), *Pollen and Pollen Tube Biology: Methods and Protocols*, Methods in Molecular Biology, vol. 2160, https://doi.org/10.1007/978-1-0716-0672-8_20,

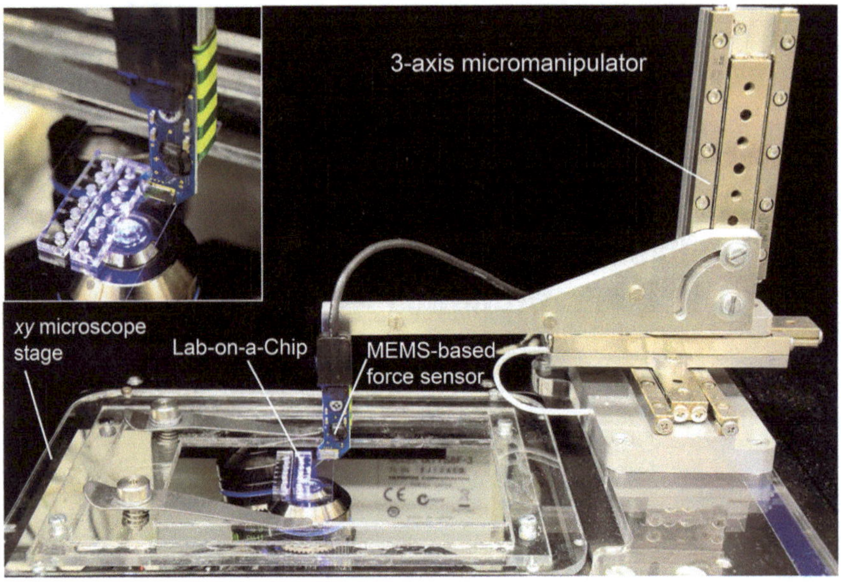

Fig. 1 System configuration to simulate the natural environment of the pollen tube in vitro with integrated force readout. It consists of a LOC device, a MEMS-based capacitive force sensor, and a 3-axis micropositioner mounted on a *xy* microscope stage of a fluorescence microscope. Close up view on the top left

To simultaneously study the forces that are involved in PT growth and their effect on cellular parameters, we developed a system that combines a microfluidics Lab-On-a-Chip (LOC) device with a microrobotics system, i.e., a MicroElectroMechanical System (MEMS)-based force sensor (Fig. 1). The LOC consists of multiple microchannel systems and serves to parallelize the growth of multiple PTs confined to the same focal plane, which greatly facilitates force measurements and microscopic observation [8]. Polydimethylsiloxane (PDMS) microchannels can be produced for a wide range of PT diameters and, if required, in many shapes (i.e., zigzag). Based on components of the Cellular Force Microscope (CFM), [9–17] the microrobotics system precisely positions the force-sensing microplate in front of the microchannels to wait for the PTs to emerge and to interact with them. With this setup, we found that PTs press against an obstacle until they reach a perceptive threshold force. Then, after a short lag phase, they reorient their growth direction [18].

To analyze the corresponding changes in the levels of physiological parameters or cytoskeletal organization in real-time, fluorescence imaging at high spatial and temporal resolution is necessary. Especially, if quantification is based on ratiometric imaging, the use of a dual view camera system, which acquires both the donor and acceptor channels simultaneously, is advisable. For ratiometric systems with two different excitation wavelengths, a fast-switchable

LED illumination system works well. Alternatively, fast filterwheels can be used in the excitation and/or emission light path.

Since individual measurements often take minutes or even hours, the resulting image stacks can contain hundreds or even thousands of frames. Manual handling of such large datasets is tedious if not impossible and, therefore, requires automatic image processing, including background subtraction and bleaching correction. If growth-related changes in tip-growing cells are to be analyzed, permanent tip tracking is of the essence.

Here, we present a protocol for quantifying the changes in the tip-based Ca^{2+} gradient when a PT grows into an obstacle, while simultaneously measuring the involved forces and changes in growth direction. For Ca^{2+} quantification we used the Förster Resonance Energy Transfer (FRET)-based ratiometric sensor Yellow Cameleon 3.60 (YC3.60) [19]. We describe custom-made image processing and analysis toolkits that we developed for the efficient analysis of large datasets. As the LOCs used for this method are not commercially available, we also provide a protocol for their fabrication.

2 Materials

2.1 Lab-on-a-Chip for High-Throughput Pollen Germination and Guided Pollen Tube Growth

1. 4 in. silicon (Si) wafer.
2. Photoresist SU-82005 and SU-83025.
3. Photomasks with the profile of the microchannels and the grain reservoirs (e.g., designed in the Siemens NX CAD software and ordered from Selba AG).
4. Spin-coater.
5. Hot plate.
6. Mask aligner.
7. Developer mr-DEV 600.
8. Polydimethylsiloxane (PDMS).
9. Desiccator.
10. Acetone, isopropanol (IPA), deionized (DI) water.
11. Heating oven at 80 °C.
12. Razor blade or box cutter.
13. 1.5 mm biopsy punch.
14. Cover glass (24 mm × 60 mm).
15. Adhesive tape.
16. O_2 plasma asher.

2.2 Lab-on-a-Chip Loading Tool

1. 5 ml syringe (or smaller) with a slip tip.
2. 200 μl plastic pipette tip.
3. 15-cm-long silicon tube with an inner diameter of 2 mm.
4. Short plastic tube (6–8 mm long) that fits tightly into the inlet holes of the LOCs (1.5 mm in diameter).

2.3 Pollen Tube Germination and Lab-on-a-Chip Loading

1. Flowering *Arabidopsis thaliana* plants. Wild-type or transformed plants expressing the molecular sensor YC3.60 (*see* **Note 1**).
2. Plant incubator or growth chamber set at 23 °C, 60% humidity.
3. *Arabidopsis* Pollen Tube Growth Medium (PTGM) based on [20] (stored at 4 °C, equilibrated to room temperature before experiment): 1.6 mM H_3BO_3, 5 mM $CaCl_2$, 5 mM KCl, 1 mM $MgSO_4$, and 10% sucrose at pH 7.5.
4. Ultra-low-gelling agarose with a melting point ≤60 °C and a gelling point ≤20 °C.
5. 1.5 ml Eppendorf tubes.
6. Thermomixer.
7. Dissecting microscope.
8. LOCs prepared under Subheading 2.1.
9. Loading tool made under Subheading 2.2.
10. Fridge at 4 °C.
11. Pipette and tips.

2.4 Microscope and Camera Setup

1. Inverted fluorescence microscope with 40× and 60× objective lenses suitable for fluorescence microscopy.
2. Dual CCD camera (alternatively: beam splitter or fast filter-wheels) set up for CFP/YFP ratio imaging.
3. Fluorescence LED illumination system equipped with suitable LED Array Modules (LAMs).
4. An *xy* microscope stage actuated by a piezomotor controller with microscopy slide holder.

2.5 Force Measurement and Force Sensor Positioning

1. A force-sensing plate with a side length of the same scale as the diameter of the PTs and the microchannels which can be placed less then 1 μm above the cover glass in front of the microchannels of the LOC. The force-sensing direction has to correspond to the direction of PT growth (e.g., parallel to the microchannels). We used a lateral force sensor from FemtoTools AG (www.femtotools.com), offering a 50 × 50 μm force-sensing plate. The standard deviation of the force in our system was 0.18 μN (*see* **Note 2**).

2. 3-axis micropositioner with submicrometer resolution composed of three orthogonal linear actuators and a control unit to either manually control the axes or, if connected to a computer, perform automated movements.
3. An adapter arm to mount the force sensor to the micropositioner (*see* **Note 3**).
4. Cable to connect the force sensor to a data acquisition device and to a computer. This connection is used for force readout and powering the sensor.

2.6 Image Processing and Analysis

1. Image processing software such as Fiji with the Bio-Formats plugin (https://imagej.net/Fiji; [21, 22]).
2. FRET-IBRA toolkit (https://github.com/gmunglani/fret-ibra).
3. TIGRMUM toolkit (https://github.com/gmunglani/TIGRMUM).

3 Methods

3.1 Lab-on-a-Chip Device Fabrication Using Soft-Lithography [8]

3.1.1 Create 2-Layer Mold (First Layer: Defines Height of Microchannels, Second Layer: Defines Height of Grain Reservoirs)

1. Clean Si-wafer (Fig. 2a) with acetone, IPA, and DI water in an ultrasonic bath.
2. Apply an even layer of SU-8 photoresist to the Si-wafer with spin-coating (Fig. 2b). The thickness of the resist defines the height of the microchannels (e.g., 7 μm for *Arabidopsis* with SU-82005, *see* **Note 4**).
3. Dry on a hot plate for 2 min 10 s at 95 °C (soft bake).
4. The design of the photomask containing the microchannels and the grain reservoirs is transferred into the photoresist using UV exposure (Fig. 2c).
5. Put on a hot plate for 3 min 10 s at 95 °C (postexposure bake).
6. Develop photoresist in two subsequent baths with developer mr-DEV 600 for 2 min 10 s and for 10 s, rinse with IPA and DI water and dry it with N_2 (Fig. 2d).
7. Put on a hot plate for 10 min at 150 °C (hard bake) to reduce stress in resist material.
8. Spin coat a second layer of SU-8 to set the height of the grain reservoirs (e.g., 30 μm for *Arabidopsis* with SU-83025, *see* **Note 4**).
9. Repeat **steps 3–5** with a photomask containing the reservoirs without the microchannels and different times and temperatures: Soft bake for 11 min 40 s at 95 °C, postexposure bake for 1 min at 65 °C and for 3 min 40 s at 95 °C, develop for 6 min, and hard bake for 10 min at 150 °C (Fig. 2f).

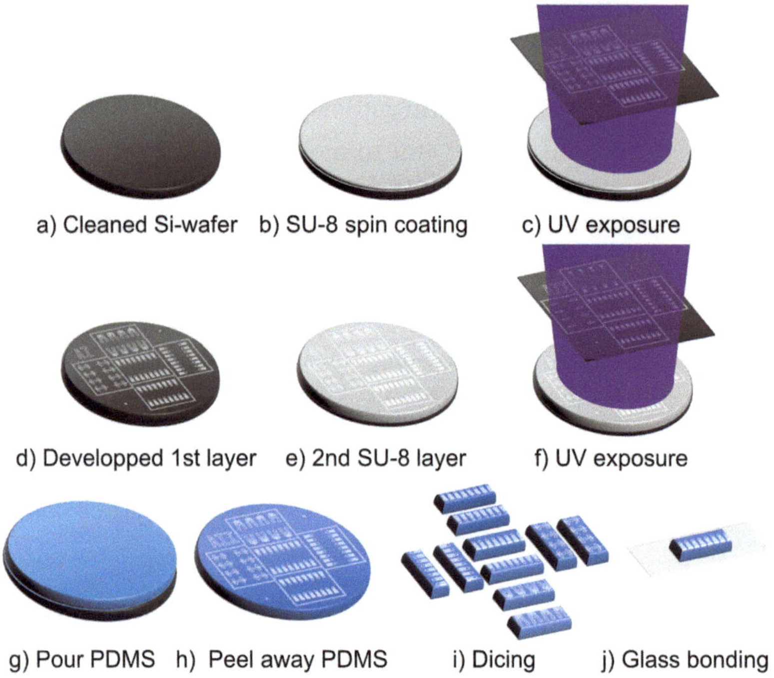

Fig. 2 Lab-on-a-chip fabrication steps. (**a–c**) First layer of the SU-8 mold, starting with a cleaned chip, which is then covered with SU-8 photoresist. The pattern of the microchannels and the grain reservoir is transferred into the resist with UV-light. (d–f) After developing the first layer, the second layer is deposited by repeating step (**b**) and (**c**) with another SU-8 resist and a mask with the design of the grain reservoirs. (**g–h**) PDMS is poured onto the mold, cured by heating, and peeled off. (**i**) The PDMS is cut into the desired shapes and (**j**) bonded to separate cover glass slides (Illustration adapted from [8])

3.1.2 PDMS Replica Molding

1. Fill ten parts of PDMS prepolymer with 1 part (in mass) of cure agent in a plastic cup. Stir the PDMS solution well until it looks cloudy and nontransparent.
2. Pour the PDMS mixture onto the mold (Fig. 2g).
3. Degas PDMS in a desiccator.
4. Cure the PDMS for 1 h in a preheated oven at 80 °C.
5. Peel the PDMS from the template and cut it into shape using razor blades or box cutters (Fig. 2h, i, *see* **Note 5**).
6. Punch inlet holes for the grain reservoirs using a biopsy puncher.
7. Clean the PDMS with adhesive tape to remove dust.
8. Expose PDMS chips and cover glasses to oxygen plasma to activate the surface.
9. Place the PDMS chips onto the cover glasses to chemically bond them (Fig. 2j).

3.2 Lab-on-a-Chip Loading Tool Assembly

1. Prepare the pipette tip: cut about 1 cm from the wide end of the 200 μl pipette tip to fit it tightly to the outlet nozzle of the syringe. Cut also about 5 mm from the pointy end to increase the opening diameter.
2. Fit prepared pipette tip to the syringe.
3. Pull silicone tube over the pipette tip as far as you can.
4. Add plastic tube into the other end of the silicone tube such that it protrudes for about 1–2 mm (*see* **Note 6**).

3.3 Pollen Germination and Growth

3.3.1 Prepare PTGM-Agarose Aliquots

1. Dissolve 0.8% (w/v) agarose in PTGM by gentle heating in a microwave. To avoid agglutination, add the agarose after the PTGM.
2. Aliquot into Eppendorf tubes (1 ml per tube).
3. Let the agarose solidify (it can be re-melted at 63 °C for later use).

3.3.2 Arabidopsis Pollen Collection and Lab-on-a-Chip Loading

1. Thaw PTGM (without agarose) or prepare freshly as in Subheading 2.3.
2. Collect 20–30 open flowers and incubate in moist chamber for ≥30 min.
3. Melt PTGM-agarose aliquot at 63 °C or prepare freshly, *see* Subheading 3.3.1.
4. Equilibrate to 23 °C (*see* **Note 7**).
5. At the same time equilibrate the loading device (syringe, tube, and whatever is used) to 23 °C (*see* **Note 8**).
6. Transfer the flowers to the prepared Eppendorf tube(s) and vortex vigorously. Additionally, shake the Eppendorf tube at maximal frequency in a thermocycler (preheated to 23 °C).
7. Spin pollen grains down at 950 × rcf for 4 min.
8. Remove as much of the floating flower debris as possible using tweezers.
9. Carefully load the LOC loading tool with the agarose-pollen mix (avoid air bubbles in the tube).
10. Load each channel system of the LOC (Fig. 3a) by fitting the protruding plastic tube of the loading tool into the inlet hole. Observe the accumulation of pollen grains in the reservoir and, when it is full, carefully remove the loading tool (*see* **Note 9**).
11. Place loaded chip in the fridge at 4 °C for 4 min (*see* **Note 10**).
12. Add liquid medium on top of the channels and in front of the channel exits (*see* **Note 11**).
13. Incubate the loaded LOCs in a moisture chamber and place it in a growth chamber at 23 °C under light (*see* **Note 12**) for 3 h (*see* **Note 13**).

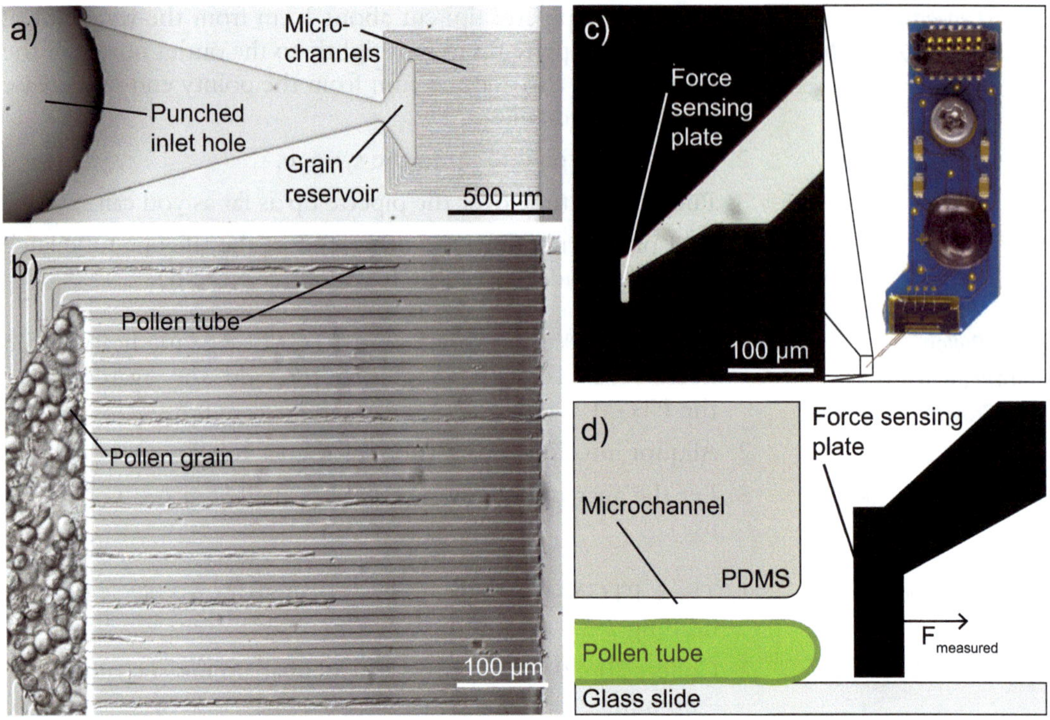

Fig. 3 Experimental configuration. (**a**) The LOC device consists of a grain reservoir and microchannels. Pollen grains are filled into the reservoir through the 1.5 mm inlet. (**b**) Pollen grains germinate inside the reservoir and the protruding PTs grow through the microchannels. (**c**) The MEMS-based capacitive force sensor provides a 50×50 µm force-sensing plate. (**d**) The force-sensing plate is placed in front of a microchannel to measure the forces during the interaction of exiting PTs with this microrobotic obstacle

14. Transfer LOC to the microscope and check for PTs in the channels to get an overview of the growing PTs and how far away they are from the channel exit (Fig. 3b).

3.4 Force Measurements and Force Sensor Positioning

1. Mount the 3-axis micropositioner onto the *xy* microscope stage.
2. Power up the microscope stage and the micropositioner.
3. Connect the sensor to the data acquisition device and mount it to the micropositioner.
4. Place the LOC on the microscope holder with the microchannels facing the force-sensing plate of the force sensor.
5. Place the force-sensing plate roughly 3 mm in front of the PDMS and above the cover glass by manually controlling the micropositioner.
6. Move the force sensor down until the force-sensing plate touches the cover glass (we refer to that as "find contact") and jump back to a safe distance of about 500 µm (*see* **Note 14**).

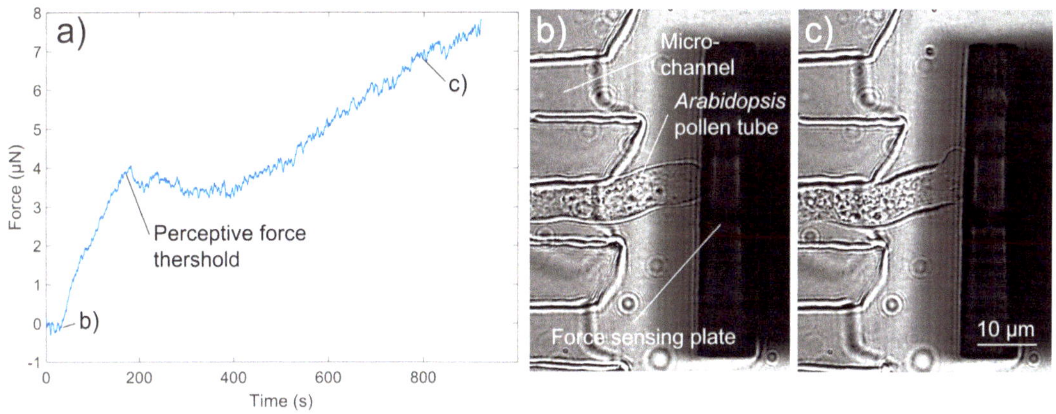

Fig. 4 Typical interaction of a pollen tube with the force-sensing plate. (**a**) Force curve of the PT-obstacle interaction, showing the initial contact and the perceptive force threshold. (**b**) Initial contact of the PT with the plate. (**c**) Change of the PT's growth behavior

7. Use the microscope stage to check the LOC for growing PTs close to the exit of a microchannel (Fig. 3b, *see* **Note 15**).
8. Move the force-sensing plate (Fig. 3c) close to the exit of the microchannel of interest (manually with the micropositioner) (*see* **Note 16**). Find contact with the cover glass and move up 1 μm (*see* **Note 14**). Then fine-adjust the position of the force-sensing plate in front of the exit manually (Fig. 3d, *see* **Note 17**).
9. Start recording the force signal (Fig. 4a) and start fluorescence imaging (*see* Subheading 3.5).
10. Wait till the PT emerges from the microchannel and interacts with the force-sensing plate (Fig. 4b).
11. Stop recording the force signal after the PT adapted its growth direction (Fig. 4c).
12. Move the sensor plate away from the microchannel and 500 μm up and repeat from **step** 7 (*see* **Note 18**).

3.5 Microscope and Camera Setup

1. Set up your imaging system for CFP/YFP ratio imaging. Excitation: use LED at 440 nm to illuminate the sample. Emission: use optical block for simultaneous imaging of CFP and YFP fluorescent signals.
2. Set focus carefully with transmitted light illumination and quickly confirm it in the fluorescence channels (*see* **Note 19**).
3. Set the LED power and exposure time to the minimum possible to reduce bleaching and phototoxic damage to the sample. Turn on LEDs only during the exposure time when images are taken or use a shutter (*see* **Note 20**).
4. Take image stacks at 1 Hz or whatever suits your needs.

3.6 Image Analysis

3.6.1 Preparation of Image Stacks for Further Analysis

The FRET-IBRA software needs two TIFF stacks as an input. Depending on your image acquisition software it may require more or less steps to get your raw image data into a useful form. Therefore, some or all of the following steps may be facultative.

1. Start Fiji and import your image stacks via the Bio-Formats Importer plugin.

   If your camera acquires both channels in a single image per time point (i.e., Orca-D2 from Hamamatsu), you have to crop each frame to get individual stacks for the donor and the acceptor channel. Tick the *Crop on import* option from the import options window and set the coordinates in the following crop options window.

   If your camera acquires individual channels per frame (i.e., Leica systems), tick *Split channels* from the import options instead.

   If you don't know the stack parameters, you can additionally tick *Display metadata* in the import options window. Later on, you will need information such as frame rate, resolution, and bit depth to configure the FRET-IBRA software. In case, your image acquisition software does not save metadata, you have to take notes of these parameters during the experiment.

2. Trim the stack if you don't want to analyze the entire stack, or if, for example, the focus is lost after some time.
3. Save the donor stack as a TIFF with the name *filename*_donor.tif. Do the same with the acceptor stack (*filename*_acceptor.tif, *see* **Note 21**).

3.6.2 Background Subtraction, Bleaching Correction, and Ratiometric Processing

1. Make yourself familiar with the FRET-IBRA software. A tutorial can be found here (https://github.com/gmunglani/fret-ibra/blob/master/examples/Tutorial.md).
2. Set the *File parameters* in the config file. An example can be found here (https://github.com/gmunglani/fret-ibra/blob/master/ibra/Config_tutorial.cfg). *File parameters* are the *input_path* (the absolute path to where your TIFF stacks are saved), the *filename*, which can be freely chosen, and the range of frames you want to process. For the *frames* parameter you can either choose a continuous range (i.e., 1:10) or individual frames (i.e., 1, 5, 10). Set the *bit_depth* parameter of your images (8, 12, or 16 bits).
3. Run the *Background subtraction* module by setting the value for the *option* parameter. The values are 0 for the acceptor channel (YFP) and 1 for the donor channel (CFP). The following steps have to be done for both channels separately.

4. Set the *Background parameters*. The size of the moving *window* must be an integer fraction of the image dimension in pixels. For an image of 1280 × 960 pixels the suggested value is 40. The *eps* value for the DBSCAN clustering algorithm should be set to 0.01 for a start. For more information check the tutorial.
5. Set the frame range (*see* Subheading 3.6.2, **step 2**). Especially for very large stacks, it is advisable to choose short ranges at the beginning and at the end of the stack (i.e., 1:20 and end-20: end) to check if the chosen *eps* values are suitable. Batches of discontinuous frames can be processed with different *eps* values, which are then automatically sorted and concatenated together. Overlapping frames are overwritten when re-processed.
6. Run FRET-IBRA with the command ./ibra.py -c <config file path> [options]. Available options can be found in the FRET-IBRA README file or by running ./ibra.py -h.
7. Check the produced video animation for appropriate background subtraction. If it didn't work for several frames, adjust the *eps* value. For an overview of all output files of FRET-IBRA *see* **Note 22**.
8. When suitable *eps* values are found, run the entire range you want to analyze.

3.6.3 Ratiometric Processing

1. Open the config file and activate the *Ratio processing* module by setting the *options* parameter to 2 and changing the values in the *Ratio parameters* section to your needs.
2. Set the *register* and *union* parameters to 1. This will align the two channels and correct for signal overlap, respectively.
3. Check the two graphs that are produced (*filename*_intensity_-nonbleach.png and *filename*_pixelcount.png, Fig. 5) for frames that did not work. They are easy to identify by sharp drops in signal intensity (*see* tutorial).
4. Correct these frames by going back and running the *Background subtraction* module on these frames for both the donor and acceptor stacks.
5. Run the *Ratio processing* module again.

3.6.4 Bleaching Correction (Optional)

1. Open the config file once more and activate the *Bleach correction* module if one or both channels showed strong bleaching (*see* **Note 23**) by setting the *options* parameter to 3.
2. Choose the frame range to which to fit the bleaching effect (*see* **Note 24**). This can be done individually for the donor and acceptor stacks.

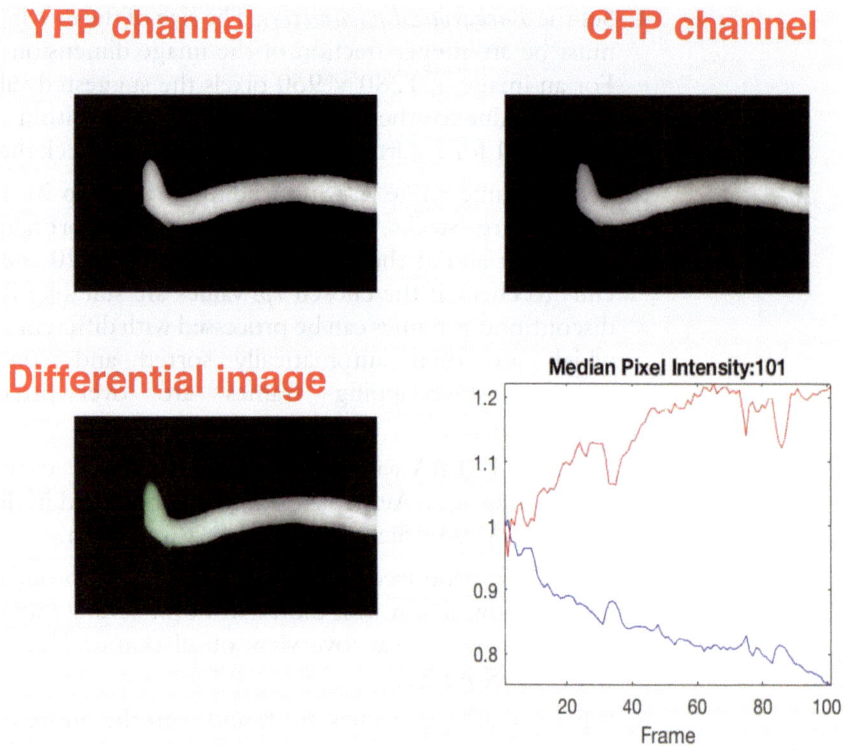

Fig. 5 Output from the FRET-IBRA software. All panels show the same frame (101) of an exemplary pollen tube. The two panels on the top show the YFP and the CFP channels after successful background subtraction. A differential image at the bottom left gives an indication of the quality of the image registration and union of the two channels. The two channels are shown in false colors (YFP: green, CFP: pink). The plot at the bottom right shows the median pixel intensity of the two channels (YFP: red curve, CFP: blue curve). Annotations in red do not appear in the original output

3. Set the *fit* parameter to either linear or exponential. This will define the type of regression that is fit to the chosen signal range.

3.6.5 Tip Tracking, Ratio Quantification and Pollen Tube Feature Analysis

1. The TIGRMUM package can be downloaded at github.com/gmunglani/TIGRMUM. It does not require any installation and can be run directly from within MATLAB.
2. The *path* and filename (*fname*) of the input files must first be set. This is followed by the start (*stp*) and end (*smp*) frame numbers of the input TIFF stack on which the algorithm should be run (*see* **Note 25**).
3. The output analysis options must then be set. This includes *tip_plot*, which outputs an animation of the frame by frame tip position named *filename*_growth.avi (Fig. 6a), *video_intensity*, which outputs a normalized pixel-intensity distribution animation named *filename*_ratio.avi (Fig. 6d), *distributions*, which displays histograms of the pixel intensities of the ROI and full

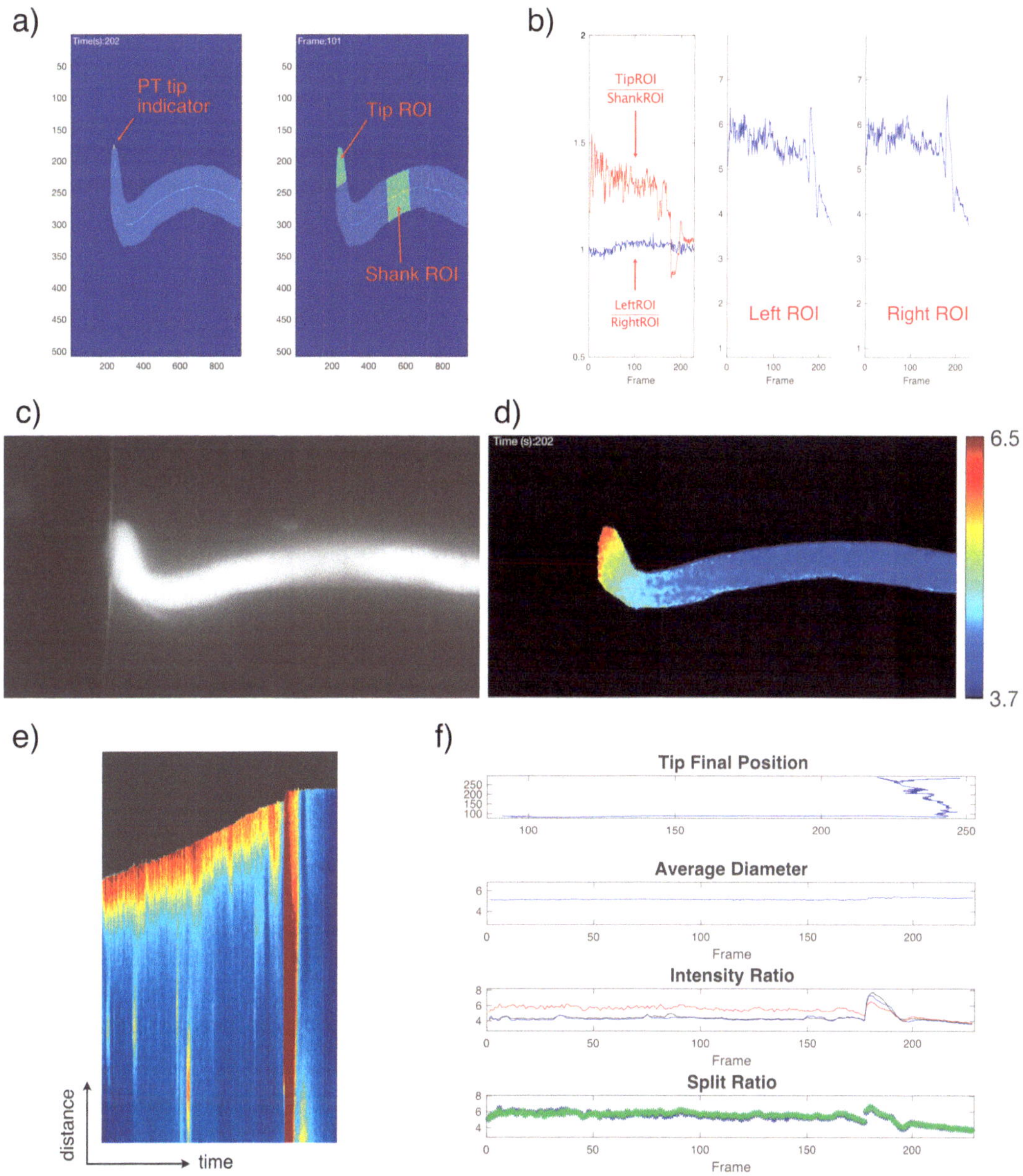

Fig. 6 Output of the TIGRMUM toolkit. (**a**) A control video to check the correct localization of the PT tip (left panel) and the correct position of the chosen Regions Of Interest (ROIs, right panel). (**b**) Quantification of the signal intensities within the ROIs. In this example, the Tip ROI was split along the central axis of the PT, such that the left and right halves can be quantified separately (middle and right panel, respectively). The left panel shows the ratio between the Tip ROI and the Shank ROI (red curve), indicating that there is a tip-focused Ca^{2+} gradient. The blue curve represents the ratio between the Left ROI and the Right ROI at the tip. A slight shift in signal intensity towards the left is seen. (**c**) The PT growing along the sensor plate after a change in growth direction (YFP channel). (**d**) The ratiometric image of the situation in (**c**) showing a shift in the tip focused Ca^{2+} gradient towards the sensor plate. (**e**) The kymograph of the ratio image along the center line of the pollen tube, again showing the tip-focused gradient and a huge Ca^{2+} burst towards the end before PT growth stops. (**d**) The toolkit delivers additional data such as tip position, the average diameter along the entire PT over time, the signal intensity of the YFP and CFP channels, and the average ratio of the whole PT (Intensity Ratio panel) as well as the ratio of the Left ROI and the Right ROI (Split Ratio). Annotations in red do not appear in the original software output

pollen tube, and *workspace*, which saves the output in a HDF5 file named *filename*_result.h5.

4. The *tip_detection* parameter then needs to be set (default: 0.5). This is a ratio parameter to aid in tip detection by weighting the ellipse method and morphological thinning (0 is only thinning and 1 is only ellipse).
5. The region of interest options then need to be stated. These include *ROI_type* (0, 1, or 2 for No ROI, Moving ROI, and Stationary ROI), *split*, which splits the ROI along its center line to measure the ratio of the intensities of right and left halves of the pollen tube, *circle*, which indicates a circle ROI as a fraction of the pollen tube diameter, *starti* and *stopi*, which prescribes the geodesic distance along the center line of the pollen tube that the ROI is created for, and *pixelsize*, which gives the pixel to μm conversion of the image.
6. Finally, the *Cmin* and *Cmax* parameters indicating the normalization scale of the pixel values for visualization need to be set, along with the averaging width (*nkymo*) of the output kymograph, *filename*_kymo.png (Fig. 6e).
7. A diameter measurement animation named *filename*_diameter.avi, the ratio of the pixel intensities between both halves of the ROI named *filename*_ROI.avi, and the average pixel intensity per frame named *filename*_pixel.avi are also included in the output.

4 Notes

1. The method described here works essentially with any kind of cytoplasmic fluorescent reporter.
2. The advantage of the lateral force sensor provided by FemtoTools AG is that the force-sensing plate is far away from the rest of the sensor. A silicon arm (around 4 mm long and 500 μm thin) connects the force-sensing plate to the sensor at an angle of 45°. In this way, the placing of the sensing plate is not hindered by the base of the sensor (including integrated circuit board) and can be as close to the exit of the microchannels as wanted. Furthermore, the risk of crashing the sensor into the cover glass or the LOC is minimized.
3. The sensor is connected to the micropositioner using an adapter arm. The design of the adapter arm can be changed according to the application. If the arm is too long (>20 cm), the position of the sensing plate slightly changes over time, which can influence the force measurement. The arm should be as short as possible to get a stable position of the sensing plate during a single PT experiment (up to 30 min). The arm can be

fabricated with a laser cutter (acrylic), with a CNC machine (metallic), or with a 3D printer.

4. The thickness varies with the type of SU-8 used and the revolutions per minute (rpm) of the spin-coater. There are look-up sheets to find the optimal combination for the desired thickness (e.g., SU-82015: https://kayakuam.com/wp-content/uploads/2019/09/SU-82000DataSheet2000_5thru2015Ver4.pdf, SU-83025, https://kayakuam.com/wp-content/uploads/2019/09/SU-8-3000-Data-Sheet.pdf).
5. The cuts in front of the microchannels have to be very straight or even a little bit tilted backwards, such that the angle between the glass slide and PDMS cut is equal or smaller than 90°. If the angle between glass slide and PDMS cut is obtuse, the top part of the PDMS will be overhanging, which makes it impossible to bring the force-sensing plate close to the exit of the microchannels. It further increases the risk of damaging the sensor, e.g., by crashing into the PDMS when moving up.
6. The plastic tube should be long enough to fit into the inlet hole but only as long as to allow the silicon tube to sit on the PDMS and seal the system tightly.
7. The equilibration (cooling) of the re-melted agarose in a thermocycler takes quite some time. Start at least 1 hour before LOC loading.
8. If two incubation time points are planned, equilibrate an empty Eppendorf tube to 25 °C and split the agarose between the two Eppendorf tubes.
9. Excess PTGM exits from the channel openings. These drops can be left to solidify and later on being carefully removed with a toothpick without disturbing the channel openings. You need to take care not to leave traces of agarose in front of the channels since this can interfere with sensor positioning.
10. Be careful. Keeping the LOCs in the fridge for too long can delay the germination for hours!
11. Medium on top of the channels needs to be largely but not completely removed for bright-field microscopy. Otherwise, the light scattering will greatly affect the quality of the images.
12. We used an incubator for *Arabidopsis* PT growth during "day time" conditions. Could also work in the dark but we have never tried it.
13. Depends on conditions and needs to be tested empirically.
14. A simple LabView™ program is used to move the sensor down at a constant velocity while the signal from the force sensor is read out. The sensor has a small cross sensitivity and slightly reacts to forces applied parallel to the force plate. When the

bottom of the force-sensing plate touches the glass slide and the signal increases above a set threshold, the motion is stopped and the micropositioner moves the sensor to a predefined position above the cover glass (e.g., 500 μm, 1 μm).

15. The micropositioner moves along with the microscope stage since it is mounted on it. Therefore, the relative position between LOC and force sensor does not change while checking the entire LOC for suitable PTs using the microscope stage, thus preventing the sensor tip from crashing into the PDMS structure.
16. For safety reasons, we move the sensor plate roughly 500 μm above the glass slide before moving it in *x* and *y* direction in order not to crash into obstacles that can lay on top of the cover glass (e.g., pollen grains). However, the sensor can still hit the PDMS structure (height ~3 mm).
17. Sometimes there is some agarose in front of the channel exit, which would influence the force measurements. By moving the plate back and forth along the PDMS chip in front of the microchannels, the agarose can be "plowed" away from the exit of the microchannels.
18. Repeat as long as there are promising PTs in the channels. Make sure that the media reservoirs on top and in front of the channels do not dry out by adding fresh medium when necessary. Once in a while, e.g., if no suitable PTs are ready for measuring, put the LOC back into the moist chamber and resupply the liquid reservoirs on top and in front of the channels.
19. Should the sample get out of focus while filming, do not try to readjust the focus because the FRET-IBRA software will have problems to deal with this later on. Instead, stop the video, readjust the focus, and start a new stack.
20. Use the CFP channel as a reference. If you can just see the signal, the FRET-IBRA software can extract it. If the software has problems discriminating between signal and background, increase exposure by either increasing LED power or exposure time.
21. Replace "*filename*" with any name of your choice, but it must be the same for the donor and acceptor stack, respectively. The "_acceptor.tif" and "_donor.tif" parts of the stack name cannot be changed.
22. Output variable names in FRET-IBRA.

    # Animation of background subtraction.

    *filename*_acceptor_frames22_28.avi (frames22_28 indicate the frame range).

    *filename*_donor_frames22_28.avi

TIFF stacks and h5 after background subtraction.
*filename*_acceptor_back.tif
*filename*_donor_back.tif
*filename*_back.h5
TIFF stack and h5 after ratio processing.
*filename*_back_ratio.h5
*filename*_back_ratio.tif
Median intensity after ratio processing.
*filename*_intensity_nonbleach.png
Pixel count after ratio processing.
*filename*_pixelcount.png
TIFF stack after bleaching (h5 is same as before).
*filename*_back_ratio_bleach.tif
Median intensity after bleaching.
*filename*_intensity_bleach.png
Log file.
filename.log.

23. If both channels bleach at the same rate, the ratio is not affected. In our hands, however, the YFP channel bleaches at a different rate than the CFP channel, which leads to an apparent decrease in the ratio. This can be corrected for by running the *bleach* module.
24. Bleaching for the YFP channel is usually pretty strong at the beginning of the exposure. Sometimes it is a good option to fit the bleach effect only to this region instead of the entire stack.
25. The input stack can be any TIFF stack produced by FRET-IBRA. If the quality is good enough, TIGRMUM works also on non-background-subtracted TIFF stacks.

References

1. Michard E, Simon AA, Tavares B et al (2017) Signaling with ions: the keystone for apical cell growth and morphogenesis in pollen tubes. Plant Physiol 173:91–111
2. Jaffe LA, Weisenseel MH, Jaffe LF (1975) Calcium accumulations within the growing tips of pollen tubes. J Cell Biol 67:488–492
3. Obermeyer G, Weisenseel MH (1991) Calcium channel blocker and calmodulin antagonists affect the gradient of free calcium ions in lily pollen tubes. Eur J Cell Biol 56:319–327
4. Pierson ES, Miller DD, Callaham DA et al (1996) Tip-localized calcium entry fluctuates during pollen tube growth. Dev Biol 174:160–173
5. Malho R, Trewavas AJ (1996) Localized apical increases of cytosolic free calcium control pollen tube orientation. Plant Cell 8:1935–1949
6. Sanati Nezhad A, Packirisamy M, Geitmann A (2014) Dynamic, high precision targeting of growth modulating agents is able to trigger pollen tube growth reorientation. Plant J 80:185–195
7. Uslu VV, Grossmann G (2016) The biosensor toolbox for plant developmental biology. Curr Opin Plant Biol 29:138–147
8. Shamsudhin N, Laeubli N, Atakan HB et al (2016) Massively parallelized pollen tube guidance and mechanical measurements on a lab-on-a-chip platform. PLOS One 11: e0168138
9. Felekis D, Muntwyler S, Vogler H et al (2011) Quantifying growth mechanics of living, growing plant cells *in situ* using microrobotics. Micro Nano Lett 6:311–316

10. Vogler H, Draeger C, Weber A et al (2013) The pollen tube: a soft shell with a hard core. Plant J 73:617–627
11. Routier-Kierzkowska AL, Weber A, Kochova P et al (2012) Cellular force microscopy for *in vivo* measurements of plant tissue mechanics. Plant Physiol 158:1514–1522
12. Felekis D, Vogler H, Grossniklaus U et al (2015) Microrobotic tools for plant biology. In: Sun Y, Liu X (eds) Micro- and nanomanipulation tools. Wiley-VCH Verlag GmbH & Co KGaA, Hoboken, NJ, pp 283–306
13. Vogler H, Felekis D, Nelson BJ et al (2015) Measuring the mechanical properties of plant cell walls. Plants (Basel) 4:167–182
14. Majda M, Sapala A, Routier-Kierzkowska AL et al (2019) Cellular force microscopy to measure mechanical forces in plant cells. In: Cvrčková F, Žárský V (eds) Plant cell morphogenesis. Humana, New York, NY, pp 215–230
15. Felekis D, Vogler H, Mecja G et al (2015) Real-time automated characterization of 3D morphology and mechanics of developing plant cells. Int J Robot Res 34:1136–1146
16. Vogler H, Shamsudhin N, Nelson BJ et al (2017) Measuring cytomechanical forces on growing pollen tubes. In: Obermeyer G, Feijó J (eds) Pollen tip growth. Springer, Cham, pp 65–85
17. Burri JT, Vogler H, Munglani G et al (2019) A microrobotic system for simultaneous measurement of turgor pressure and cell-wall elasticity of individual growing plant cells. IEEE Robot Automat Lett 4:641–646
18. Burri JT, Vogler H, Läubli NF et al (2018) Feeling the force: how pollen tubes deal with obstacles. New Phytol 220:187–195
19. Nagai T, Yamada S, Tominaga T et al (2004) Expanded dynamic range of fluorescent indicators for Ca^{2+} by circularly permuted yellow fluorescent proteins. Proc Natl Acad Sci U S A 101:10554–10559
20. Boavida LC, McCormick S (2007) Temperature as a determinant factor for increased and reproducible *in vitro* pollen germination in *Arabidopsis thaliana*. Plant J 52:570–582
21. Linkert M, Rueden CT, Allan C et al (2010) Metadata matters: access to image data in the real world. J Cell Biol 189:777–782
22. Schindelin J, Arganda-Carreras I, Frise E et al (2012) Fiji: an open-source platform for biological-image analysis. Nature Methods 9:676–682

Chapter 21

Measuring Exocytosis Rate in Arabidopsis Pollen Tubes Using Corrected Fluorescence Recovery After Photoconversion (cFRAPc) Technique

Jingzhe Guo and Zhenbiao Yang

Abstract

Exocytosis is a fundamental process essential for many cellular functions by targeting signal peptides, proteins, and cell wall components to the plasma membrane (PM) or extracellular matrix. Conventional methods, such as FRAP, often underestimate the exocytosis rate of a specific molecule, because retrieval of the molecules from the PM by endocytosis can impact the measurement. To overcome this issue, we have previously established a novel method, corrected fluorescence recovery after photoconversion (cFRAPc), which allows us to accurately measure the exocytosis rate by monitoring both exocytosis-dependent and exocytosis-independent events. In this chapter, we provide detailed procedures for the cFRAPc method to measure the exocytosis rate of Arabidopsis receptor-like kinase PRK1 in pollen tubes. This method should be widely applicable to various cell types.

Key words Pollen tube, Exocytosis rate, Confocal microscopy, Fluorescence recovery after photobleaching, Photoconversion, PRK1-Dendra2

1 Introduction

The rapid tip growth of the pollen tube requires robust exocytosis at the tip region to provide materials for the PM and cell wall [1–4], and coordinate the signaling events during elongation and guidance [5, 6]. Therefore, the pollen tube has been one of the most studied cell models for investigating the mechanism of exocytosis. Various methods have been developed to visualize and measure the exocytosis in pollen tubes, these include observing secretion vesicles by electron microscopy [7] or differential interference contrast (DIC) microscopy [8], monitoring exocytosis activity indirectly by quantifying the change of thickness of the apical cell wall [9], labeling the exocytic vesicles with fluorescent dyes or vesicle proteins with fluorescent protein tags and observing with various advanced microscopy techniques [3, 8, 10–12]. Among these methods, a fluorescence recovery after photobleaching

Anja Geitmann (ed.), *Pollen and Pollen Tube Biology: Methods and Protocols*, Methods in Molecular Biology, vol. 2160,
https://doi.org/10.1007/978-1-0716-0672-8_21, © Springer Science+Business Media, LLC, part of Springer Nature 2020

(FRAP) assay has been widely used due to its ability to quantitatively analyze the exocytosis level of a specific cargo or a population of labeled vesicles [3, 13–17]. In this type of FRAP assay, fluorescently labeled molecules that were secreted to the PM or extracellular matrix through exocytosis were firstly bleached in a specific region of interest (ROI), and the recovery of the fluorescence in the ROI was then recorded. Because the molecules were delivered primarily through exocytosis, the rate of fluorescence is directly correlated with the rate of exocytosis. However, this method probably underestimates the exocytosis rate, considering that the recovery of fluorescence in the ROI may be counteracted by endocytosis-mediated membrane internalization [2] or dilution due to insertion of new PM, both of which could dampen the recovery of fluorescence.

We have recently developed a novel FRAP-based method, corrected fluorescence recovery after photoconversion (cFRAPc) [18], to accurately determine the relative exocytosis rate of a specific protein-of-interest by correcting the loss of fluorescence recovery due to exocytosis-independent events. In the cFRAPc method, we fuse Dendra2, a green-to-red photoconvertible green fluorescent protein derived from *Dendronephthya sp.* [19, 20], with the protein-of-interest that is delivered to the PM through exocytosis. The fused Dendra2 in a specific ROI on the PM is irreversibly converted from green fluorescence to red fluorescence by local irradiation with an ultraviolet laser, after which the change of green and red fluorescence in the photoconverted ROI is monitored simultaneously. The intensity of the red fluorescence in the ROI decreases with time due to endocytosis and dilution. Because the internalized red Dendra2 will be rapidly diluted into the much larger cytoplasmic space, where the intensity of the red fluorescence remains close to zero during the cFRAPc experiment [18], we consider that exocytosis will not contribute any red fluorescence Dendra2 to the PM. Therefore, the reduction rate of the red fluorescence intensity in the photoconverted ROI directly represents the rate of exocytosis-independent events. Meanwhile, green fluorescence in the ROI is gradually recovered by two joint processes: delivery of unconverted Dendra2 fused molecules to the ROI by exocytosis, and reduction of Dendra2 on the PM by the removal or dilution via exocytosis-independent events. The reduction rate of Dendra2 on the PM by exocytosis-independent events and the rate of green fluorescence recovery are both measurable from the time-lapse images in the red and green channel, respectively; thus the rate of exocytosis can be calculated accordingly. However, our cFRAPc method is limited to those membrane proteins with low lateral diffusion rate, due to the difficulty of calibrating the signal reduction in the photoconverted region if the protein diffuses laterally. Therefore, the lateral diffusion of your protein-of-interest needs to be examined before cFRAPc analysis.

In this chapter, taking Dendra2 fused Arabidopsis transmembrane receptor-like kinase PRK1 (At5g35390) [5, 14, 18] as an example, we describe the detailed protocol of the cFRAPc method for measuring the exocytosis rate of PRK1 on the apical PM of the Arabidopsis pollen tube. This method is readily applicable to study the exocytosis rate of other PM proteins in various types of cells.

2 Materials

2.1 Germinating Pollen Tubes

1. Pollen Germination Medium for Arabidopsis: 0.01% (w/v) Boric acid, 1 mM $CaCl_2$, 1 mM Ca $(NO_3)_2$, 1 mM $MgSO_4$ (*see* **Note 1**), 18% Sucrose, 0.5% Noble agar (Difco), adjust pH to 7.0 with 0.5 M KOH.
2. 50-mm-diameter petri dishes.
3. 10 mL Luer lock syringe.
4. 0.22 μm sterile syringe filters.
5. 10 mL serological pipette.
6. Fine forceps.
7. Empty tip box.

2.2 Mounting of Pollen Tubes for Observation

1. Glass microscope slides (75 × 25 mm).
2. Square cover glasses (22 × 22 mm, No. 1.5).
3. Lanolin (loaded into a 10 mL syringe, *see* **Note 2**).
4. Mini spatula.

2.3 Imaging of Pollen Tubes

1. Zeiss upright LSM 880 confocal laser scanning microscope (*see* **Note 3**) equipped with at least two detection channels and three laser lines (405 nm, for photoconversion of Dendra2; 488 nm, for imaging green colored Dendra2 before photoconversion; 561 nm or 543 nm, for imaging red colored Dendra2 after photoconversion).
2. 40× water immersion objective lens (Zeiss LD LCI Plan-Apochromat 40×/1.2 Imm AutoCorr DIC M27).

2.4 Software for Imaging and Data Analysis

1. Zeiss Zen (black edition) microscope software.
2. Fiji (https://fiji.sc/) [21].
3. Microsoft Office Excel.

3 Methods

1. To prepare 100 mL PGM, 18 g sucrose is first dissolved into 85 mL milli-Q water in a clean glass bottle (*see* **Note 4**).

3.1 Preparation of Solid Pollen Germination Medium (PGM)

2. Add 1 mL of each 100× concentrated stock solution of boric acid, $CaCl_2$, Ca $(NO_3)_2$, and $MgSO_4$ (*see* **Note 1**).
3. Adjust the pH of the PGM to 7.0 with a few μL of 0.5 M KOH (*see* **Note 5**).
4. (Optional) Filter sterilize the PGM through a 0.22 μm syringe filter (*see* **Note 6**).
5. Add 0.5 g Noble agar, adjust the final volume to 100 mL (*see* **Note 4**).
6. Put the bottle with the PGM in a boiling water bath and swirl occasionally until agar dissolves completely (*see* **Note** 7).
7. Transfer 2.5 mL of the PGM onto each petri dish with a 10 mL serological pipette.
8. Solidified PGM can be stored at 4 °C refrigerator for up to 2 months.

3.2 Germinating Arabidopsis Pollen Tubes for Imaging

In this protocol, we use pollen tubes germinated from pollen grains of *Arabidopsisthaliana* (Col.0) expressing PRK1 (AT5G35390) fused with Dendra2 driven by the pollen specific promoter Lat52 (Lat52::PRK1-Dendra2) [14, 18].

1. Grow Arabidopsis plant in a greenhouse at 23 °C under 16 h-light/8 h-dark condition.
2. We may start to germinate pollen tubes after the plants start to flower and produce elongated siliques.
3. Fill one-third of the empty tip box with tap water, and warm up the tip box by keeping it inside a 28 °C incubator at least 1 day before germinating pollen tubes.
4. Detach a freshly opened flower from the plant, hold the pedicel in one hand, while carefully removing the sepals and petals with a pair of fine forceps.
5. Detach the filament and brush the pollen grains onto the surface of the PGM until a 3 × 3 mm region is evenly covered with pollen grains.
6. Put back the cover of the petri dish and transfer into the prewarmed tip box at 28 °C incubator.

3.3 Mounting of Pollen Tubes for Confocal Microscopic Observation

1. After incubation at 28 °C for 3–4 h, take out the petri dish and examine the germination of pollen grains under a bright-field optical microscope. If there are enough pollen tubes, you may proceed to mount the pollen tubes for imaging. If there are few or no pollen tubes germinated, you need to optimize the PGM and germinate the pollen grains again (*see* **Note 5**).
2. Apply four drops of lanolin by pushing it through the syringe onto each corner of the square cover glass.

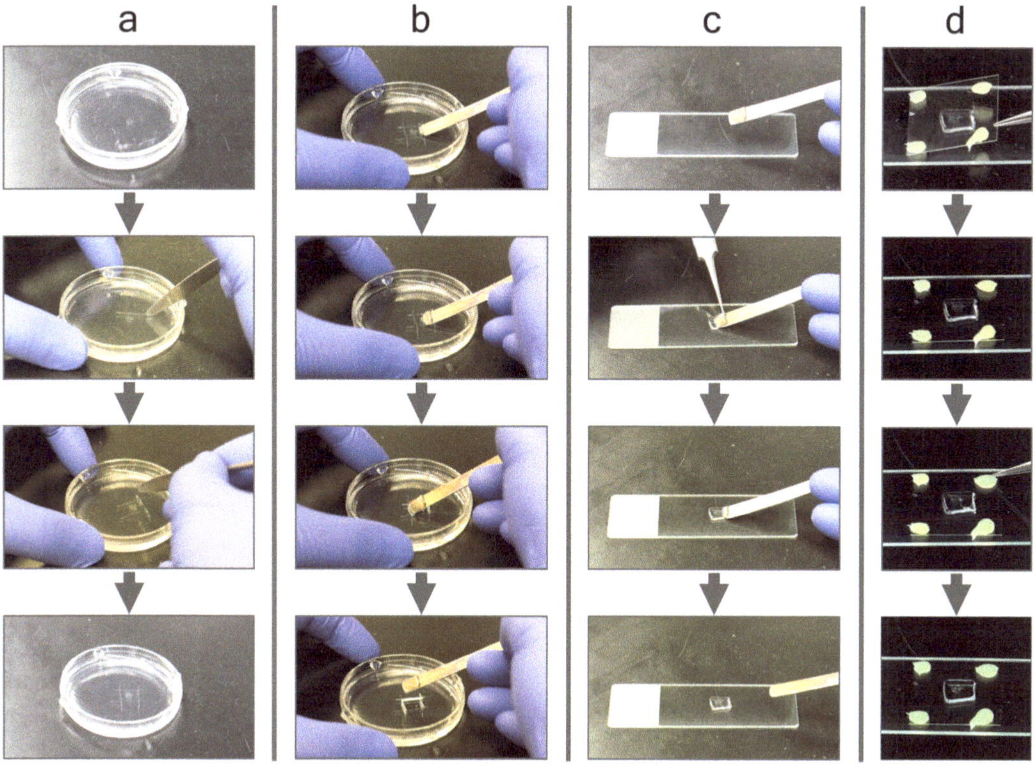

Fig. 1 Mounting of pollen tubes for confocal microscopic observation. (**a**) Split the PGM hosting pollen tubes by cutting with a mini spatula. (**b**) Dig out the PGM pad with pollen tube. (**c**) Place the isolated PGM pad onto the surface of a glass slide. (**d**) Cover with lanolin drop supported cover glass, and press gently until it touched the surface of the PGM pad

3. Split the PGM with a mini spatula around the region with pollen tubes (*see* Fig. 1a).
4. Dig out the square-shaped PGM pad that accommodates the pollen tubes on top using the mini spatula, and transfer it to the surface of a clean glass microscope slide, while keeping the pollen tubes on the top (*see* Fig. 1b, c).
5. Cover the PGM using the cover glass with lanolin drops, then gently push down the cover glass until it contacts the surface of PGM (*see* Fig. 1d). Now the pollen tubes are ready to be observed.

3.4 cFRAPc Assay

This example cFRAPc experiment was performed on a Zeiss upright LSM 880 confocal microscope. Other confocal microscopes with similar imaging capability shall also work, but require proper optimization on the imaging parameters.

3.4.1 Imaging Setup

1. Start the Zeiss upright LSM 880 confocal microscope, open Zeiss Zen black microscope software and turn on 405, 488, and 561 nm laser lines.
2. Mount the slide with the cover glass facing objective lens onto the microscope.
3. Locate the pollen tubes under bright field with a 10× objective lens.
4. Apply a drop of milli-Q water on top of the cover glass, switch to the 40× water-immersion objective lens and refocus.
5. Switch to the acquisition mode on the Zeiss Zen black software, click on *Time Series*, *Bleaching*, and *Regions* at the multidimensional acquisition panel. Use the following setting for cFRAPc assay:

   Imaging parameters:

   Frame size: 512 × 512 pixels.

   Zoom: 4.0.

   Pixel dwell time: 8.19 μs.

   Pinhole: 1 Airy unit.

   Averaging: 1.

   Bi-directional scan.

   Time series: 30 cycles of continuous imaging (interval = 0 s, about 5.2 s/frame).

   Use sequential scan mode to avoid crosstalk between the converted (red) and unconverted (green) Dendra2:

   Track 1: 0.6% 561 nm laser; Ch2 GaAsP detector: 566–635 nm, gain 750–800 (Imaging red colored Dendra2 after photoconversion).

   Track 2: 0.25% 488 nm laser; Ch1 PMT detector: 493–550 nm, gain 900–1000 (Imaging green colored Dendra2 before photoconversion), T-PMT, gain 350–400 (Transmission light image).

   Bleaching parameters:

   Start bleaching after five scans (pre-bleaching).

   Bleaching laser: 2% 405 nm laser (*see* **Note 8**).

   Bleaching time: set *Different scan speed* as 2 (pixel dwell time: 65.54 μs) (*see* **Note 9**).

3.4.2 Estimating the Photobleaching Rate

1. Pollen grains can be used for estimating the photobleaching rate of both green and red colored Dendra2, therefore, pollen grains of Lat52::PRK1-Dendra2 transgenic Arabidopsis are mounted on slides after they are brushed on the PGM following **steps 2–5** in Subheading 3.3.

2. Mount the slide on the confocal microscope and locate the pollen grains under 40× water immersion lens.
3. Use the imaging setting described in **step 5** of Subheading 3.4.1.
4. Click *Live* and adjust position of the scan area so that two or more pollen grains are visualized in the imaging window.
5. Select one whole pollen grain by drawing a polyline along its contour using the polyline tool under the *Regions* tool tab (*see* Fig. 2a).
6. Click *Start Experiment* to start imaging. Unselected pollen grain will be used to estimate photobleaching rate of the green colored Dendra2, while the selected pollen grain will be photoconverted and then used to estimate photobleaching rate of the red colored Dendra2 (*see* Fig. 2b).
7. Import the image file into Fiji for analysis.
8. Select the unconverted pollen grain with the Polygon selections tool in the green channel (Track 2: 493–550 nm) image stack, and analyze the intensity vs time by clicking *Image > Stacks > Plot Z-axis Profile.*
9. In the new plot window, copy the quantification data by clicking *Data > Copy All Data.*
10. Paste above data into a new Excel file, delete first five rows of pre-photoconversion data and keep only the data after photoconversion. Normalize the intensity by dividing all intensity data with that of the first frame post-photoconversion.
11. Repeat above measurement for another nine unconverted pollen grains.
12. Plot the average intensity vs time of the above ten pollen grains, and fit the curve by $I(t) = e^{-kt}$ (exponential fitting method, *see* Fig. 2c). Here, I is the normalized fluorescence intensity, t is the time, and k is the decay (photobleaching) constant of green colored Dendra2. We assign this calculated k as k_{green}, which will be used later for correcting the photobleached fluorescence during the time-lapseimaging.
13. Use the same approach mentioned above (**step 7–12**), the decay constant (k_{red}) of the red colored Dendra2 can be estimated by selecting the photoconverted pollen grain in the red channel (Track 1: 566–635 nm) image stack.

3.4.3 Recording Fluorescence Recovery After Photoconversion

1. Prepare and mount the pollen tubes to the Zeiss upright confocal microscope.
2. Keep all settings identical to that of the experiment for measuring the photobleaching rate, click *Live* and adjust scan area so that a single pollen tube is in the center of the field of view.

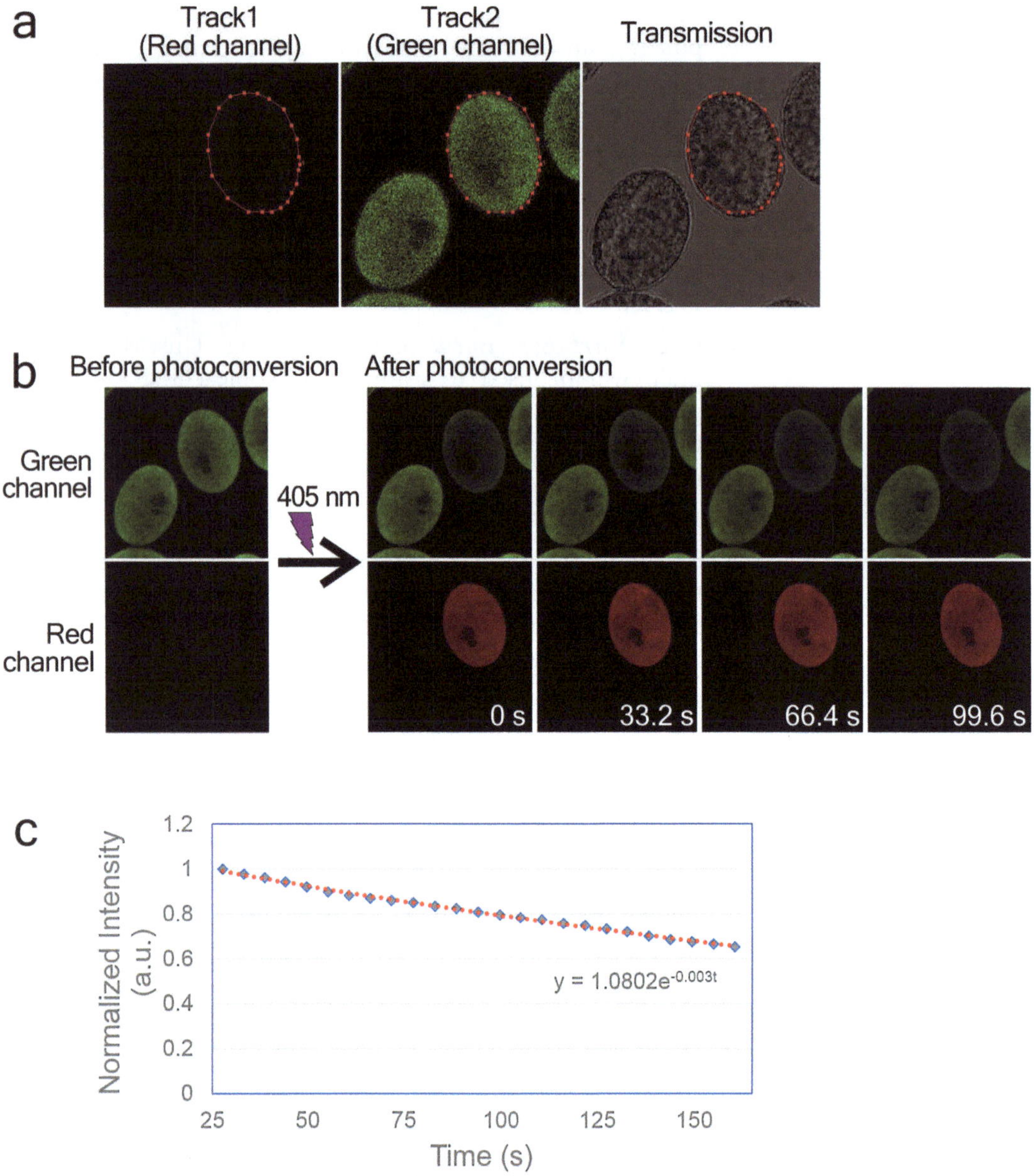

Fig. 2 Estimating the photobleaching rate. (**a**) Use *Polyline* tool to select one pollen grain (highlighted by the red dotted line) for photoconversion. (**b**) Representative images of a photoconversion experiment using the pollen grains from A. (**c**) A photobleaching curve (blue dots) and a fitting curve (red dotted line, fitted with the exponential fitting method) of green colored PRK1-Dendra2

3. Rotate the scan area until the longitudinal axis of the pollen tube is aligned with the vertical or horizontal direction of the imaging window.
4. Stop *Live* mode, use *Rectangle* tool on the *Regions* tool tab to draw a rectangle shaped region of interest (ROI) covering the

apical PM region (*see* the ROI highlighted by the yellow rectangle in Fig. 3a).

5. Click *Start Experiment* to start imaging (*see* **Note 10**).

3.5 Data Analysis

3.5.1 Measurement of the Intensity

Open the imaging file of cFRAPc experiment in Fiji and measure the following parameters (*see* Fig. 3b):

1. Background signal of green (g_{mem0}) and red (r_{mem0}) PRK1-Dendra2 on the PM.

   In the first frame (t_0) after photoconversion in the green channel image stack, draw a segmented line over the photoconverted PM region, click *Analyze* > *Measure* to acquire the mean gray value as g_{mem0}.

   In the last frame ($t - 1$) before photoconversion in the red channel image stack, draw a segmented line over the photoconverted PM region, click *Analyze* > *Measure* to acquire the mean gray value as r_{mem0}.

2. Length (L_t) of the photoconverted region over time in the post-conversion image sequence.

   In every frame ($t_0 - t_{24}$) after photoconversion in the red channel image stack, draw a segmented line over the PM region that resides red PRK1-Dendra2, click *Analyze* > *Measure* to acquire the value of area as the corresponding length for each frame (L_{t0} to L_{t24}).

3. The average intensity of green ($g_{\text{mem}(t)}$) and red ($r_{\text{mem}(t)}$) PRK1-Dendra2 on the photoconverted PM region over time in the post-conversion image sequence.

   In every frame ($t_0 - t_{24}$) after photoconversion in the red channel image stack, draw a segmented line over the PM region that resides red PRK1-Dendra2, click *Analyze* > *Measure* to acquire the mean gray value as the corresponding red PRK1-Dendra2 intensity for each frame ($r_{\text{mem}(t0)}$ to $r_{\text{mem}(t24)}$).

   Similarly, green PRK1-Dendra2 intensity for each frame ($g_{\text{mem}(t0)}$ to $g_{\text{mem}(t24)}$) can be measured in every frame ($t_0 - t_{24}$) after photoconversion in the green channel image stack. Please note that $g_{\text{mem}(t0)}$ is the same as background signal (g_{mem0}) for green PRK1-Dendra2.

4. The average intensity of green PRK1-Dendra2 in the cytoplasm (g_{cyt}).

   In the first frame (t_0) after photoconversion in the green channel image stack, select all cytoplasm region of the pollen tube using *polygon selection* tool, click *Analyze* > *Measure* to acquire the mean gray value as g_{cyt}.

3.5.2 Correction of Photobleaching

Photobleaching causes a significant drop in fluorescence of Dendra2 (Fig. 1c), therefore, the average intensity of green ($g_{\text{mem}(t)}$) and red ($r_{\text{mem}(t)}$) colored PRK1-Dendra2 on the photoconverted

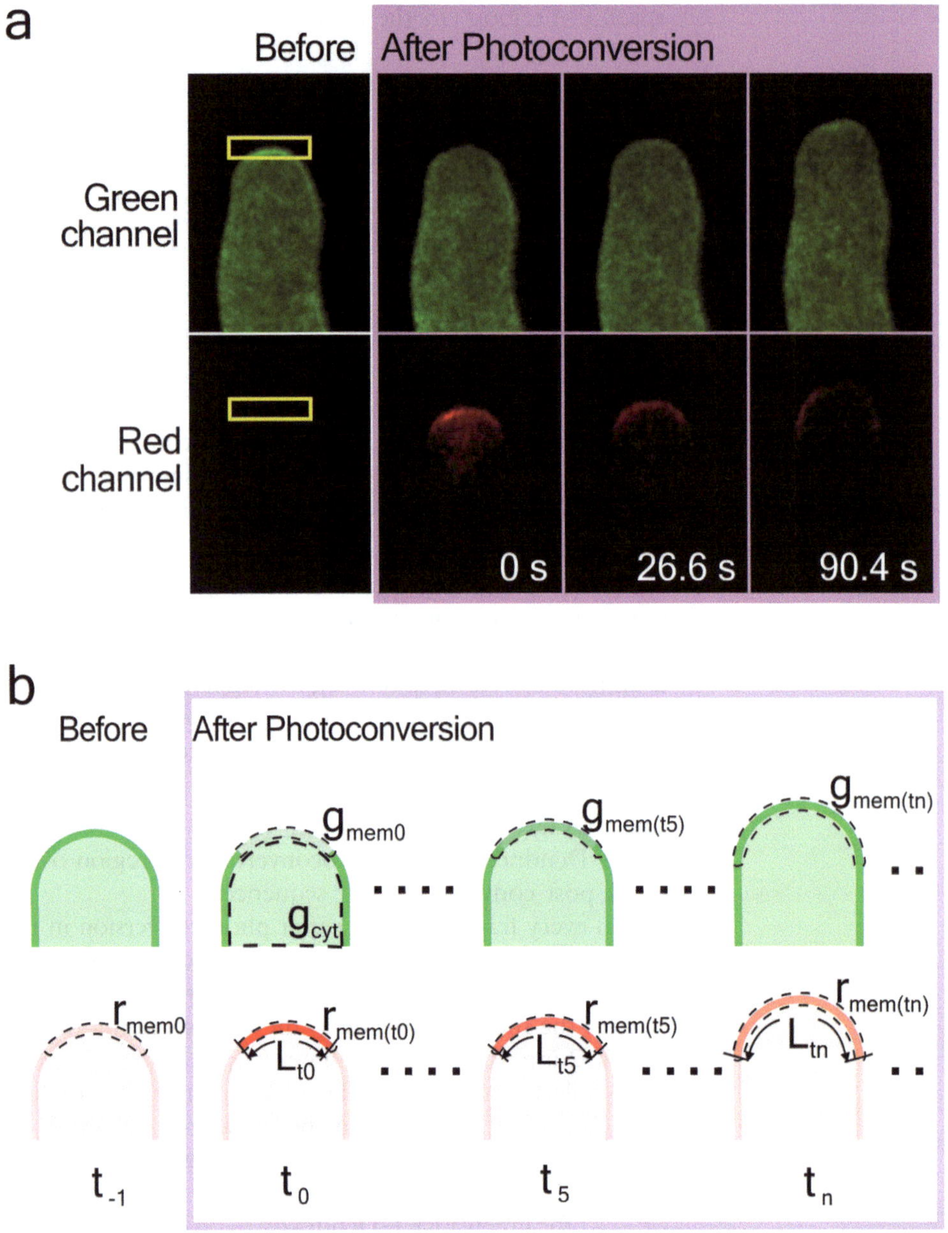

Fig. 3 Acquisition and measurement of PRK1-Dendra2 fluorescent recovery after photoconversion. (**a**) Representative images of a typical cFRAPc assay. PRK1-Dendra2 localized at the apical plasma membrane (yellow rectangle indicated region) was photoconverted using ROI bleaching function on the confocal microscope. (**b**) The length (L), average green (g_{mem}), and red (r_{mem}) Dendra2 intensity on the plasma membrane of the photoconverted region over time after photoconversion is measured as illustrated on the figure. The average intensity of green Dendra2 in the cytosolic region (g_{cyt}) of the pollen tube at the first frame post photoconversion is also measured

PM region over time must be corrected before further analysis. Correction can be made by the following equation: Corrected intensity $= I_t/\exp^{-k\ t}$. Here, I_t is the average intensity of the green or red PRK1-Dendra2 at a given time t, k is the decay constant that has been estimated in Subheading 3.4.2. Use k_{green} for correcting the average intensity of green PRK1-Dendra2 ($g_{\text{mem}(t)}$), and k_{red} for correcting that of red PRK1-Dendra2 ($r_{\text{mem}(t)}$). Now, we substitute $g_{\text{mem}(t)}$ and $r_{\text{mem}(t)}$ with the corrected value.

3.5.3 Calculating Exocytosis Rate

1. In an Excel data sheet, use the following equations to calculate the total green signal (g_{t}) and the total red signal (r_{t}) at each time point (t).

$$g_{\text{t}} = \left[g_{\text{mem}(t)} - g_{\text{mem0}}\right] L_t.$$

$$r_{\text{t}} = \left[r_{\text{mem}(t)} - r_{\text{mem0}}\right] L_t.$$

2. Plot the total red signal r_{t} versus time t in the Excel data sheet. Add a linear trendline to the data points nearby $t0$ in Excel, and obtain the slope, $k_{r(t0)}$, of the curve at time point t0.
3. Use equation $R_{\text{red}(t0)} = -k_{r(t0)}/[r_{\text{mem}(t0)}L_{t0}]$ to calculate the signal reduction rate at time point $t0$.
4. Plot the total green signal g_{t} versus time t in the Excel data sheet. Add a linear trendline to the data points nearby $t0$ in Excel, and obtain the slope, $k_{g(t0)}$, of the curve at time point $t0$.
5. The corrected exocytosis rate $R_{\text{ex}(t0)}$ (*see* **Note 11**) at time point $t0$ can be calculated as:

$$R_{\text{ex}(t0)} = [k_{\text{g}(t0)}/L_{t0} + R_{\text{red}(t0)}g_{\text{mem}(t0)}]/g_{\text{cyt}}.$$

4 Notes

1. We usually prepare 100 mL of 100× concentrated stock solutions for each of the 0.01% (w/v) boric acid, 1 mM $CaCl_2$, 1 mM Ca $(NO_3)_2$, and 1 mM $MgSO_4$. Each stock solution is aliquoted to 1.5 mL volume inside 2 mL microcentrifuge tubes, which were then stored at −20 °C freezer for up to 3 years.
2. Vaseline can be used instead of lanolin. We routinely load lanolin into a 10 mL syringe for easier application onto the cover glass. To load lanolin, we first melt it to liquid in a 50 °C oven, remove the plunger of the syringe, then pour the melted lanolin into the syringe from the wide opening on the flange side. After lanolin fully solidified, insert the plunger into the syringe and push forward until lanolin just emerges out of the tip.

3. Confocal laser scanning microscopes from other manufacturers can be used, but optimization of imaging setup is required. You can also use an inverted confocal microscope, but remember to mount the slide upside down so that cover glass is placed between the pollen tubes and the objective lens.
4. In our experience, if you dissolve everything into 85 mL milli-Q water according to our protocol, the final volume is approximately 100 mL, therefore there is no need to adjust the final volume.
5. Properly adjusted pH of the PGM is important for good pollen germination.
6. PGM prepared without filter sterilization can only be stored for 2–3 weeks, after which growth of fungi from contaminated spores usually appears on the PGM. We found that PGM prepared with this extra sterilization step can be stored at 4 °C in the refrigerator for up to 2 months without any fungal growth, and still give good pollen germination rate.
7. Alternatively, you can heat up the PGM faster with microwave oven than using the water bath, but you must be careful not to boil the medium, which will spill and change the composition of the medium.
8. The 405 nm laser light is harmful to cells. If your sample shows signs of damage after photoconversion using 405 nm laser irradiation, you may consider primed photoconversion using 488 nm and 633 nm dual laser irradiation [22, 23].
9. Other confocal systems may use the same speed during bleaching and imaging; in such case, you may need to increase the number of bleaching scans to achieve an optimal photoconversion of Dendra2.
10. The apical PM of pollen tube moves significantly between each frame. To compensate for the movement of the pollen tube, the ROI position can be adjusted to properly cover the apical PM region of pollen tube during the initial 5 pre-bleaching scans. To adjust the ROI position during scanning, you can change the value of *Center X* or *Center Y* in the *Regions* tool tab.
11. The exocytosis rate measured by the cFRAPc method is a relative value, it is most suitable for comparing the exocytosis rate in different regions of the PM, or between different chemical treatments, or between different genetic backgrounds.

Acknowledgments

This work is supported by the U.S. National Institute of General Medical Sciences (GM100130).

References

1. Chebli Y, Kaneda M, Zerzour R, Geitmann A (2012) The cell wall of the Arabidopsis pollen tube—spatial distribution, recycling, and network formation of polysaccharides. Plant Physiol 160(4):1940–1955. https://doi.org/10.1104/pp.112.199729
2. Grebnev G, Ntefidou M, Kost B (2017) Secretion and endocytosis in pollen tubes: models of tip growth in the spot light. Front Plant Sci 8:154. https://doi.org/10.3389/fpls.2017.00154
3. Wang H, Zhuang X, Cai Y, Cheung AY, Jiang L (2013) Apical F-actin-regulated exocytic targeting of NtPPME1 is essential for construction and rigidity of the pollen tube cell wall. Plant J 76(3):367–379. https://doi.org/10.1111/tpj.12300
4. Cheung AY, Wu HM (2008) Structural and signaling networks for the polar cell growth machinery in pollen tubes. Annu Rev Plant Biol 59:547–572. https://doi.org/10.1146/annurev.arplant.59.032607.092921
5. Luo N, Yan A, Liu G, Guo J, Rong D, Kanaoka MM, Xiao Z, Xu G, Higashiyama T, Cui X, Yang Z (2017) Exocytosis-coordinated mechanisms for tip growth underlie pollen tube growth guidance. Nat Commun 8(1):1687. https://doi.org/10.1038/s41467-017-01452-0
6. Hwang JU, Vernoud V, Szumlanski A, Nielsen E, Yang Z (2008) A tip-localized RhoGAP controls cell polarity by globally inhibiting Rho GTPase at the cell apex. Curr Biol 18(24):1907–1916. https://doi.org/10.1016/j.cub.2008.11.057
7. Ketelaar T, Galway ME, Mulder BM, Emons AM (2008) Rates of exocytosis and endocytosis in Arabidopsis root hairs and pollen tubes. J Microsc 231(2):265–273. https://doi.org/10.1111/j.1365-2818.2008.02031.x
8. Zonia L, Munnik T (2008) Vesicle trafficking dynamics and visualization of zones of exocytosis and endocytosis in tobacco pollen tubes. J Exp Bot 59(4):861–873. https://doi.org/10.1093/jxb/ern007
9. McKenna ST, Kunkel JG, Bosch M, Rounds CM, Vidali L, Winship LJ, Hepler PK (2009) Exocytosis precedes and predicts the increase in growth in oscillating pollen tubes. Plant Cell 21(10):3026–3040. https://doi.org/10.1105/tpc.109.069260
10. Wang X, Teng Y, Wang Q, Li X, Sheng X, Zheng M, Samaj J, Baluska F, Lin J (2006) Imaging of dynamic secretory vesicles in living pollen tubes of Picea meyeri using evanescent wave microscopy. Plant Physiol 141(4):1591–1603
11. Parton RM, Fischer-Parton S, Watahiki MK, Trewavas AJ (2001) Dynamics of the apical vesicle accumulation and the rate of growth are related in individual pollen tubes. J Cell Sci 114(Pt 14):2685–2695
12. Camacho L, Malho R (2003) Endo/exocytosis in the pollen tube apex is differentially regulated by Ca2+ and GTPases. J Exp Bot 54(380):83–92. https://doi.org/10.1093/jxb/54.380.83
13. Bove J, Vaillancourt B, Kroeger J, Hepler PK, Wiseman PW, Geitmann A (2008) Magnitude and direction of vesicle dynamics in growing pollen tubes using spatiotemporal image correlation spectroscopy and fluorescence recovery after photobleaching. Plant Physiol 147(4):1646–1658. https://doi.org/10.1104/pp.108.120212
14. Lee YJ, Szumlanski A, Nielsen E, Yang Z (2008) Rho-GTPase-dependent filamentous actin dynamics coordinate vesicle targeting and exocytosis during tip growth. J Cell Biol 181(7):1155–1168. https://doi.org/10.1083/jcb.200801086
15. Moscatelli A, Idilli AI, Rodighiero S, Caccianiga M (2012) Inhibition of actin polymerisation by low concentration Latrunculin B affects endocytosis and alters exocytosis in shank and tip of tobacco pollen tubes. Plant Biol 14(5):770–782. https://doi.org/10.1111/j.1438-8677.2011.00547.x
16. Yan A, Yang Z (2012) FRAP-based analysis of Rho GTPase-dependent polar exocytosis in pollen tubes. Methods Mol Biol 827:393–401. https://doi.org/10.1007/978-1-61779-442-1_26
17. Parton RM, Fischer-Parton S, Trewavas AJ, Watahiki MK (2003) Pollen tubes exhibit regular periodic membrane trafficking events in the absence of apical extension. J Cell Sci 116

(Pt 13):2707–2719. https://doi.org/10.1242/jcs.00468
18. Luo N, Yan A, Yang Z (2016) Measuring exocytosis rate using corrected fluorescence recovery after photoconversion. Traffic 17 (5):554–564. https://doi.org/10.1111/tra.12380
19. Chudakov DM, Lukyanov S, Lukyanov KA (2007) Tracking intracellular protein movements using photoswitchable fluorescent proteins PS-CFP2 and Dendra2. Nat Protoc 2 (8):2024–2032. https://doi.org/10.1038/nprot.2007.291
20. Chudakov DM, Lukyanov S, Lukyanov KA (2007) Using photoactivatable fluorescent protein Dendra2 to track protein movement. Biotechniques 42(5):553–557. https://doi.org/10.2144/000112470
21. Schindelin J, Arganda-Carreras I, Frise E, Kaynig V, Longair M, Pietzsch T, Preibisch S, Rueden C, Saalfeld S, Schmid B, Tinevez JY, White DJ, Hartenstein V, Eliceiri K, Tomancak P, Cardona A (2012) Fiji: an open-source platform for biological-image analysis. Nat Methods 9(7):676–682. https://doi.org/10.1038/nmeth.2019
22. Klementieva NV, Lukyanov KA, Markina NM, Lukyanov SA, Zagaynova EV, Mishin AS (2016) Green-to-red primed conversion of Dendra2 using blue and red lasers. Chem Commun 52(89):13144–13146. https://doi.org/10.1039/c6cc05599k
23. Mohr MA, Argast P, Pantazis P (2016) Labeling cellular structures in vivo using confined primed conversion of photoconvertible fluorescent proteins. Nat Protoc 11 (12):2419–2431. https://doi.org/10.1038/nprot.2016.134

Chapter 22

Testing Pollen Tube Proteins for In Vivo Binding to Phosphatidic Acid by n-Butanol Treatment and Confocal Microscopy

Carolin Fritz and Benedikt Kost

Abstract

The general role of cellular membranes is to provide a barrier and to generate separate reaction spaces. However, additional functions of membrane domains enriched in certain classes of lipids have been discovered, which represent an important area of ongoing research. Such membrane domains can be found in cells at different size scales (e.g., nanodomains, microdomains), represent membrane regions with special physical properties and play important roles in the direct or indirect propagation of signaling processes. Domain formation within the plasma membrane (PM) does not only involve the accumulation of specific lipids, but also the recruitment of specific transmembrane or PM-associated peripheral proteins. Phosphatidic acid (PA) is increasingly recognized as an important signaling lipid and component of PM domains. This lipid is involved in the regulation not only of biotic or abiotic stress responses, but also of pollen tube tip growth and of other forms of polar cell expansion. Although many PA-binding proteins have been characterized, a conserved PA interaction motif could not be identified in these proteins. Consequently, protein binding to PA cannot be predicted based on sequence analysis, but has to be biochemically tested using lipid strip or liposome assays. Although these assays are often informative, they are generally based on the use of artificial model membranes, which compared to natural membranes contain fewer lipid types often at non-physiological concentrations. In this chapter, we describe an alternative in vivo assay that can be employed to analyze protein binding to PA at the PM of normally elongating tobacco pollen tubes. This assay is based on the use of n-butanol (n-ButOH), which inhibits phospholipase D (PLD) and thereby blocks a major biosynthetic pathway that generates PA within the PM from substrates like phosphatidylcholine (PC) or phosphatidylethanolamine (PE). PLD inhibition reduces the PA content of the PM and consequently the level of PM association of PA-binding proteins, which can be analyzed using fluorescence microscopy. Methods enabling n-ButOH treatment of cultured tobacco pollen tubes expressing YFP-tagged PA-binding proteins as well as the quantitative determination of the PM association of these proteins are described.

Key words PA-binding proteins, YFP-tagging, PM/cytoplasm ratio of fluorescence intensity, Pollen tubes, *Nicotiana tabacum*, Confocal microscopy, Stable transformation, Transient transformation, n-ButOH

Anja Geitmann (ed.), *Pollen and Pollen Tube Biology: Methods and Protocols*, Methods in Molecular Biology, vol. 2160, https://doi.org/10.1007/978-1-0716-0672-8_22,

1 Introduction

Pollen tubes rapidly expand exclusively in one direction based on tip growth. The polarity of this process is maintained by specialized cytoskeletal structures, ion gradients, and different regulatory proteins, which specifically associate within distinct, clearly defined plasma membrane (PM) domains [1–3]. These proteins include Rac/Rop GTPases, the Rho GTPases of plants, as well as upstream regulators (Rac/RopGEFs, Rac/RopGAPs) and downstream effectors of these proteins [4]. Important effectors of pollen tube-specific Rac/Rop GTPases are phosphoinositol-4-phosphatekinases, which phosphorylate phosphoinositol-4-phosphate (PI4P) to produce phosphoinositol-4,5-bisphosphate PI4,5P (PIP_2) [5, 6]. PIP_2 and other phosphoinositols (PIs) are well-characterized signaling lipids, which accumulate in different organelles or PM domains and play important roles in the control of vesicle trafficking [7]. Delivery of material necessary for cell wall synthesis at the tip of growing pollen tubes depends on secretion, vesicle trafficking, and the endomembrane system. In order to provide sufficient cell wall material, secretory vesicles need to fuse with the PM at an about 10× higher rate than required for PM expansion [8, 9]. This results in the incorporation of excess material into the PM, which needs to be compensated by endocytosis. Thus, pollen tube growth depends on precisely coordinated interplay between exo- and endocytosis [10]. These two processes are thought to occur in distinct PM domains at the tip, between which PM components are subjected to drift [10]. Sites of exo- and endocytosis are characterized by the accumulation of specific sets of proteins and lipids, which determine the properties of these PM domains. For instance, curvature depends on membrane composition and can support vesicle fission as well as vesicle fusion [11, 12], processes that are mediated by complex interactions between different groups of proteins, whose PM association at least in part depends on the presence of specific lipids [13]. To achieve a better understanding of the cellular mechanism and regulatory processes underlying vesicle trafficking during tip growth, it is important to characterize protein–lipid interaction within distinct domains of the pollen tube PM.

A particularly interesting lipid is phosphatidic acid (PA), which constitutes only 1–2% of total phospholipids, but has important signaling functions [14, 15]. PA is involved not only in abiotic and biotic stress reactions (induced by wounding, pathogens or changes in temperature, osmotic, and oxidative conditions) [14], but also plays an important role in the control of pollen tube growth [16]. PA levels need to rapidly increase or decrease in response to different triggers and therefore must be tightly regulated. PA can be generated based on three different biosynthetic pathways. One of them is compartmentalized to the endoplasmatic reticulum (ER) and generates PA as a precursor for the production of structural lipids. The two other pathways generate signaling pools of PA within the PM [17], and either depend on phospholipase D (PLD), which cleaves off the head groups of structural lipids including phosphatidylcholine (PC) or phosphatidylethanolamine (PE) [18], or on phospholipase C (PLC), which cleaves PIP_2 to generate diacylglycerol (DAG) and phosphoinositol-3-phosphate (IP_3) [2, 15]. DAG produced by PLC activity is subsequently phosphorylated by diacylglycerolkinase (DGK) to generate PA [15, 19].

Both pathways producing signaling pools of PA can be pharmacologically blocked. Whereas R059022 is an inhibitor of DGKs, PLDs can be specifically blocked by primary alcohols including n-butanol (n-ButOH). PLD catalyzes a transphosphatidylation reaction, which in the absence of primary alcohols uses H_2O as nucleophilic acceptor resulting in the release of PC or PE head groups and in PA generation. Primary alcohols such as n-ButOH can serve as alternative nucleophilic acceptors, causing the production of phosphatidylbutanol or related molecules instead of PA. As phosphatidylbutanol represents a metabolic dead end [14], treatment with n-ButOH and other primary alcohols effectively reduces the PA content of the PM. By contrast, the closely related tertiary alcohol t-ButOH is ineffective as a PLD inhibitor in pollen tubes [16].

The phosphatidic acid binding domain (PABD) of the yeast snare Spo20p fused to a fluorescent protein (e.g., YFP or mRFP) has been established as in vivo marker for PA, which specifically associates with a clearly defined lateral PM domain in tobacco pollen tubes [1]. Although PA is known to be essential for the normal expansion of tobacco pollen tubes, the exact function of this signaling lipid during tip growth is not well understood [1, 16]. Increased membrane curvature resulting from high PA levels may support vesicle fission from the lateral pollen tube PM, where a major site of endocytosis appears to be located [10, 13, 20]. In addition, it is conceivable that PA recruits specific proteins to the PM and/or modulates the activity of PM-associated proteins [21].

The association of the Spo20p-based PA marker with the lateral PM of tobacco pollen tubes was shown to be sensitive to n-ButOH [1]. Similarly, treatment with n-ButOH can be used to investigate whether in vivo binding of other proteins to PA is necessary for their recruitment to the PM. Decreased PA levels caused by n-ButOH treatment should result in reduced protein association with the lateral pollen tube PM, if this association requires PA binding. Any protein of interest fused to a fluorescent protein can be tested for in vivo PA binding by quantitatively analyzing its association with the lateral PM of tobacco pollen tubes before and after n-ButOH application. Treatment with t-ButOH instead serves as an excellent control in such experiments. Obviously, this assay also allows quantitative analysis of point mutations that potentially affect the PA binding of investigated proteins.

2 Materials

2.1 Plant Material

Nicotiana tabacum L. cv. Petit Havana SR1 are grown for 2 months (*see* **Notes 1** and **2**) under long-day conditions (16 h light/8 h dark; 20 °C) in a growth chamber or a greenhouse. Depending on the setup of the experiment, wild-type plants providing pollen for transient transformation or transgenic plants already containing appropriate expression constructs can be used (*see***Note 3**).

2.2 Preparation of PTNT Medium and Plates

1. 50-mL plastic tubes.
2. 5.5-cm diameter plastic petri dish.
3. 5-mL serological pipette.
4. Pipette boy.
5. Sterile hood.
6. pH meter.
7. KOH for adjusting the pH.
8. 10× salt stock solution: 1.0 mM $CaCl_2$, 1.0 mM KCl, 0.8 mM $MgSO_4$, 1.6 mM H_3BO_4, 30 μM $CuSO_4$. Store at −20 °C (*see* **Note 4**).
9. 1.5% casein acid hydrolysate ("amicase") stock solution. Store at −20 °C (*see* **Note 5**).
10. Sucrose.
11. MES hydrate.
12. PEG-6000.
13. 20 mg/mL rifampicin in methanol. Store at −20 °C.
14. 0.5% phytagel.
15. Microwave oven.
16. 300-mL beaker.
17. 2× PTNT medium: 10% (w/v) sucrose, 0.6% (w/v) MES hydrate, 0.06% (v/v) casein acid hydrolysate (1.5% stock solution), 2× salt stock solution, 20 mg/mL rifampicin; adjust pH with KOH to 5.9. 25% (w/v) PEG-6000. Filter sterilize and store at 4 °C (*see* **Note 6**).

2.3 Pollen Preparation and Plating

1. 50-mL plastic tube.
2. 2× PTNT medium.
3. 0.5% phytagel.

4. Sterile water (H_2O_{bidest}).
5. n-Butanol.
6. t-Butanol.
7. 10 Anthers (transient transformation) or 5 anthers (transgenic plants) with pollen from tobacco flowers.
8. Tweezers.
9. Filter sieve.
10. Vortex mixer.
11. Polyamide filter membrane (pore size 0.45 μm; diameter 47 mm).
12. Vacuum pump.
13. 5.5-cm diameter petri dish.
14. 500-mL Büchner flask (vacuum stable) (Fig. 2f).
15. GV 050/0/02 glass frit filter with lower glass base (Fig. 2d).
16. Conical Büchner flask adapter to seal the flask (Fig. 2d).
17. Whatman PTFE-centering ring GV 050/0/05 (Fig. 2e).
18. Ceramic filter fitted into the PTFE-centering ring (Fig. 2e).
19. Parafilm.

2.4 Transient Pollen Transformation

1. 0.5-mL PCR tubes.
2. Microcentrifuge.
3. Absolute ethanol.
4. 60 mg/mL gold particles in 50% glycerol (*see* **Note 7**).
5. Midi or maxi preparation of pDNA containing the gene of interest encoding a fluorescent fusion protein: concentration adjusted to 1 μg/μL (*see* **Note 8**).
6. 2.5 M $CaCl_2$. Filter sterilize and store at −20 °C.
7. 0.1 M spermidine or 1 mg/mL protamine. Store at −20 °C.
8. Vortex mixer.
9. PDS-10000-He™ Biolistic Particle Delivery System.
10. Rupture disk 1100 psi.
11. Biolistic macrocarrier.
12. Stopping screen.
13. Parafilm.

2.5 Imaging and Analysis

1. Slides and cover slips (22 × 22 mm; *see* **Note 9**).
2. Scalpel.
3. Timer.

4. Confocal microscope equipped with a 20× multi-immersion objective and a 63× oil or water immersion objective, with lasers allowing excitation of the fluorescent protein used (515 nm for YFP, 488 nm for GFP). Detection windows or emission filters should cover wavelength regions of 497–526 nm for GFP and 520–560 nm for YFP.
5. Software to acquire images.
6. Software to analyze fluorescence intensities, e.g., Fiji/ ImageJ [22].

3 Methods

3.1 Plasmid Design for Transient and Stable Transformation

Plasmids containing expression cassettes encoding the protein of interest tagged with a fluorescent protein need to be constructed. These expression cassettes should contain in the indicated order a promotor strongly active in tobacco pollen tubes (typically *LAT52*, rarely *UBI10*), the coding sequence of the protein of interest fused to a sequence encoding a fluorescent protein (typically GFP or YFP), and a terminator (e.g., the *NOS* terminator). It is essential to ensure in frame fusion of the sequences encoding the protein of interest and the fluorescent protein (Fig. 1) (*see* **Note 10**).

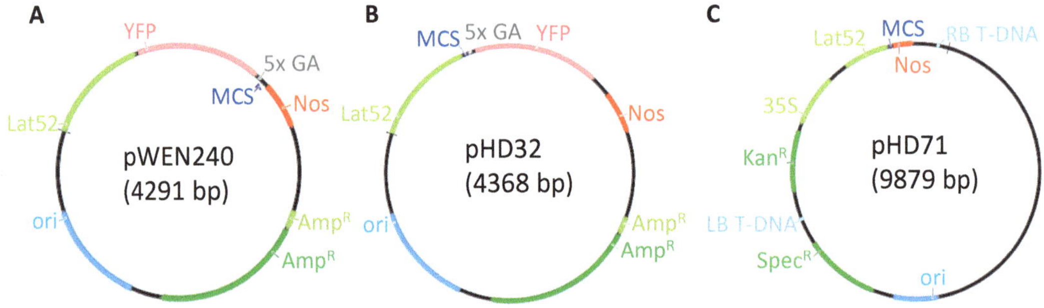

Fig. 1 Expression vectors suitable for transient and stable expression of fluorescent fusion protein in tobacco pollen tubes. (**a**, **b**) Physical maps of vectors suitable for transient expression in pollen tubes of proteins of interest fused to YFP either at the *N*- or *C*-terminus (pWEN240 and pHD32). The CDS of the gene coding for the protein of interest can be cloned into the multiple cloning sites (MCS) of these vectors. Between the MCS and the YFP cDNA, both vectors contain a sequence coding for a 5× GA (glycine, alanine) linker, which promotes correct folding and intracellular targeting of fluorescent fusion proteins. Both vectors contain bacterial genes conferring ampicillin (Amp^R; ß-lactamase). (**c**) Physical map of a binary vector containing left and right border sequences (LB and RB) required for genome integration, which is suitable for stable expression in pollen tubes of proteins of interest fused to YFP (pHD71). Sequences encoding YFP-tagged proteins of interest can be inserted into the MCS of this plasmid, which in addition to the pollen tube expression cassette also contains a plant kanamycin resistance gene under the control of the constitutively active *35S* promotor, as well as a bacterial gene conferring spectinomycin resistance (Spe^R). The pollen tube expression cassette in all three plasmids contains the strong pollen tube-specific *Lat52* promotor along with a *Nos* terminator. An origin of replication (ori) required for amplification in bacteria is also present in all three plasmids

1. Clone *N*- and *C*-terminal fusions of the protein of interest and the fluorescent tag (Fig. 1a, b), as the attached tag may interfere with protein folding and/or intracellular targeting depending on its position.
2. A flexible spacer, e.g., a 5× GA linker, between the protein of interest and the fluorescent tag enhances the probability of correct protein folding and intracellular targeting (Fig. 1a, b).
3. In addition to the expression cassette described above, to enable amplification and maintenance in bacteria plasmids need to contain a bacterial origin of replication and a bacterial antibiotic resistance gene (Fig. 1).
4. Vectors suitable for the construction of plasmids enabling transient expression of proteins of interest tagged with a fluorescent protein in tobacco pollen tubes are e.g., pWEN240 [3] and pHD32 [23] (Fig. 1 a, b).

Binary vectors required for stable transformation are in general designed as described above, but need to contain the additional elements indicated below. Prior to using binary vectors for stable transformation, it is advisable to test expression of the protein of interest tagged with a fluorescent protein based on transient transformation (*see* Subheading 3.3).

1. For stable insertion into the plant genome, left and right border sequences (LB and RB) flanking introduced plant expression cassettes are required (Fig. 1c).
2. In addition to the expression cassette encoding the protein of interest tagged with a fluorescent protein, a marker gene allowing the selection of transformed plants has to be inserted between LB and RB sequences (*see* **Note 11**). The expression of this marker gene is typically under control of the strong *35S* promotor, which is constitutively active throughout the whole transformed plants. Marker genes providing resistance to BASTA, kanamycin (Kan^R), spectinomycin ($Spec^R$) or hygromycin (Hyg^R) can be used (Fig. 1c).
3. A binary vector suitable for the construction of plasmids enabling stable expression of the protein of interest tagged with a fluorescent protein in tobacco pollen tubes is e.g., pHD71 (based on pPZP212 [24]), which confers $Spec^R$ to bacteria and Kan^R to plants.

3.2 Preparation of PTNT Plates and Plating of Pollen Grains

Before plating pollen grains for transient transformation, first prepare macrocarriers (*see* Subheading 3.3). Particle bombardment as early as possible after plating pollen grains enhances the transient transformation efficiency.

1. To prepare PTNT plates dilute 2× PTNT medium 1:1 (v/v) with 0.5% (w/v) phytagel (*see* **Note 12**).

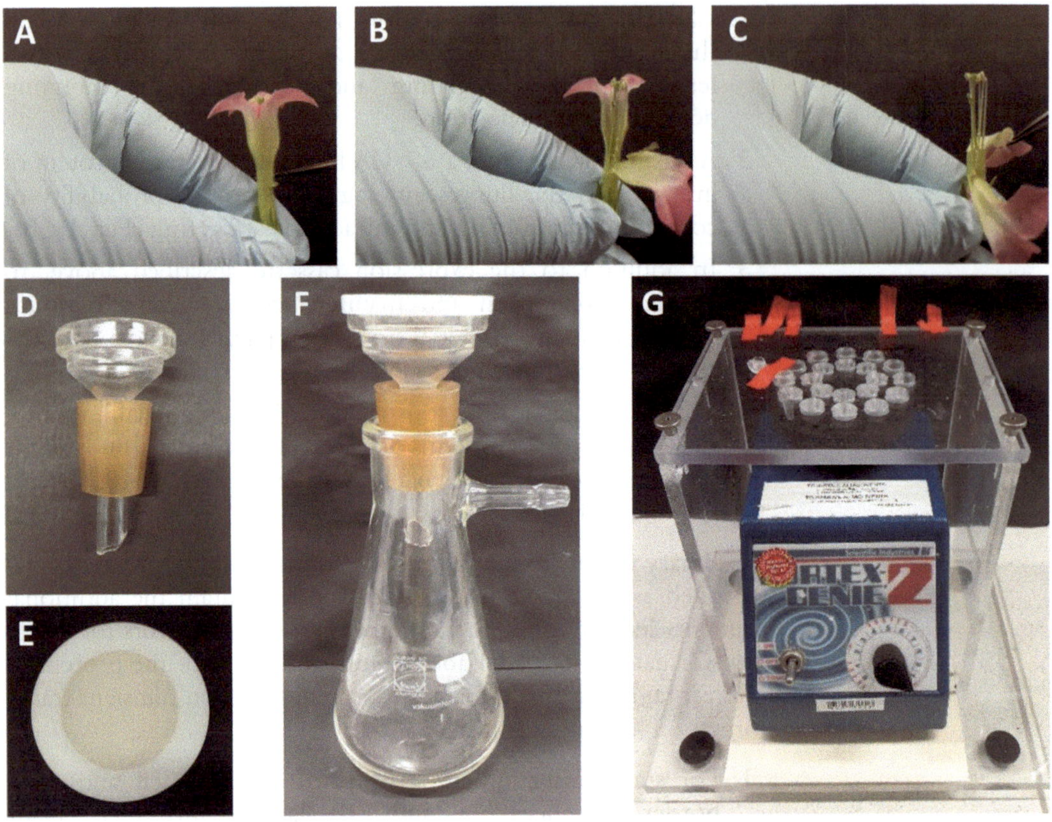

Fig. 2 Harvesting of anthers/devices used for pollen plating and for gold particle preparation. (**a**–**c**) Harvesting of tobacco anthers: use tweezers to cut two opposite sides of the corolla of collected flowers from the bottom to the top. Remove the corolla and collect the anthers without the style. (**d**) Conical Büchner flask adapter attached to a GV050/0/02 glass frit with lower glass base. (**e**) Top view of a PTFE-centering ring with ceramic filter. (**f**) Büchner flask holding the assembled conical Büchner flask adapter (**d**) and the PTFE-centering ring with the ceramic filter (**e**). (**g**) Plastic rack that can hold multiple 0.5-mL tubes on a vortex mixer to enable the preparation of several gold particle suspensions at the same time under continuous vortexing. Adhesive tape is used to fix open 0.5-mL tubes on the surface of the rack

2. Pipette 3.5 mL of the PTNT/phytagel mixture into a 5.5-cm diameter plastic petri dish under a sterile hood (*see* **Note 13**). Wait until the medium has solidified and store plates at 4 °C (*see* **Note 14**).
3. Collect flowers (cut from plants) to harvest anthers and pollen. To minimize loss of pollen, remove corolla with tweezers after cutting it open on two opposite sides from the bottom to the top (Fig. 2a–c). Harvest the anthers using tweezers and transfer them into a 50-mL plastic tube containing the appropriate volume of PTNT medium (2× PTNT diluted 1:1 (v/v) with sterile H_2O_{bidest}) (*see* **Note 15**). For transient transformation, ten anthers (two flowers) per 5 mL PTNT medium are harvested, in case of stably transformed plants five anthers (one

flower) per 5 or 10 mL PTNT medium are sufficient. For experiments involving the treatment of stably transformed pollen with n-/t-BuOH, it is advisable to plate pollen grains at relatively low density to avoid detoxification effects. For transient transformation experiments, chill the tube containing pollen grains on ice to prevent early germination.

4. Vortex 50-mL tubes containing anthers within PTNT for 3 min to release pollen grains.
5. Remove anthers by filtering the mixture through a sieve into another tube. Alternatively, it is possible to remove anthers from the suspension using ethanol sterilized tweezers.
6. To plate pollen grains on solid medium, vacuum filtrate 5 mL pollen suspension through a pre-wetted polyamide filter membrane (pollen grains are retained on the surface of the membrane), which has been positioned on top of an assembled filter system (Fig. 2f).
7. Place the membrane upside down on 3.5 mL solid PTNT medium in a petri dish and remove the filter immediately (pollen grains remain on the surface of the culture medium).
8. For transient transformation, bombard pollen grains immediately after plating. Seal plates containing freshly bombarded or transgenic pollen grains with parafilm and incubate them at 21 °C in the dark.

3.3 Transient Pollen Transformation by Biolistics

1. Prepare the gold particle suspension required for two macrocarriers in a 0.5-mL tube under continuous vortexing. If available, a rack that can hold multiple 0.5-mL tubes on a vortex mixer enables the preparation of several suspensions at the same time (Fig. 2g).
2. Add in this order to the 0.5-mL tube: 25 μL gold particles (suspended in glycerol), 2.5–5 μL plasmid DNA (1 μg/μL), 25 μL 2.5 M $CaCl_2$, and 10 μL 1 mg/mL protamine or 0.1 M spermidine (*see* **Note 16**).
3. Keep vortexing for another 3 min.
4. Spin gold particles down for 30 s and remove supernatant (*see* **Note 17**).
5. Add 200 μL absolute ethanol and vortex extensively, at least 3 min.
6. Spin gold particles down for 30 s and remove supernatant with a pipette, or using a pipette tip attached to a tube connected to a vacuum pump.
7. Resuspend the gold particles in 16 μL absolute ethanol (*see* **Note 18**).

8. Equally distribute the gold particle suspension onto two microcarriers.
9. Wait until the ethanol has evaporated.
10. Prepare a plate containing solid medium covered with pollen grains (*see* Subheading 3.2).
11. Set the parameters for the particle gun according to the instruction manual (*see* **Note 19**) and bombard pollen grains.
12. Seal plates with Parafilm and incubate them at 21 °C in the dark.

3.4 n-/t-Butanol Treatment

1. It is advisable to perform a dose–response experiment to determine the n-ButOH concentration that shows the desired effects (we recommend using the concentration inducing a half-maximal response). Cultures treated with t-ButOH or not treated at all should be used as negative controls (*see* **Note 20**). As n- and t-ButOH can be directly dissolved in PTNT medium, no additional solvent control is required [16]. If experiments with other pharmacologically active substances requiring a solvent (e.g., DMSO or methanol) are performed, PTNT medium containing just the solvent should be used as a negative control.
2. 0.5% n-ButOH has been reported to inhibit both pollen germination and pollen tube elongation without completely blocking either of these two processes [1]. This concentration appears generally suitable for the analysis of n-ButOH effects on the PA-dependent PM association of fluorescent fusion proteins. Higher concentrations may affect the PM association of such proteins more strongly, but are also more likely to have additional indirect effects resulting from the complete disruption of tip growth.
3. Different methods can be employed to apply 0.5% n- or t-ButOH to cultured pollen tubes:
   - 0.5% n- or t-ButOH can be directly added to the culture medium when preparing the PTNT plates that are subsequently used for pollen plating. As PTNT medium supplemented with phytagel is rather viscous, it is important to mix well after the addition of n- or t-ButOH to this medium (*see* **Notes 21** and **22**). Using this method, growing pollen tubes are constantly exposed to n- or t-ButOH starting with pollen germination.
   - Alternatively, pollen tubes can be grown on solid PTNT medium for variable periods of time before n- or t-ButOH is applied (*see* **Note 20**). To this end, a square (e.g., 1.5 × 1.5 cm) of solid culture medium with growing pollen tubes is cut out from culture plates and transferred to a new

petri dish using a scalpel. Subsequently, liquid PTNT medium containing 0.5% n- or t-ButOH is added until it thinly but completely covers the solid medium. Pollen tubes treated in this manner are incubated for 10 min at RT before they are transferred onto a slide for microscopic imaging (*see* **Note 23**). This method allows the analysis of n-ButOH effects early after application, before they can be affected by potential adaptation.

- 0.5% n- or t-ButOH can be added to pollen grains or pollen tubes cultured in liquid PTNT medium any time before or after pollen germination. To avoid physically damaging pollen tube cultures, it is important to gently mix these cultures after ButOH application. After 10 min of incubation at RT, pollen tubes treated in this manner can be transferred onto a slide for microscopic imaging (*see* **Note 24**). A disadvantage of this method is that when suspended in liquid PTNT medium pollen tubes are rarely positioned with their long axis parallel to the coverslip, which is required for the imaging of medial confocal optical sections.
- If available, a flow chamber enabling medium exchange during image acquisition (*see* Chapter 15 or 19) can be used to image fast and live responses of cultured pollen tubes to ButOH application (*see* **Note 25**).

3.5 Imaging

1. Switch on the confocal microscope and activate the 488 nm or 514 nm lasers for GFP or YFP imaging, respectively (*see* **Note 26**). Set the detection window for GFP or YFP imaging to 497–526 or 520–560 nm, respectively. Use emission filters with similar characteristics when working with a filter-based confocal microscope. If available, also activate the bright field channel.
2. Transfer a square (e.g., 1.5 × 1.5 cm) of solid culture medium covered with growing pollen tubes from a culture plate to a microscope slide using a scalpel. Carefully place a coverslip directly on top of the pollen tubes with the help of tweezers.
3. Screen for pollen tube tips suitable for imaging using a low magnification (e.g., 20×) objective and switch to a high magnification (e.g., 63×) objective for the acquisition of high quality images required for image analysis. When working with stably transformed pollen tubes, a high magnification (e.g., 63×) objective may be used for both screening and image acquisition.
4. To enable quantitative comparison of ButOH treated and control pollen tubes analyzed in different experiments, key imaging settings (i.e., laser intensity, detector gain, image resolution, zoom factor, averaging) need to be kept constant.

Untreated control pollen tubes displaying strong PM associated fluorescence should be used to define optimal settings for high-quality imaging at high (e.g., 63×) magnification. Generally, 1024 × 1024 resolution, 3× line averaging and a zoom factor of 2.5 are recommended.

5. To enable measurement of the growth rate of imaged pollen tubes, after acquiring a first image used to analyze PM association of fluorescent fusion proteins, a second image of the same pollen tube needs to be taken 2 min later (*see* **Note 27**) (Fig. 3a, b).

3.6 Quantitative Analysis of Pollen Tube Growth Rates and of Relative Levels of PM-Associated and Cytoplasmic Fluorescence (PM/Cytoplasm Fluorescence Ratio)

Even stably transformed pollen tubes express fluorescent fusion proteins at variable levels, which may affect the PM/cytoplasm fluorescence ratio of these proteins as well as the pollen tube growth rate. We therefore recommend measuring not only the PM/cytoplasm fluorescence ratio in each analyzed pollen tube, but also the growth rate of these cells. Differences in the PM/cytoplasm fluorescence ratio displayed by analyzed fluorescent fusion proteins in ButOH-treated and in control pollen tubes may not be obvious when looking at raw data. Therefore, quantitative analysis of these data is generally necessary. To measure pollen tube growth rates and PM/cytoplasm fluorescence ratios, Fiji/ImageJ 1.52i [22] or other software packages with comparable features can be employed.

1. It is advisable to save images, which were acquired as described above (*see* Subheading 3.5), in the original format determined by the imaging software. In addition, they need to be exported as .tif files using a gray scale lookup table, which can be imported into Fiji.
2. To measure pollen tube growth rate, use Fiji to open the two fluorescence images of each analyzed pollen tube, which were recorded at a 2 min interval (*see* Subheading 3.5, **step 5**).
3. Open the function: Fiji → Image → Color → Merge Channel. Define the first image as C1 (red) and the second image as C2 (green) and click OK.
4. In the merged image generated as a result of **step 3**, the distance between the positions of the tip of the imaged pollen tube at the two different time points can be measured with the straight-line tool (*see* **Note 28**) (Fig. 3c). This distance divided by 2 equals the growth rate of the analyzed pollen tube in μm/min.
5. To measure the PM/cytoplasm fluorescence ratio, open just the first image of each pollen tube in Fiji (recorded as described in **step 2**).

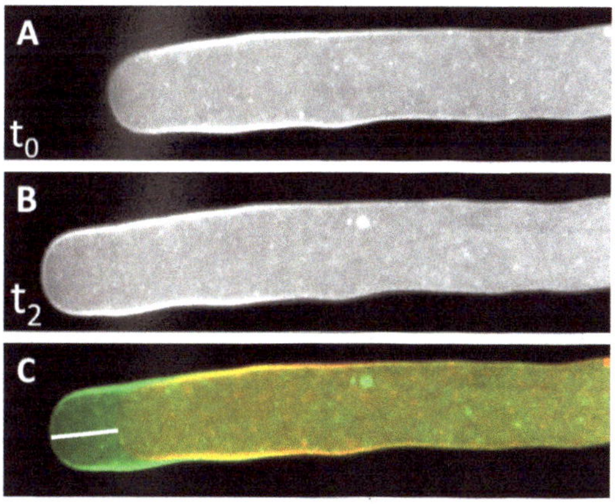

Fig. 3 Determination of the pollen tube growth rate. (**a**) Image of a tobacco pollen tube taken at t_0 to analyze PM association of a fluorescent fusion proteins. (**b**) Image of the same pollen tube 2 min later (t_2), which was taken to enable determination of the growth rate. (**c**) Overlay of the t_0 and t_2 images generated using the "merge channels" function of Fiji. The length of the white line, which can be drawn and measured with the help of the Fiji straight-line tool, divided by 2 corresponds to the pollen tube growth rate in μm/min

6. Open the function: Fiji → Image → Transform → Rotate. Rotate the image such that the longitudinal axis of the analyzed pollen tube is vertical with the tip pointing downwards.
7. Use the straight-line tool to draw a horizontal line across the analyzed pollen tube in a region, which displays PM association of the expressed fluorescent fusion protein (Fig. 4a, c). The line needs to be longer than the width of the pollen tube.
8. Open the ROI Manager: Fiji → Analyze → Tools → ROI Manager. Add the horizontal line drawn as described above (**step 7**). Repeat **step 7** and add two additional horizontal lines to the ROI Manager such that all three horizontal lines are positioned at different distances from the apex. These three lines indicate the positions at which fluorescence intensity profiles will be plotted (Fig. 4a, c). In addition, draw a fourth vertical line perpendicularly across the three horizontal line such that one end of this line is positioned exactly at the extreme pollen tube apex. This vertical line defines the distance from the apex at which the three horizontal lines are placed (Fig. 4a, c). Combine all four lines and save them as one single ROI. To this end, mark all four lines in the ROI Manager, click the right mouse button, choose "OR (Combine)" and add the combination to the ROI Manager. Finally, save not only the combined ROI, but also each of the three individual horizontal lines as single ROIs (*see* **Note 29**).
9. Use Fiji to open a pollen tube image and to position the combined ROI (all four lines) as described above (**step 8**).

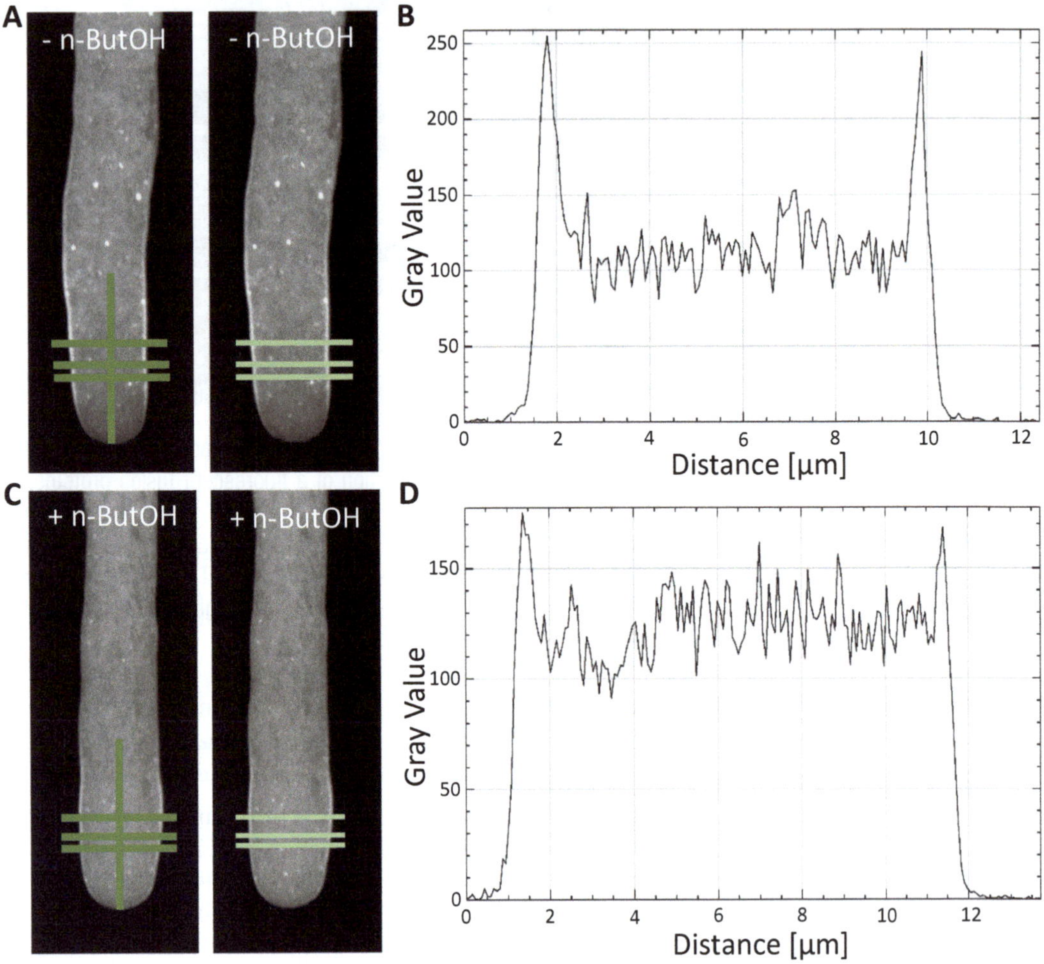

Fig. 4 Quantification of relative levels of PM-associated and cytoplasmic fluorescence (PM/cytoplasm fluorescence ratio) displayed by fluorescent fusion proteins in tobacco pollen tubes after n-ButOH treatment. (**a**, **b**) Untreated control pollen tube. (**c**, **d**) Pollen tube grown in the presence of n-ButOH. (**a**, **c**) Combined ROI (left images: indicated in dark green) used for the standardized positioning of three single horizontal line ROIs (right images: indicated in light green) across the tips of analyzed pollen tubes. The lower end of the single vertical line that is part of the combined ROI was positioned to exactly overlap with the extreme pollen tube apex. Along each of the three single horizontal line ROIs (light green) fluorescence intensity plots were generated (**b**, **d**). Based on these plots, levels of PM-associated (average of the two peak values) and cytoplasmic fluorescence (average intensity in a 4 μm wide region in the center of the plots between the two peaks) were determined. Image analysis was performed using the Fiji/ImageJ software package

Add the three individually saved single line ROIs as well and arrange them on top of the horizontal lines of the combined ROI. Remove the combined ROI. Click on one horizontal line and use the function: Fiji → Analyze → Plot Profile (short cut: ctrl + k) to plot the fluorescence intensity along this line (Fig. 4b, d). Click on "list" in the plot window to get access to all intensity values along the analyzed line. Compute the average of the two peak intensity values representing the level of PM associated fluorescence on both flanks of the analyzed pollen tube (in case of low PM/cytoplasm fluorescence ratio: chose the outer-most peak on both sides). Also, compute the average intensity value in a 4 μm wide region in the center of the plot between the two peaks, which represents the level of cytoplasmic fluorescence. Determine the PM/cytoplasm fluorescence ratio for the analyzed line. Finally, repeat the same procedure for the other two horizontal lines defined by the saved ROIs (**step 8**). Determine the average PM/cytoplasm fluorescence ratio for the analyzed pollen tube (average of all three analyzed line plots).

10. Repeat **steps 5**, **6**, and **9** with all pollen tube images to be analyzed and determine the average PM/cytoplasm fluorescence ratio. If this ratio is reduced in the presence of n-ButOH, the analyzed fluorescent fusion protein is likely to bind to PA in living tobacco pollen tubes. A fluorescent Spo20p fusion protein, which has been reported to display this behavior [1], can be used as a positive control.

4 Notes

1. The development of the tobacco plants including the time point of flower induction varies substantially depending on cultivation conditions (e.g., temperature). This is particularly evident if plants are grown in a green house, in which these conditions fluctuate considerably.
2. Lifetime as well as flowering period of tobacco plants can be substantially prolonged by cutting off senescent flowers and maturing capsules (fruits), as well as by adding fertilizer once a week.
3. Stably transformed pollen tubes are preferred, because all of them express the protein of interest and a sufficiently large number of pollen tubes (at least 30) for meaningful quantification can be analyzed within a relatively short time. In addition, the preparation time before imaging is minimal, since the pollen grains do not need to be transformed. However, it takes about half a year to obtain the first stably transformed flowering plants.
4. Considering the low salt concentrations even in the stock solution, it is advisable to prepare a large volume of this solution (e.g., 1 L, freeze 50 mL aliquots). For 500 mL 2× PTNT medium 100 mL 10× salt stock solution is required.
5. Prepare and freeze 10 mL aliquots. For 500 mL 2× PTNT medium, 20 mL 1.5% casein acid hydrolysate are required.

6. Weigh sucrose and MES hydrate into a 1-L beaker. Thaw 1.5% casein acid hydrolysate and 10× salt stock solution in a water bath or in another beaker filled with water heated in a microwave oven and add both stock solutions to the sucrose and MES hydrate in the 1-L beaker. Fill up to half of the total volume with H_2O_{bidest} and add rifampicin as required. Adjust the pH to 5.9 with KOH and add PEG-6000. It takes a while until the PEG-6000 is completely dissolved (heating speeds this process up).
7. To sterilize 60 mg of 1.6 μm gold particles add 1 mL absolute ethanol and vortex for 3 min, spin down for 1 min, wash two times with sterile H_2O_{bidest}, and resuspend in 1 mL 50% (v/v) glycerol (sterile).
8. Mini preparation plasmid isolation does not yield DNA of the required quantity and purity.
9. Use cover slips with optimal thickness for confocal microscopy (generally 0.17 mm).
10. Virtual cloning using a software package like Geneious (R 7.1.8) [25] before starting the physical cloning is advisable to test and confirm the cloning strategy.
11. Be aware that binary plasmids contain a bacterial as well as a plant gene conferring resistance to different antibiotics or herbicides. Make sure to use the appropriate antibiotic or herbicide to select transformed bacteria or plants.
12. Heat water in a 300-mL beaker in the microwave oven until the water boils and put a 50-mL tube containing 2× PTNT in the beaker to pre-warm this viscous medium. Heat the phytagel in a microwave oven until the first bubbles appear, add to the pre-warmed 2× PTNT medium and mix by inverting the tube (gently to avoid bubbles) until the medium is homogenous. Keep the tube in hot water to prevent premature solidification of the medium.
13. Considering the high viscosity and temperature of the PTNT/phytagel mixture, it is advisable to pipette up and down a few times to equilibrate the pipette. Avoid releasing the medium too fast to make sure the pipette empties completely. Gently shake plates to evenly distribute the medium. 3.5 mL medium per plate is optimal for biolistic transformation. Less medium can be used if stably transformed pollen tubes are to be cultured, although thinner layers of solid medium are more difficult to transfer to slides for microscopy.
14. PTNT plates can be stored for up to 4 weeks.
15. Handle plants and especially harvested flowers with care to avoid loss of pollen. Collect pollen only from fully opened

flowers showing light to dark pink color, not from flowers that are turning purple, which indicates beginning senescence.

16. As gold particles sediment rapidly even in 50% glycerol, vortex the particle suspension at least 3 min before taking an aliquot from the center of the tube to ensure an average particle concentration.
17. Remove the supernatant with a pipette after tilting the tube: try to avoid removing gold particles that have not been pelleted and form a layer on top of the supernatant.
18. Resuspend by vortexing and pipetting up and down. Scratch along the inner wall of the tube with the tip of the pipette to bring gold particles sticking to this surface back into suspension.
19. Appropriate parameters for the BioRad PDS-10000-He™ Biolistic Particle Delivery System are: vacuum 28 inches Hg, pressure 1000–1100 psi (rupture disk 1100 psi, tank pressure 1300 psi), plates containing cultured pollen are placed on a tray inserted into the second slot below the macrocarrier holder.
20. The protein of interest should be most strongly associated with the PM in untreated control pollen tubes. Note that in transiently transformed pollen tubes the Spo20p:YFP marker, for example, only displays clearly detectable PM association 6–7 h after particle bombardment.
21. Make sure that the medium is not too hot when adding n- or t-ButOH to prevent evaporation. Prepare and use PTNT plates with and without n-ButOH separately. Plates containing n-ButOH cannot be stored and need to prepared freshly. n- or t-ButOH plates need to be incubated in an airtight box separately from each other and from untreated controls.
22. This method works best with stably transformed pollen tubes. As 0.5% n-ButOH substantially affects pollen germination, it critically reduces the efficiency of transient transformation.
23. This is the most suitable method to analyze transiently transformed pollen tubes treated with drugs that strongly inhibit pollen germination, as it allows normal pollen germination before drug application. In case pollen tubes are required to grow for more than 5 h before drug application, it is recommended to place a thin film of liquid PTNT medium over the grains immediately after particle bombardment. This prevents pollen tubes growing on the surface of solid medium from developing a tendency to burst upon treatment with liquid medium. This method is of course also applicable for transgenic pollen tubes.
24. This method can only be applied to stably transformed pollen tubes, because particle bombardment requires pollen grains to

be cultured on solid medium. This method facilitates the application of drugs at exactly defined concentrations, but is not optimal for microscopic imaging.

25. This method is also just recommended for stably transformed pollen tubes, which all express the analyzed fluorescent fusion protein. The small size of the flow chamber prevents effective screening for transiently transformed pollen tubes.
26. For quantitative analysis of fluorescence intensity, some lasers (e.g., argon lasers) need to be started at least 20 min prior to imaging to guarantee constant laser intensity.
27. To ensure that pollen tubes were viable and normally growing at the time images were taken to analyze PM association of fluorescent fusion proteins, it is important to determine the pollen tube growth rate after recording these images. For this purpose, it is best to take a single second image 2 min later, as constant time-lapse imaging may result in phototoxicity, which potentially reduces pollen tube growth rates. To take two consecutive images, pollen tube tips should be positioned in the center of the area selected to record the first image, such that they do not grow out of the this area within 2 min. In general, it is recommended to position the longitudinal pollen tube axis either horizontally or vertically in the area selected for imaging to facilitate not only the recording of two consecutive images, but also subsequent image analysis and presentation.
28. To facilitate growth rate quantification, tips of analyzed pollen tubes can be enlarged by zooming-in using the "up" curser of the computer keyboard. In general, pollen tubes displaying a growth rate of at least 3.0 μm/min are selected for further analysis, as they are considered unaffected by fluorescent fusion protein expression and imaging.
29. The width of horizontal lines drawn using straight-line tool can be increased to e.g., 3 pixels to enhance data averaging, which may produce more reliable results.

Acknowledgments

This work was supported by the German Research Foundation (DFG) through the "Major Equipment Grants" INST90/1074-1FUGG (SP8 DIVE-FALCON microscope) and INST90/1025-1FUGG (plant growth chamber facility for tobacco), as well as within the framework of the Research Training Group (RTG) 1962.

References

1. Potocky M, Pleskot R, Pejchar P et al (2014) Live-cell imaging of phosphatidic acid dynamics in pollen tubes visualized by Spo20p-derived biosensor. New Phytol 203 (2):483–494
2. Helling D, Possart A, Cottier S et al (2006) Pollen tube tip growth depends on plasma membrane polarization mediated by tobacco PLC3 activity and endocytic membrane recycling. Plant Cell 18(12):3519–3534
3. Klahre U, Kost B (2006) Tobacco RhoGTPase ACTIVATING PROTEIN1 spatially restricts signaling of RAC/Rop to the apex of pollen tubes. Plant Cell 18(11):3033–3046
4. Kost B (2008) Spatial control of Rho (Rac-Rop) signaling in tip-growing plant cells. Trends Cell Biol 18(3):119–127
5. Ischebeck T, Stenzel I, Heilmann I (2008) Type B phosphatidylinositol-4-phosphate 5-kinases mediate Arabidopsis and *Nicotiana tabacum* pollen tube growth by regulating apical pectin secretion. Plant Cell 20 (12):3312–3330
6. Kost B, Lemichez E, Spielhofer P et al (1999) Rac homologues and compartmentalized phosphatidylinositol 4, 5-bisphosphate act in a common pathway to regulate polar pollen tube growth. J Cell Biol 145(2):317–330
7. Noack LC, Jaillais Y (2017) Precision targeting by phosphoinositides: how PIs direct endomembrane trafficking in plants. Curr Opin Plant Biol 40:22–33
8. Picton JM, Steer MW (1983) Membrane recycling and the control of secretory activity in pollen tubes. J Cell Sci 63:303–310
9. Derksen J, Rutten T, Lichtscheidl IK et al (1995) Quantitative analysis of the distribution of organelles in tobacco pollen tubes: implications for exocytosis and endocytosis. Protoplasma 188(3):267–276
10. Grebnev G, Ntefidou M, Kost B (2017) Secretion and endocytosis in pollen tubes: models of tip growth in the spot light. Front Plant Sci 8:154
11. Weigert R, Silletta MG, Spano S et al (1999) CtBP/BARS induces fission of Golgi membranes by acylating lysophosphatidic acid. Nature 402(6760):429–433
12. Roth MG, Bi K, Ktistakis NT et al (1999) Phospholipase D as an effector for ADP-ribosylation factor in the regulation of vesicular traffic. Chem Phys Lipids 98 (1):141–152
13. Kaneda M, van Oostende-Triplet C, Chebli Y et al (2019) Plant AP180 N-terminal homolog proteins are involved in clathrin-dependent endocytosis during pollen tube growth in *Arabidopsis thaliana*. Plant Cell Physiol 60 (6):1316–1330. https://doi.org/10.1093/pcp/pcz036
14. Munnik T (2001) Phosphatidic acid: an emerging plant lipid second messenger. Trends Plant Sci 6(5):227–233
15. Munnik T, Irvine RF, Musgrave A (1998) Phospholipid signalling in plants. Biochim Biophys Acta 1389(3):222–272. https://doi.org/10.1016/s0005-2760(97)00158-6
16. Potocky M, Elias M, Profotova B et al (2003) Phosphatidic acid produced by phospholipase D is required for tobacco pollen tube growth. Planta 217(1):122–130
17. Arisz SA, Munnik T (2013) Distinguishing phosphatidic acid pools from de novo synthesis, PLD, and DGK. Meth Mol Biol (Clifton, NJ) 1009:55–62
18. Pappan K, Austin-Brown S, Chapman KD et al (1998) Substrate selectivities and lipid modulation of plant phospholipase D alpha, -beta, and -gamma. Arch Biochem Biophys 353 (1):131–140
19. Arisz SA, Testerink C, Munnik T (2009) Plant PA signaling via diacylglycerol kinase. Biochim Biophys Acta 1791(9):869–875
20. Kooijman EE, Chupin V, de Kruijff B et al (2003) Modulation of membrane curvature by phosphatidic acid and lysophosphatidic acid. Traffic (Copenhagen, Denmark) 4 (3):162–174
21. Testerink C, Munnik T (2005) Phosphatidic acid: a multifunctional stress signaling lipid in plants. Trends Plant Sci 10(8):368–375
22. Schindelin J, Arganda-Carreras I, Frise E et al (2012) Fiji: an open-source platform for biological-image analysis. Nat Methods 9:676
23. van Engelen FA, Molthoff JW, Conner AJ et al (1995) pBINPLUS: an improved plant transformation vector based on pBIN19. Transgenic Res 4(4):288–290
24. Hajdukiewicz P, Svab Z, Maliga P (1994) The small, versatile pPZP family of Agrobacterium binary vectors for plant transformation. Plant Mol Biol 25(6):989–994
25. Kearse M, Moir R, Wilson A et al (2012) Geneious basic: an integrated and extendable desktop software platform for the organization and analysis of sequence data. Bioinformatics 28 (12):1647–1649

Index

Anja Geitmann (ed.), *Pollen and Pollen Tube Biology: Methods and Protocols*, Methods in Molecular Biology, vol. 2160, https://doi.org/10.1007/978-1-0716-0672-8, © Springer Science+Business Media, LLC, part of Springer Nature 2020

Q

R

S

MIX
Papier aus verantwortungsvollen Quellen
Paper from responsible sources
FSC® C105338

If you have any concerns about our products,
you can contact us on
ProductSafety@springernature.com

In case Publisher is established outside the EU,
the EU authorized representative is:
Springer Nature Customer Service Center GmbH
Europaplatz 3, 69115 Heidelberg, Germany

Printed by Libri Plureos GmbH
in Hamburg, Germany